THE
MINISTRY
FOR THE
FUTURE

By Kim Stanley Robinson

Icehenge

The Memory of Whiteness

THREE CALIFORNIAS

The Wild Shore
The Gold Coast
Pacific Edge

The Planet on the Table

Escape from Kathmandu

A Short, Sharp Shock

Remaking History

THE MARS TRILOGY

Red Mars
Green Mars
Blue Mars

The Martians

Antarctica

The Years of Rice and Salt

SCIENCE IN THE CAPITAL

Forty Signs of Rain
Fifty Degrees Below
Sixty Days and Counting
combined as *Green Earth*

Galileo's Dream

2312

Shaman

Aurora

New York 2140

Red Moon

The Ministry for the Future

THE MINISTRY FOR THE FUTURE

KIM STANLEY ROBINSON

www.orbitbooks.net

Copyright © 2020 by Kim Stanley Robinson

Cover design by Lauren Panepinto
Cover images by Trevillion and Shutterstock
Cover copyright © 2020 by Hachette Book Group, Inc.

Orbit
Hachette Book Group
1290 Avenue of the Americas
New York, NY 10104
www.orbitbooks.net

First Edition: October 2020
Simultaneously published in Great Britain by Orbit

Orbit is an imprint of Hachette Book Group.
The Orbit name and logo are trademarks of Little, Brown Book Group Limited.

The Hachette Speakers Bureau provides a wide range of authors for speaking events. To find out more, go to www.hachettespeakersbureau.com or call (866) 376-6591.

Library of Congress Cataloging-in-Publication Data
Names: Robinson, Kim Stanley, author.
Title: The ministry for the future / Kim Stanley Robinson.
Description: First edition. | New York, NY : Orbit, 2020.
Identifiers: LCCN 2020014375 | ISBN 9780316300131 (hardcover) |
    ISBN 9780316300162 (ebook) | ISBN 9780316300124
Subjects: GSAFD: Science fiction.
Classification: LCC PS3568.O2893 M56 2020 | DDC 813/.54—dc23
LC record available at https://lccn.loc.gov/2020014375

ISBNs: 978-0-316-30013-1 (hardcover), 978-0-316-59170-6 (signed edition), 978-0-316-59169-0 (BarnesAndNoble.com signed edition), 978-0-316-30016-2 (ebook)

Printed in the United States of America

LSC-C

10 9 8 7 6 5 4 3 2 1

For Fredric Jameson

THE
MINISTRY
FOR THE
FUTURE

1

I t was getting hotter.

Frank May got off his mat and padded over to look out the window. Umber stucco walls and tiles, the color of the local clay. Square apartment blocks like the one he was in, rooftop patios occupied by residents who had moved up there in the night, it being too hot to sleep inside. Now quite a few of them were standing behind their chest-high walls looking east. Sky the color of the buildings, mixed with white where the sun would soon rise. Frank took a deep breath. It reminded him of the air in a sauna. This the coolest part of the day. In his entire life he had spent less than five minutes in saunas, he didn't like the sensation. Hot water, maybe; hot humid air, no. He didn't see why anyone would seek out such a stifling sweaty feeling.

Here there was no escaping it. He wouldn't have agreed to come here if he had thought it through. It was his home town's sister city, but there were other sister cities, other aid organizations. He could have worked in Alaska. Instead sweat was dripping into his eyes and stinging. He was wet, wearing only a pair of shorts, those too were wet; there were wet patches on his mat where he had tried to sleep. He was thirsty and the jug by his bedside was empty. All over town the stressed hum of windowbox air conditioner fans buzzed like giant mosquitoes.

And then the sun cracked the eastern horizon. It blazed like an atomic bomb, which of course it was. The fields and buildings underneath that brilliant chip of light went dark, then darker still as the chip flowed to the sides in a burning line that then bulged to a crescent he couldn't look at. The heat coming from it was palpable, a slap to the face. Solar radiation heating the skin of his face, making him blink. Stinging eyes flowing, he couldn't see much. Everything was tan and beige and a brilliant, unbearable white. Ordinary town in Uttar Pradesh, 6 AM. He looked at his phone: 38 degrees. In Fahrenheit that was—he tapped—103 degrees. Humidity about 35 percent. The combination was the thing. A few years ago it would have been among the hottest wet-bulb temperatures ever recorded. Now just a Wednesday morning.

Wails of dismay cut the air, coming from the rooftop across the street. Cries of distress, a pair of young women leaning over the wall calling down to the street. Someone on that roof was not waking up. Frank tapped at his phone and called the police. No answer. He couldn't tell if the call had gone through or not. Sirens now cut the air, sounding distant and as if somehow submerged. With the dawn, people were discovering sleepers in distress, finding those who would never wake up from the long hot night. Calling for help. The sirens seemed to indicate some of the calls had worked. Frank checked his phone again. Charged; showing a connection. But no reply at the police station he had had occasion to call several times in his four months here. Two months to go. Fifty-eight days, way too long. July 12, monsoon not yet arrived. Focus on getting through today. One day at a time. Then home to Jacksonville, comically cool after this. He would have stories to tell. But the poor people on the rooftop across the way.

Then the sound of the air conditioners cut off. More cries of distress. His phone no longer showed any bars. Electricity gone. Brownout, or blackout. Sirens like the wails of gods and goddesses, the whole Hindu pantheon in distress.

Generators were already firing up, loud two-stroke engines. Illegal gas, diesel, kerosene, saved for situations like these, when the law requiring use of liquid natural gas gave way to necessity. The air, already bad, would soon be a blanket of exhaust. Like breathing from the exhaust pipe of an old bus.

Frank coughed at the thought of it, tried again to drink from the jug by his bed. It was still empty. He took it downstairs with him, filled it from their filtered tank in the refrigerator in the closet there. Still cold even with power off, and now in his thermos jug, where it would stay cold for a good long while. He dropped an iodine pill in the jug for good measure, sealed it tight. The weight of it was reassuring.

The foundation had a couple of generators here in the closet, and some cans of gasoline, enough to keep the generators going for two or three days. Something to keep in mind.

His colleagues came piling in the door. Hans, Azalee, Heather, all red-eyed and flustered. "Come on," they said, "we have to go."

"What do you mean?" Frank asked, confused.

"We need to go get help, the whole district has lost power, we have to tell them in Lucknow. We have to get doctors here."

"What doctors?" Frank asked.

"We have to try!"

"I'm not leaving," Frank said.

They stared at him, looked at each other.

"Leave the satellite phone," he said. "Go get help. I'll stay and tell people you're coming."

Uneasily they nodded, then rushed out.

Frank put on a white shirt that quickly soaked up his sweat. He walked out into the street. Sound of generators, rumbling exhaust into the super-heated air, powering air conditioners he presumed. He suppressed a cough. It was too hot to cough; sucking back in air was like breathing in a furnace, so that one coughed again. Between the intake of steamy air and the effort of coughing, one ended up hotter than ever. People came up to him asking for help. He said it would be coming soon. Two in the afternoon, he told people. Come to the clinic then. For now, take the old ones and the little ones into rooms with air conditioning. The schools would have A/C, the government house. Go to those places. Follow the sound of generators.

Every building had a clutch of desperate mourners in its entryway, waiting for ambulance or hearse. As with coughing, it was too hot to wail very much. It felt dangerous even to talk, one would overheat. And what was there to say anyway? It was too hot to think. Still people approached him. Please sir, help sir.

Go to my clinic at two, Frank said. For now, get to the school. Get inside, find some A/C somewhere. Get the old ones and the little ones out of this.

But there's nowhere!

Then it came to him. "Go to the lake! Get in the water!"

This didn't seem to register. Like Kumbh Mela, during which people went to Varanasi and bathed in the Ganges, he told them the best he could. "You can stay cool," he told them. "The water will keep you more cool."

A man shook his head. "That water is in the sun. It's as hot as a bath. It's worse than the air."

Curious, alarmed, feeling himself breathing hard, Frank walked down streets toward the lake. People were outside buildings, clustered in doorways. Some eyed him, most didn't, distracted by their own issues. Round-eyed with distress and fear, red-eyed from the heat and exhaust smoke, the dust. Metal surfaces in the sun burned to the touch, he could see heat waves bouncing over them like air over a barbeque. His muscles were jellied, a wire of dread running down his spinal cord was the only thing keeping him upright. It was impossible to hurry, but he wanted to. He walked in the shade as much as possible. This early in the morning one side of the street was usually shaded. Moving into sunlight was like getting pushed toward a bonfire. One lurched toward the next patch of shade, impelled by the blast.

He came to the lake and was unsurprised to see people in it already, neck deep. Brown faces flushed red with heat. A thick talcum of light hung over the water. He went to the curving concrete road that bordered the lake on this side, crouched and stuck his arm in up to the elbow. It was indeed as warm as a bath, or almost. He kept his arm in, trying to decide if the water was cooler or hotter than his body. In the cooking air it was hard to tell. After a time he concluded the water at the surface was approximately the same temperature as his blood. Which meant it was considerably cooler than the air. But if it was a little warmer than body temperature...well, it would still be cooler than the air. It was strangely hard to tell. He looked at the people in the lake. Only a narrow stretch of water was still in the morning shade of buildings and trees, and that stretch would be gone soon. After that the entire lake would be lying there in the sun, until the late afternoon brought shadows on the other side. That was bad. Umbrellas, though; everyone had an umbrella. It was an open question how many of the townspeople could fit in the lake. Not enough. It was said the town's population was two hundred thousand. Surrounded by fields and small hills, other towns a few or several kilometers away, in every direction. An ancient arrangement.

He went back to the compound, into the clinic on the ground floor.

Up to his room on the next floor, huffing and puffing. It would be easiest to lie there and wait it out. He tapped in the combination on his safe and pulled open its door, took out the satellite phone. He turned it on. Battery fully charged.

He called headquarters in Delhi. "We need help," he said to the woman who answered. "The power has gone out."

"Power is out here too," Preeti said. "It's out everywhere."

"Everywhere?"

"Most of Delhi, Uttar Pradesh, Jharkhand, Bengal. Parts of the west too, in Gujarat, Rajasthan…"

"What should we do?"

"Wait for help."

"From where?"

"I don't know."

"What's the forecast?"

"The heat wave is supposed to last awhile longer. The rising air over the land might pull in cooler air off the ocean."

"When?"

"No one knows. The high pressure cell is huge. It's caught against the Himalaya."

"Is it better to be in water than in air?"

"Sure. If it's cooler than body temperature."

He turned off the phone, returned it to the safe. He checked the particulate meter on the wall: 1300 ppm. This for fine particulates, 25 nanometers and smaller. He went out onto the street again, staying in the shade of buildings. Everyone was doing that; no one stood in the sun now. Gray air lay on the town like smoke. It was too hot to have a smell, there was just a scorched sensation, a smell like heat itself, like flame.

He returned inside, went downstairs and opened the safe again, took out the keys to the closet, opened the closet and pulled out one of the generators and a jerrycan of gas. He tried to fill the generator's gas tank and found it was already full. He put the can of gas back in the closet, took the generator to the corner of the room where the window with the air conditioner was. The windowbox A/C had a short cord and was plugged into the wall socket under the window. But it wouldn't do to run a generator in a room, because of the exhaust. But it also wouldn't do to run the generator out on the street below the window; it would surely be snatched. People were desperate. So...He went back to the closet, rooted around, found an extension cord. Up to the building's roof, which had a patio surrounded by a rampart and was four floors off the street. Extension cord only reached down to the floor below it. He went down and took the A/C unit out of the window on the second floor, hefted it up the stairs, gasping and sweating. For a moment he felt faint, then sweat stung his eyes and a surge of energy coursed through him. He opened the fourth-floor office window, got the A/C unit balanced on the ledge and closed the window on it, pulled out the plastic sidepanels that closed off the parts of the window still open. Up to the rooftop terrace, start the generator, listen to it choke and rattle up to its two-stroke percussion. Initial puff of smoke, after that its exhaust wasn't visible. It was loud though, people would hear it. He could hear others around the town. Plug in the extension cord, down the stairs to the upper office, plug in the A/C unit, turn it on. Grating hum of the A/C. Inrush of air, ah God, the unit wasn't working. No, it was. Lowering the temperature of the outer air by 10 or 20 degrees—that left it at about 85 degrees as he thought of it, maybe more. In the shade that was fine, people could do that, even with the humidity. Just rest and be easy. And the cooler air would fall down the stairs and fill the whole place.

Downstairs he tried to close the window where the A/C unit had been, found it was stuck. He slammed it downward with his fists, almost breaking the glass. Finally it gave a jerk and came down. Out onto the street, closing the door. Off to the nearest school. A little shop nearby sold food and drinks to students and their parents. The school was closed, the shop too, but people were there, and he recognized some. "I've got air conditioning going at the clinic," he said to them. "Come on over."

Silently a group followed him. Seven or eight families, including the shop owners, locking their door after them. They tried to stay in the shade but now there was little shade to find. Men preceded wives who herded children and tried to induce their single file to stay in the shade. Conversations were in Awadhi, Frank thought, or Bhojpuri. He only spoke a little Hindi, as they knew; they would speak in that language to him if they wanted to talk to him, or confer with someone who would speak to him in English. He had never gotten used to trying to help people he couldn't talk to. Embarrassed, ashamed, he blasted past his reluctance to reveal his bad Hindi and asked them how they felt, where their families were, whether they had anyplace they could go. If indeed he had said those things. They looked at him curiously.

At the clinic he opened up and people filed in. Without instruction they went upstairs to the room where the A/C was running, sat down on the floor. Quickly the room was full. He went back downstairs and stood outside the door and welcomed people in if they showed any interest. Soon the whole building was as full as it could be. After that he locked the door.

People sat sweltering in the relative cool of the rooms. Frank checked the desk computer; temperature on the ground floor 38 degrees. Perhaps cooler in the room with the A/C unit. Humidity now 60 percent. Bad to have both high heat and high humidity, unusual; in the dry season on the Gangetic plain, January through March, it was cooler and drier; then it grew hot, but was still dry; then with the soaking of the monsoon came cooler temperatures, and omnipresent clouds that gave relief from direct sunlight. This heat wave was different. Cloudless heat and yet high humidity. A terrible combination.

The clinic had two bathrooms. At some point the toilets stopped working. Presumably the sewers led to a wastewater treatment plant somewhere that ran on electricity, of course, and might not have the generator capacity to keep working, although that was hard to believe. Anyway it had happened. Now Frank let people out as needed so they could go in the alleys somewhere, as in hill villages in Nepal where there were no toilets at any time. He had been shocked the first time he saw that. Now he took nothing for granted.

Sometimes people began crying and little crowds surrounded them; elders in distress, little children in distress. Quite a few accidents of excretion. He put buckets in the bathrooms and when they were full he took them out into the streets and poured them into the gutters, took them back. An old man died; Frank helped some younger men carry the body up to the rooftop patio, where they wrapped the old one in a thin sheet, maybe a sari. Much worse came later that night, when they did the same thing for an infant. Everyone in every room cried as they carried the little body up to the roof. Frank saw the generator was running out of gas and went down to the closet and got the fuel can and refilled it.

His water jug was empty. The taps had stopped running. There were two big water cans in the refrigerator, but he didn't talk about those. He refilled his jug from one of them, in the dark; the water was still a bit cool. He went back to work.

Four more people died that night. In the morning the sun again rose like the blazing furnace of heat that it was, blasting the rooftop and its sad cargo of wrapped bodies. Every rooftop and, looking down at the town, every sidewalk too was now a morgue. The town was a morgue, and it was as hot as ever, maybe hotter. The thermometer now said 42 degrees, humidity 60 percent. Frank looked at the screens dully. He had slept about three hours, in snatches. The generator was still chuntering along in its irregular two-stroke, the A/C box was still vibrating like the bad fan it was. The sound of other generators and air conditioners still filled the air. But it wasn't going to make any difference.

He went downstairs and opened the safe and called Preeti again on the satellite phone. After twenty or forty tries, she picked up. "What is it?"

"Look, we need help here," he said. "We're dying here."

"What do you think?" she said furiously. "Do you think you're the only ones?"

"No, but we need help."

"We all need help!" she cried.

Frank paused to ponder this. It was hard to think. Preeti was in Delhi.

"Are you okay there?" he asked.

No answer. Preeti had hung up.

His eyes were stinging again. He wiped them clear, went back upstairs to get the buckets in the bathroom. They were filling more slowly now; people were emptied out. Without a water supply, they would have to move soon, one way or the other.

When he came back from the street and opened his door there was a rush and he was knocked inside. Three young men held him down on the floor, one with a squared-off black handgun as big as his head. He pointed the gun and Frank looked at the round circle of the barrel end pointed at him, the only round part of a squared-off thing of black metal. The whole world contracted to that little circle. His blood pounded through him and he felt his body go rigid. Sweat poured from his face and palms.

"Don't move," one of the other men said. "Move and you die."

Cries from upstairs tracked the intruders' progress. The muffled sounds of the generator and A/C cut off. The more general mumble of the town came wafting in the open doorway. People passing by stared curiously and moved on. There weren't very many of them. Frank tried to breathe as shallowly as possible. The stinging in his right eye was ferocious, but he only clamped the eye shut and with the other stared resolutely away. He felt he should resist, but he wanted to live. It was as if he were watching the whole scene from halfway up the stairs, well outside his body and any feelings it might be feeling. All except the stinging in his eye.

The gang of young men clomped downstairs with generator and A/C unit. Out they went into the street. The men holding Frank down let him go. "We need this more than you do," one of them explained.

The man with the gun scowled as he heard this. He pointed the gun at Frank one last time. "You did this," he said, and then they slammed the door on him and were gone.

Frank stood, rubbed his arms where the men had grasped him. His heart was still racing. He felt sick to his stomach. Some people from upstairs came down and asked how he was. They were worried about him, they were concerned he had been hurt. This solicitude pierced him, and suddenly he felt more than he could afford to feel. He sat on the lowest stair and hid his face in his hands, racked by a sudden paroxysm. His tears made his eyes sting less.

Finally he stood up. "We have to go to the lake," he said. "There's water there, and it will be cooler. Cooler in the water and on the sidewalk."

Several of the women were looking unhappy at this, and one of them said, "You may be right, but there will be too much sun. We should wait until dark."

Frank nodded. "That makes sense."

He went back to the little store with its owner, feeling jittery and light-headed and weak. The sauna feeling hammered him and it was hard to carry a sack of food and canned and bottled drinks back to the clinic. Nevertheless he helped ferry over six loads of supplies. Bad as he felt, it seemed as if he was stronger than many of the others in their little group. Although at times he wondered if some of them could in fact just keep dragging along like this all day. But none of them spoke as they walked, nor even met eyes.

"We can get more later," the shop owner finally declared.

The day passed. Wails of grief were now muffled to groans. People were too hot and thirsty to make any fuss, even when their children died. Red eyes in brown faces, staring at Frank as he stumbled among them, trying to help get corpses of family members up onto the roof, where they scorched in the sun. Bodies would be rotting, but maybe they would anneal and dry out before that, it was so hot. No odors could survive in this heat, only the smell of scorched steamy air itself. Or maybe not: sudden smell of rotting meat. No one lingered up here now. Frank counted fourteen wrapped bodies, adult and child. Glancing across that rooftop level of the town he saw that other people were similarly engaged, silent, withdrawn, down-gazing, hurrying. No one he could see looked around as he was looking around.

Downstairs the food and drink were already gone. Frank made a count, which he found hard. Something like fifty-two people in the clinic. He sat on the stairs for a while, then went in the closet and stared at its contents. He refilled his water jug, drank deeply, refilled it again. No longer cool, but not hot. There was the can of gas; they could burn the bodies if they had to. There was another generator, but there was nothing to power with it that would do any good. The satellite phone was still charged, but there was no one to call. He wondered if he should call his mom. Hi Mom, I'm dying. No.

The day crawled second by second to its last hour, and then Frank conferred with the store owner and his friends. In murmurs they all agreed; time to go to the lake. They roused the people, explained the plan, helped those who needed it to stand, to get down the stairs. A few couldn't do it; that presented a quandary. A few old men said they would stay behind as long as they were needed, then come along to the lake. They said goodbye to the people leaving as if things were normal, but their eyes gave it all away. Many wept as they left the clinic.

They made their way in the afternoon shadows to the lake. Hotter than ever. No one on the streets and sidewalks. No wailing from the buildings. Still some generators grumbling, some fans grinding. Sound seemed stunted in the livid air.

At the lake they found a desperate scene. There were many, many people in the lake, heads dotted the surface everywhere around the shores, and out where it was presumably deeper there were still heads, people semi-submerged as they lay on impromptu rafts of one sort or another. But not all of these people were alive. The surface of the lake seemed to have a low miasma rising out of it, and now the stink of death, of rotting meat, could be discerned in one's torched nostrils.

They agreed it might be best to start by sitting on the low lakeshore walkway or corniche and put their legs in the water. Down at the end of the walkway there was still room to do that, and they trudged down together and sat as a group, in a line. The concrete under them was still radiating the day's heat. They were all sweating, except for some who weren't, who were redder than the rest, incandescent in the shadows of the late afternoon. As twilight fell they propped these people up and helped them to die. The water of the lake was as hot as bath water, clearly hotter than body temperature, Frank thought; hotter than the last time he had tested it. It only made sense. He had read that if all the sun's energy that hit Earth were captured by it rather than some bouncing away, temperatures would rise until the seas boiled. He could well imagine what that would be like. The lake felt only a few degrees from boiling.

And yet sometime after sunset, as the quick twilight passed and darkness fell, they all got in the water. It just felt better. Their bodies told them to do it. They could sit on the shallowest part of the lake bottom, heads out of water, and try to endure.

Sitting next to Frank was a young man he had seen playing the part of Karna in one of the plays at the local mela, and Frank felt his blankness pierced again, as when the people had shown concern for him, by the memory of the young man at the moment Arjuna had rendered Karna helpless with a spoken curse and was about to kill him; at that point the young man had shouted triumphantly, "It's only fate!" and managed to take one last swing before going down under Arjuna's impervious sword. Now the young man was sipping the water of the lake, round-eyed with dread and sorrow. Frank had to look the other way.

The heat began to go to his head. His body crawled with the desire to get out of this too-hot bath, run like one would from a sauna into the icy lake that ought to accompany all such saunas, feel that blessed shock of cold smacking the breath out of his lungs as he had felt it once in Finland. People there spoke of trying to maximize the temperature differential, shift a hundred degrees in a second and see what that felt like.

But this train of thought was like scratching an itch and thereby making it worse. He tasted the hot lake water, tasted how foul it was, filled with organics and who knew what. Still he had a thirst that couldn't be slaked. Hot water in one's stomach meant there was no refuge anywhere, the world both inside and outside well higher than human body temperature ought to be. They were being poached. Surreptitiously he uncapped his water jug and drank. Its water was now tepid, but not hot, and it was clean. His body craved it and he couldn't stop himself, he drank it all down.

People were dying faster than ever. There was no coolness to be had. All the children were dead, all the old people were dead. People murmured what should have been screams of grief; those who could still move shoved bodies out of the lake, or out toward the middle where they floated like logs, or sank.

Frank shut his eyes and tried not to listen to the voices around him. He was fully immersed in the shallows, and could rest his head back against the concrete edge of the walkway and the mud just under it. Sink himself until he was stuck in mud and only half his head exposed to the burning air.

The night passed. Only the very brightest stars were visible, blurs swimming overhead. A moonless night. Satellites passing overhead, east to west, west to east, even once north to south. People were watching, they knew what was happening. They knew but they didn't act. Couldn't act. Didn't act. Nothing to do, nothing to say. Many years passed for Frank that night. When the sky lightened, at first to a gray that looked like clouds, but then was revealed to be only a clear and empty sky, he stirred. His fingertips were all pruney. He had been poached, slow-boiled, he was a cooked thing. It was hard to raise his head even an inch. Possibly he would drown here. The thought caused him to exert himself. He dug his elbows in, raised himself up. His limbs were like cooked spaghetti draping his bones, but his bones moved of their own accord. He sat up. The air was still hotter than the water. He watched sunlight strike the tops of the trees on the other side of the lake; it looked like they were bursting into flame. Balancing his head carefully on his spine, he surveyed the scene. Everyone was dead.

2

I am a god and I am not a god. Either way, you are my creatures. I keep you alive.

Inside I am hot beyond all telling, and yet my outside is even hotter. At my touch you burn, though I spin outside the sky. As I breathe my big slow breaths, you freeze and burn, freeze and burn.

Someday I will eat you. For now, I feed you. Beware my regard. Never look at me.

3

Article 14 of the Paris Agreement Under the United Nations Framework Convention on Climate Change called for a periodic taking stock of all the signatory nations' carbon emissions, which meant in effect the total global carbon burn for the year in question. The first "global stocktake" was scheduled for 2023, and then every five years after that.

That first global stocktake didn't go well. Reporting was inconsistent and incomplete, and yet still it was very clear that carbon emissions were far higher than the Parties to the Agreement had promised each other they would be, despite the 2020 dip. Very few nations had hit the targets they had set for themselves, even though they had set soft targets. Aware of the shortfall even before the 2023 stocktake, 108 countries had promised to strengthen their pledges; but these were smaller countries, amounting together to about 15 percent of global total emissions.

So at the annual Conference of the Parties the following year, some delegations pointed out that the Agreement's Article 16, clause 4, specified that the COP "shall make the decisions necessary to promote the Agreement's effective implementation by establishing such subsidiary bodies as are deemed necessary for the implementation of the Agreement." They also pointed to Article 18, clause 1, which allowed the COP to create new "Subsidiary Bodies for Implementation of the Agreement." These subsidiary bodies had previously been understood to mean committees that met only during the annual COP gatherings, but now some delegates argued that given the general failure of the Agreement so far, a new subsidiary body with permanent duties, and the resources to pursue them, was clearly needed to help push the process forward.

So at COP29, held in Bogotá, Colombia, the Parties to the Agreement created a new Subsidiary Body for Implementation of the Agreement, as

authorized by Articles 16 and 18, to be funded using the funding protocols outlined in Article 8, which bound all Parties to the methods outlined in the Warsaw International Mechanism for Loss and Damage. The
announcement said:

"Be it resolved that a Subsidiary Body authorized by this twenty-ninth
Conference of the Parties serving as the meeting of the parties to the Paris
Climate Agreement (CMA) is hereby established, to work with the Inter-
governmental Panel on Climate Change, and all the agencies of the United
Nations, and all the governments signatory to the Paris Agreement, to
advocate for the world's future generations of citizens, whose rights, as
defined in the Universal Declaration of Human Rights, are as valid as our
own. This new Subsidiary Body is furthermore charged with defending
all living creatures present and future who cannot speak for themselves, by
promoting their legal standing and physical protection."

Someone in the press named this new agency "the Ministry for the
Future," and the name stuck and spread, and became what the new agency
was usually called. It was established in Zurich, Switzerland, in January of
2025.

Not long after that, the big heat wave struck India.

4

Above the campus of the Eidgenössische Technische Hochschule, Zurich rises to a forest park on top of the Zuriberg, the hill that forms the eastern flank of the city. Most of the city covers the banks of the river Limmat, which begins as the outlet of the Zurichsee and drains northward between two hills, the Zuriberg on the east and the Uetliberg to the west. Between these two hills the land is pretty flat, for Switzerland anyway, and here almost a quarter of all the Swiss have congregated to make a compact handsome city. Those lucky enough to live on the rise of the Zuriberg often think they have the best location, with a view over the roofs of the downtown, and out to the big lake to the south, and sometimes a glimpse of the Alps. In the late afternoon sun a sense of luminous calm can radiate up from this mixed vista of human and natural features. A good place. Visitors often call it boring, but the locals don't complain.

At the Kirche Fluntern tram stop, about halfway up the Zuriberg, you can get off one of the blue trams and walk north along Hochstrasse, past an old church with a steeple that has a big clockface on it, with a bell that rings the hours. Next door to it is where one finds the offices of the Paris Agreement's Ministry for the Future. They're within easy walking distance of the ETH, with all its geotechnical expertise, and not far above offices of the big Swiss banks, with their immense amounts of capital, all out of proportion to Switzerland's small size. These proximities were not accidental; for centuries now the Swiss have pursued a national policy of creating maximum safety for Switzerland by helping to increase peace and prosperity worldwide. "No one is safe until all are secure" seems to be their principle here, and in that project, geotechnical expertise and lots of money are both very useful.

That being the case, and since Geneva already hosted the headquarters of the World Health Organization and several other UN agencies, when the Paris Agreement established this new agency of theirs, Zurich forcefully made the case that Geneva was already too crowded with agencies, and expensive as a result, and after some vigorous inter-cantonal tussling, they won the bid to host the new agency. Offering rent-free the compound on Hochstrasse, and several nearby ETH buildings as well, was no doubt one of many reasons their bid succeeded.

Now the head of the ministry—Mary Murphy, an Irish woman of about forty-five years of age, ex–minister of foreign affairs in the government of the Irish Republic, and before that a union lawyer—walked into her office to find a crisis that didn't surprise her one bit. Everyone had been transfixed with horror at the news of the deadly heat wave in India; there were sure to be immediate ramifications. Now the first of these had come.

Her chief of staff, a short slight man named Badim Bahadur, followed Mary into her office saying, "You must have heard that the Indian government is beginning a solar radiation management action."

"Yes, I just saw it this morning," she said. "Have they given us the details of their plan?"

"They came half an hour ago. Our geoengineering people are saying that if they do it as planned, it will equate to about the same as the Pinatubo volcanic eruption of 1991. That lowered global temperature by about a degree Fahrenheit, for a year or two. That was from the sulfur dioxide in the ash cloud that the volcano shot into the stratosphere. It will take the Indians several months to replicate that boost of sulfur dioxide, our people say."

"Do they have the capacity to do it?"

"Their air force can probably do it, yes. They can certainly try, they've got the necessary aircraft and equipment. A lot of it will simply reconfigure aerial refueling technology. And planes dump fuel all the time, so that part won't be so hard. The main problem will be getting up as high as possible, and then it's just a matter of quantity, the number of missions needed. Thousands of flights, for sure."

Mary pulled her phone from her pocket, tapped the screen for Chandra. Head of India's delegation to the Paris Agreement, she was well known to Mary. It would be late in Delhi, but this was when they usually talked.

When she answered, Mary said, "Chandra, it's Mary, can you talk for a minute?"

"For a minute, yes," Chandra said. "It's very busy here."

"I'm sure. What's this about your air force doing a Pinatubo?"

"Or a double Pinatubo, yes. This is what our academy of sciences is recommending, and the prime minister has ordered it."

"But the Agreement," Mary said, sitting down on her chair and focusing on her colleague's voice. "You know what it says. No atmospheric interventions without consultation and agreement."

"We are breaking the Agreement," Chandra said flatly.

"But no one knows what the effects will be!"

"They will be like Pinatubo, or hopefully double that. Which is what we need."

"You can't be sure that there won't be other effects—"

"Mary!" Chandra exclaimed. "Stop it right now. I know what you are going to say even before you say it. Here's what we are sure of in India: millions of people have just died. We'll never even know how many died, there are too many to count. It could be twenty million people. Do you understand what that means?"

"Yes."

"No. You don't understand. I invite you to come see it in person. Really you should, just so you know."

Mary found herself short of breath. She swallowed. "I will if you want me to."

A long silence followed. Finally Chandra spoke, her voice tight and choked. "Thank you for that, but maybe there is too much trouble here now for us to handle such a visit. You can see in the reports. I will send you some we are making. What you need to know now is that we are scared here, and angry too. It was Europe and America and China who caused this heat wave, not us. I know we have burned a lot of coal in the last few decades, but it's nothing compared to the West. And yet we signed the Agreement to do our part. Which we have done. But no one else is fulfilling commitments, no one is paying the developing nations, and now we have this heat wave. And another one could happen next week! Conditions are much the same!"

"I know."

"Yes, you know. Everyone knows, but no one acts. So we are taking matters into our own hands. We'll lower global temperatures for a few years, everyone will benefit. And perhaps we'll dodge another massacre like this one."

"All right."

"We do not need your permission!" Chandra shouted.

"I didn't mean that," Mary said. But the line had gone dead.

5

We drove in with a fuel truck, water truck, all that. It was like going out into nothing. With the electricity out, pumps weren't working, nothing was working. We set to work on the power plants before we did anything about all the dead. In any case there was nothing we could do about them, the bodies were sleeping where they fell. This wasn't just the people, but all the cattle too. Seeing all the bodies, cows, people, dogs, someone said something about the way Tibetans bury their dead, called sky burial—let the vultures eat the bodies. And there were some vultures doing that, yes. Clouds of vultures and crows. They must have flown in afterward. Sometimes the stink was horrible, but then we would move on or the wind would shift, and it went away. It seemed like it was too hot for smells, the air was cooked. The main smell was of burning. And things were burning, yes. Once the power came back on, there were some downed lines east of Lucknow, and brush fires started from them. Next day a wind came and the fire spread and got into the towns and we had to fight the fire before anything else. We got particulate readings of 1500 ppm.

There was a lake we could pump from, next to one town near Lucknow. The lake was filled with dead bodies, it was awful, but we threw the pump intake out into the lake anyway, because we needed the water. We were downwind of a brush fire, it was coming at us fast. So when the pump started filling our water trucks we were relieved.

Then I heard a noise, at first I thought it was something in the pump line, a kind of squeak it was. But then it seemed to be coming from the lakeshore, where there was a sidewalk running around the edge of the water. So I went over to look. I don't know. I guess it sounded alive.

He was lying against a building across the sidewalk from the lake. He had a shirt draped over his head. I saw him move and shouted to the others and went to him. He was a firangi, with brown hair and skin that was all peeling off. He looked like he had been burned, or boiled, I don't know— he looked dead but he was moving. His eyes were almost swollen shut, but I could see he was looking at me. Once we started helping him he never said a thing, never made another sound. His lips were cracked bloody. I thought maybe his voice was gone, that he was too cooked to talk. We gave him water by the spoonful. We were afraid to give him too much at once. Once we got the word to team command, the medicos were with us pretty quickly. They took over and gave him infusions. He watched them do it. He looked around at us, and back at the lake, but he never said a thing. His eyes were just slits, and so red. He looked completely mad. Like a different kind of being entirely.

6

Following the great Indian heat wave, the emergency meeting of the Paris Agreement signatories was fraught indeed. The Indian delegation arrived in force, and their leader Chandra Mukajee was excoriating in her denunciation of the international community and its almost complete failure to adhere to the terms of the agreement that every nation on Earth had signed. Reductions in emissions ignored, payment into investment funds that were to be spent on decarbonization not paid—in every way the Agreement had been ignored and abrogated. A performance without substance, a joke, a lie. And now India had paid the price. More people had died in this heat wave than in the entirety of the First World War, and all in a single week and in a single region of the world. The stain of such a crime would never go away, it would remain forever.

No one had the heart to point out that India had also failed to meet its emission reduction targets. And of course if total emissions over historical time were totted up, India would come in far behind all of the developed nations of the Western world, as everyone knew. In dealing with the poverty that still plagued so much of the Indian populace, the Indian government had had to create electricity as fast as they could, and also, since they existed in a world run by the market, as cheaply as they could. Otherwise outside investors would not invest, because the rate of return would not be high enough. So they had burned coal, yes. Like everyone else had up until just a few years before. Now India was being told not to burn coal, when everyone else had finished burning enough of it to build up the capital to afford to shift to cleaner sources of power. India had been told to get better without any financial help to do so whatsoever. Told to tighten the belt and embrace austerity, and be the working class for the bourgeoisie of the developed world, and suffer in silence until better times came—but the better times could never come, that plan was shot. The deck had been stacked, the game was over. And now twenty million people were dead.

The people in the big room in the center of Zurich's Kongresshall sat there in silence. This was not the same silence as the earlier memorial moment, a ritual period of silence to honor the memory of the dead which had stretched on for minute after minute. Now it was the silence of shame, confusion, dismay, guilt. The Indian delegation was done talking, they had nothing more they cared to say. Time for a response, an answer to them; but there was no answer. Nothing could be said. It was what it was: history, the nightmare from which they could not wake.

Finally that year's president of the Paris Agreement organization, a woman from Zimbabwe, stood up and went to the podium. Briefly she embraced Chandra, nodded to the other Indians on stage, and went to the microphone.

"Obviously we have to do better," she said. "The Paris Agreement was created to avoid tragedies like this one. We are all in a single global village now. We share the same air and water, and so this disaster has happened to all of us. Since we can't undo it, we have to turn it to the good somehow, or two things will happen; the crimes in it will go unatoned, and more such disasters will happen. So we have to act. At long last, we have to take the climate situation seriously, as the reality that overrides everything else. We have to act on what we know."

Everyone nodded. They could not applaud, not now, but they could nod. They could raise their hands, some of them with their fists clenched, and commit themselves to action.

That was all very well. It was a moment, maybe even a moment to remember. But very soon they were back to the usual horse-trading of national interests and commitments. The disaster had happened in India, in a part of India where few foreigners ever went, a place said to be very hot, very crowded, very poor. Probably more such events in the future would mostly happen in those nations located between the Tropics of Capricorn and Cancer, and the latitudes just to the north and south of these lines. Between thirty north and thirty south: meaning the poorest parts of the world. North and south of these latitudes, fatal heat waves might occur from time to time, but not so frequently, and not so fatally. So this was in some senses a regional problem. And every place had its regional problems. So when the funerals and the gestures of deep sympathy were done with, many people around the world, and their governments, went back to business as usual. And all around the world, the CO_2 emissions continued.

For a while, therefore, it looked like the great heat wave would be like mass shootings in the United States—mourned by all, deplored by all, and then immediately forgotten or superseded by the next one, until they came in a daily drumbeat and became the new normal. It looked quite possible that the same thing would happen with this event, the worst week in human history. How long would that stay true, about being the worst week? And what could anyone do about it? Easier to imagine the end of the world than the end of capitalism: the old saying had grown teeth and was taking on a literal, vicious accuracy.

But not in India. Elections were held and the nationalist nativist BJP party was thrown out of office as insufficient to the task, and partly responsible for the disaster, having sold the country to outside interests and burned coal and trashed the landscape in the pursuit of ever-growing inequality. The RSS disgraced and discredited at last as an evil force in Indian life. A new party was voted in, a composite party composed of all kinds of Indians, every religion and caste, urban poor, rural poor, the educated, all banded together by the disaster and determined to make something change. The ruling elite lost legitimacy and hegemony, and the inchoate fractured resistance of victims coalesced in a party called Avasthana, Sanskrit for survival. The world's biggest democracy, taking a new way. India's electrical power companies were nationalized where they weren't already, and a vast force was put to work shutting down coal-fired power plants and building wind and solar plants, and free-river hydro, and non–battery electrical storage systems to supplement the growing power of battery storage. All kinds of things began to change. Efforts were renewed to dismantle the worst effects of the caste system—these efforts had been made before, but now it was made a national priority, the new reality, and enough Indians were now ready to work for it. All over India, governments at all levels began to implement these changes.

Lastly, though this was regretted by many, some more radical portion of this new Indian polity sent a message out to the world: change with us, change now, or suffer the wrath of Kali. No more cheap Indian labor, no more sell-out deals; no deals of any kind, unless changes were made. If changes weren't made by the countries that had signed the Paris Agreement—and every nation had signed it—then this portion of India was now their enemy, and would break off diplomatic relations and do everything short of declaring military war. But economic war—yes, economic war. The world would see what this particular one-sixth of its population, formerly the working class for the world, could do. Time for the long post-colonial subalternity to end. Time for India to step onto the world stage, as it had at the start of history, and demand a better world. And then help to make it real.

Whether that kind of aggressive stance would be revealed as a true national position or the posturing of a radical faction remained to be seen. It depended, some thought, on how far India's new national government was willing to go to back up this Kali group's threats—to in effect unleash them. War in the age of the internet, the age of the global village, the age of drones, the age of synthetic biology and artificial pandemics—this was not the same as war in the past. If they were serious, it could get ugly. In fact, if even just the Kali faction of the Indian polity was serious, it could get very ugly.

But two could play at those games, indeed everyone could play those games—not just the 195 nations that had signed the Paris Agreement, but all the various kinds of non-state actors, right down to individuals.

And so came a time of troubles.

7

He was having panic attacks whenever he got hot, and then the panic attacks made him hotter still. Feedback loop for sure. When he was stabilized enough to move him, we flew him to Glasgow. He had spent a year abroad there, he said, and we thought that familiarity might help. He didn't want to go home to the States. So we took him to Glasgow and kept him cool, and took walks with him around the neighborhood at night. It was October and so the usual rain and raw sea air. That seemed to comfort him.

One night I was out there walking the streets with him, letting him take the lead. He hardly ever said a word, and I let him be. On this night he was a little more talkative. He pointed out to me where he had gone to school, theaters he had frequented. Apparently he had taken an interest in theater, done some work backstage with lighting and sets and costumes. Then when we found ourselves on Clyde Street, he wanted to walk out onto the pedestrian bridge that ran out over the river to the south bank.

Out there in the dark the city looked foursquare and massive. It's low for a city, not much different than it must have appeared a century or two ago. A little uncanny somehow, like a city in some dark fantasy. He stood there and looked down at the black water, elbows on the railing.

We talked about various things. At one point I asked him again if he would be going home.

No, he said sharply. I'm never going back there. It was the blackest look I ever saw on him. *Never*, he said.

I let it go. I didn't want to ask. We stood there leaning against the railing. It looked like the city was slowly floating in toward the hills.

So why did I survive? he said all of a sudden. Why just me, out of all those people?

I didn't know what to say. You just did, I said. Probably you were the healthiest person there. Maybe one of the biggest, I don't know. You aren't that big, but maybe bigger than most Indians.

He shrugged. Not really.

Even a bit more body mass would help. You have to keep your core temperature under about 104. A few pounds could help with that. And a lifetime of better food and medical care. And you're a runner, right?

I was a swimmer.

That probably helped. Stronger heart, thinner blood. That sort of thing. Ultimately I think it just means you were the strongest person there, and only the strongest survived.

I don't think I was the strongest person there.

Well, maybe you were better hydrated? Or you stayed in the water more? They said they found you by the lake.

Yes, he said. Something I said had troubled him. He said, I did stay submerged as much as I could. Just my face up there to breathe, all night long. But a lot of people were doing that.

It added up to survival, I said. You made it. You were lucky.

Don't say that.

I don't mean lucky. It was chance. I mean there's always an element of chance.

He looked at the dark low city, spangled with its night lights. It's just fate, he said. He put his forehead on the railing.

I put a hand on his shoulder. Fate, I agreed.

8

Humans are burning about 40 gigatons (a gigaton is a billion tons) of fossil carbon per year. Scientists have calculated that we can burn about 500 more gigatons of fossil carbon before we push the average global temperature over 2 degrees Celsius higher than it was when the industrial revolution began; this is as high as we can push it, they calculate, before really dangerous effects will follow for most of Earth's bioregions, meaning also food production for people.

Some used to question how dangerous the effects would be. But already more of the sun's energy stays in the Earth system than leaves it by about 0.7 of a watt per square meter of the Earth's surface. This means an inexorable rise in average temperatures. And a wet-bulb temperature of 35 will kill humans, even if unclothed and sitting in the shade; the combination of heat and humidity prevents sweating from dissipating heat, and death by hyperthermia soon results. And wet-bulb temperatures of 34 have been recorded since the year 1990, once in Chicago. So the danger seems evident enough.

Thus, 500 gigatons; but meanwhile, the fossil fuels industry has already located at least 3,000 gigatons of fossil carbon in the ground. All these concentrations of carbon are listed as assets by the corporations that have located them, and they are regarded as national resources by the nation-states in which they have been found. Only about a quarter of this carbon is owned by private companies; the rest is in the possession of various nation-states. The notional value of the 2,500 gigatons of carbon that should be left in the ground, calculated by using the current price of oil, is on the order of 1,500 trillion US dollars.

It seems quite possible that these 2,500 gigatons of carbon might eventually come to be regarded as a kind of stranded asset, but in the meantime, some people will be trying to sell and burn the portion of it they own or control, while they still can. Just enough to make a trillion or two, they'll be saying to themselves—not the crucial portion, not the burn that pushes us over the edge, just one last little taking. People need it.

The nineteen largest organizations doing this will be, in order of size from biggest to smallest: Saudi Aramco, Chevron, Gazprom, Exxon-Mobil, National Iranian Oil Company, BP, Royal Dutch Shell, Pemex, Petróleos de Venezuela, PetroChina, Peabody Energy, ConocoPhillips, Abu Dhabi National Oil Company, Kuwait Petroleum Corporation, Iraq National Oil Company, Total SA, Sonatrach, BHP Billiton, and Petrobras.

Executive decisions for these organizations' actions will be made by about five hundred people. They will be good people. Patriotic politicians, concerned for the fate of their beloved nation's citizens; conscientious hard-working corporate executives, fulfilling their obligations to their board and their shareholders. Men, for the most part; family men for the most part: well-educated, well-meaning. Pillars of the community. Givers to charity. When they go to the concert hall of an evening, their hearts will stir at the somber majesty of Brahms's Fourth Symphony. They will want the best for their children.

9

Down in Zurich's Niederdorf, the old medieval district bordering the east side of the Limmat under the tower of the Grossmünster, Zwingli's austere warehouse of a cathedral, there were still some little bars tucked here and there, too stodgy to attract many tourists. Not that Zurich got many tourists in November. Rain was turning into sleet, and the old black cobbles in their pattern of overlapping fans were getting slippery. Mary Murphy glanced down a broader street that led to the river; there stood the construction crane that wasn't really a construction crane but rather a work of art, a sculptor's joke at the ubiquity of cranes in Zurich. The city was always rebuilding itself.

In one of the smallest bars she sat down with Badim Bahadur, her chief of staff, who was hunched over a whisky reading his phone. He nodded at her in a morose greeting, pushed the ice around in his glass.

"What's the word from Delhi?" she said as she sat across from him.

"It's to start tomorrow."

She nodded at the waiter, pointed at Badim's drink. Another whisky. "What's the reaction?"

"Bad." He shrugged. "Maybe Pakistan will bomb us, and we'll retaliate, and that will start a nuclear winter. That will cool the planet quite nicely!"

"I should think the Pakistanis would want this as much as anyone, or even more. A heat wave like the one that just happened could kill everyone there."

"They know that. They're just piling on. China is doing it too. We are now the pariah of the world, all for doing the needful. We're getting killed for getting killed."

"It's always that way."

"Is it?" He glanced out the window. "I don't notice Europe hurting too badly."

"This is Switzerland, not Europe. The Swiss stay out of shit like this, they always have. That's what you're seeing here."

"Is it so different in the rest of Europe?"

"They killed Greece for getting killed, remember? And the rest of southern Europe isn't doing much better. Ireland neither for that matter. We got killed by the Brits for centuries. Something like a quarter of all the Irish died in the famine, and about as many left the island. That was something."

"Post-colonials," Badim said.

"Yes. And of the same empire too. It's funny how England never seemed to pay too much of a price for its crimes."

"No one does. You pay for being the victim, not the criminal."

Her whisky arrived and she downed half of it. "We're going to have to figure out how to change that."

"If there is a way."

"Justice?"

Badim made a skeptical face. "What is that?"

"Come on, don't be cynical."

"No, I mean it. Consider the Greek goddess of justice. Bronze woman in a toga, with a blindfold covering her eyes to make her be fair. Her scales held up to measure the balance of crime and punishment, no consideration given to individual influence. But nothing ever really balances in those scales. If it's an eye for an eye, maybe. That will balance out. But if someone is killed, no. The murderer gets fined, or jailed for life—is that a real balance? No."

"And thus capital punishment."

"Which everyone agrees is barbaric. Because if killing is wrong, two wrongs don't make a right. And violence begets violence. So you try to find some equivalent, and nothing is equivalent. So the scales are never balanced. Particularly if one nation murders another nation for three centuries, takes all its goods and then says Oh, sorry—bad idea. We'll stop and all is well. But all is not well."

"Maybe India can get England to pay for this casting of dust."

He shrugged. "It costs like ten euros. I don't see why everyone isn't supporting it one hundred percent. The effect will only last three or four years at most, and during that time we can see what it does, and decide whether we should keep doing it or not."

"Lots of people think it will have knock-on effects."

"Like what?"

"You know them as well as I. If doing this stops the monsoon, you'll have doubled your own misery."

"So we decided to risk it! After that it's no one else's business."

"But it'll be a worldwide effect."

"Everyone wants the temperatures lowered."

"Not Russia."

"I'm not so sure. The sea ice is melting and the permafrost is thawing, that's half their country. If their rivers don't freeze, Siberia has no roads for nine months of the year. They're made for the cold there, they know that."

"There's cold and cold," Mary said.

"But it's colder than ever there, sometimes! You know that. No. They're just piling on, like everyone else. Someone takes the bull by the horns, grabs the wolf by the ears, and everyone takes that opportunity to stick knives in his back. I'm sick of it."

She took another sip. "Welcome to the world," she said.

"Well I don't like it." He downed his drink. "So what are we going to do? We're the Ministry for the Future. We have to take a stand on this."

"I know. We'll have to see what our scientists say about it first."

He gave her a look. "They will prevaricate."

"Well, they don't know enough now to make a considered judgment. So they'll say it's a good experiment, that we should run it and wait a decade and see what happens."

"As usual!"

"But that's science, right?"

"But we have to do more than the usual!"

"We'll say that. And I'm sure we'll end up backing India."

"With money?"

"Ten euros, sure! Cash on the barrel."

He laughed despite himself. But quickly his expression darkened.

"It isn't enough," he said. "What we're doing with this ministry. I'm telling you, it isn't enough."

Mary regarded him closely. This was a reproach. And he wasn't meeting her eye.

"Let's go for a walk," she suggested. "I've been sitting all day."

He didn't object. They polished off their drinks, paid and walked out into the twilight. Down to the crane statue and then upstream by the Limmat in its stone channel, the black surface of water sheeting past them, cracking the light reflected from the other side. Past the old stone cube of the Rathaus; as always, Mary marvelled that the entire city government could have been stuffed into such a small building. Then past the Odeon and across the big bridge spanning the lake outlet, to the tiny park on the other side, where the statue of Ganymede stood, his uplifted hand seeming to hold up the moon, low over the Zurichsee. This was a place she often came to; something in the statue, the lake, the Alps far to the south, combined in a way she found stirring, she couldn't say why. Zurich—life—she couldn't say. The world seemed a big place when she was here.

"Listen," she said to Badim. "Maybe you're right. Maybe there's no such thing as justice, in the sense of some kind of real reparation of a wrong. No eye for an eye, no matter what. Especially historical justice, or climate justice. But over the long haul, in some rough sense, that's what we have to try for. That's what our ministry is about. We're trying to set things up so that in the future, over the long haul, something like justice will get created. Some long-term ledger of more good than bad. Bending the arc and all that. No matter what happened before, that's what we can do now."

She pointed at Ganymede, holding his back hand aloft. The moon lay right there in it, as if he were about to throw it across the sky.

Badim sighed. "I know," he said. "I'm here to try." And the look in his eye—distant, intense, calculating, cold—told Mary that he would. It made her shiver to see it.

More relaxing for Mary, even entertaining, were her meetings with Tatiana Voznesenskaya, head of the ministry's legal division. They were in the habit of meeting on some mornings at the Utoquai schwimmbad, and if it was warm enough, changing into their bathing suits and swimming out into the lake, freestyling in tandem and then chatting as they did the breaststroke for a while, circling out there, looking at the city from that strange low offshore angle; then back in to shower and sit in the schwimmbad café over hot drinks. Tatiana was tall and dark, dramatic in that Russian way of pale blue eyes and fashion model cheekbones, of grim high spirits and fuligin black humor. She had gotten pretty high in the Russian state department before running afoul of some part of the power structure there and deciding she would be better off in an international agency. Her expertise in Russia had been international treaty law, which she now brought to bear in working to find allies and legal means to advance the cause of defending the generations to come. This she felt was mostly a matter of establishing situations where these generations to come were given legal standing, such that their currently existing lawyers could file suits and be heard by courts. Not easy, given the reluctance of any court to grant standing to anyone or anything outside the magic circle of the law as written. But Tatiana had experience with most of the already-existing international courts, and was now working with the Network of Institutions for Future Generations, and the Children's Trust, and many other groups, all to leverage the power given to the ministry by its origins in the Paris Agreement. Mary often felt that it was really Tatiana who should have been made the head of the ministry, that Mary's experience in Ireland and the UN had been rather lightweight compared to Tatiana's tough career.

Tatiana had waved this off when Mary once mentioned the thought over drinks. "No you are perfect! Nice Irish girl, everyone loves you! I would wreck everything at once, bashing around like a KGB thug. Which I am," she added with a dangerous glint in her eye.

"Not really," Mary said.

"No, not really. But I would wreck things. We need you at the top, getting us in the door. It's similar to legal standing, really. Less formal but just as important. You have to get people to listen to you before you can make your case. That's what you do—people listen to you. Then we can go to work."

"I hope so. Do you really think we can get significant legal standing for people who don't exist yet?"

"I'm not sure. On the one hand, the circle of inclusion has been growing over historical time, which is a kind of precedent. More kinds of people given standing, even ecologies given standing, as in Ecuador. It sets a pattern, and logically it holds water. But even if we succeed in that part, we have a second problem, maybe bigger, in the weakness of international courts generally."

"Do you think they're weak?"

Tatiana gave Mary a sharp look, as if to say Please be serious. "Nations agree to them only if they like their judgments. But judgments always side with one side or other, so the losing side is never pleased. And there is no sheriff for the world. So, the US does what it wants, and the rest of us also do what we want. The courts only work when some petty war criminal gets caught and everyone decides to look virtuous."

Mary nodded unhappily. The Indians' flouting of the Paris Agreement with their geoengineering, not much different legally than the general disregard for the Agreement's emission reduction targets, was just the latest example of this kind of behavior. "So what do you think we can do to improve that situation?"

Tatiana shrugged. "Rule of law is all we've got," she said darkly. "We tell people that and then try to make them believe it."

"How do we do that?"

"If the world blows up they'll believe it. That's why we got the international order we got after World War Two."

"Not good enough?" Mary suggested.

"No, but nothing is ever good enough. We just make do." Tatiana brightened, although Mary saw the sly look that indicated a joke: "We make a new religion! Some kind of Earth religion, everyone family, universal brotherhood."

"Universal sisterhood," Mary said. "An Earth mother religion."

"Exactly," Tatiana said, and laughed. "As it should be, right?"

They toasted the idea. "Write up the laws for that," Mary said. "Have them ready for when the time comes."

"Of course," Tatiana said. "I have entire constitution already, in here." And she tapped her forehead.

10

We took off from Bihta and Darbhanga and INS Garuda and Gandhi-nagar, mostly in Ilyushin IL-78s, bought long ago from the Soviet Union. We had some Boeing and Airbus refuelers too. They were old planes, and it was very cold inside them. Our suits were old too, they were hard to move in, and hardly anything as insulation. We got very cold up there, but the flights were relatively short.

We flew to sixty thousand feet, as high as the planes could get. Higher would have been better but we couldn't do it. It took a couple of hours, as we always carried a maximum load. Two planes got caught in the so-called coffin corner and stalled catastrophically, and one of the crews didn't get out.

Once up there we deployed the fuel lines and pumped the aerosols into the air. The plumes looked like dumped fuel at first, but they were really aerosol particulates, we were told mostly sulfur dioxide and then some other chemicals, like from a volcano, but there wasn't ash like in a volcanic explosion, it was a mix made to stay up there and reflect sunlight. Manufactured at Bhopal and elsewhere in India.

We flew most of our missions over the Arabian Sea, so the prevailing winds of late summer would carry the stuff over India before anywhere else. We wanted that, it was for us we were doing it, and some felt we might also avoid some criticism by doing it that way. But soon enough what we released would get carried by the winds all over the stratosphere, mostly in the northern hemisphere but eventually everywhere. There it would be deflecting some sunlight.

Even in India you could hardly see any difference in the sky. For all our lives we were living under the ABC, the Asian Brown Cloud, so we were used to dusty skies. Our operation only made things a little whiter by day, and the sunsets were sometimes more red than before. Quite beautiful on certain days. But mostly things looked the same. The sunlight we deflected to space was said to be about a fifth of one percent of the total incoming. Very important crucial stuff, but it's not really possible to see a difference that small.

Global effect was said to be like Pinatubo's eruption in 1991, or some said a double Pinatubo. The total release was taken to the stratosphere in several thousand individual missions. We had a fleet of only two hundred planes, so we each went up scores and scores of times, spread out over seven months. That was a lot of work. Of course it was a pretty small effort as these things go. And if it helped to prevent another heat wave, it was worth doing.

We knew the Chinese hated the idea, and Pakistan of course, and although we flew only when the jet streams were running toward the east or northeast, there were times when those countries lay in the path of dispersion. And all over the world people pointed out that the ozone layer would get hurt, which would be bad for everyone. Once a heat-seeking missile flew right by our plane, Vikram dodged it at the last minute, the plane squealed like a cat. No one ever found out who shot it at us. But we didn't care. We did what we were told, we were happy to do it. Everyone had lost someone they knew in the heat wave. Even if they hadn't, it was India. And it could happen again, anywhere in India and really anywhere in the world. As our officials told people, over and over. Even farther north a heat wave could strike. Europe once suffered one that killed seventy thousand people, even though Europe is so far north. Well more than half the land on Earth is at risk. So we did it.

Day after day for seven months. And round-the-clock, what with maintenance and refueling, and the filling of the tanks. It was a routine that took many thousands of people working together. We got tired, exhausted, but also we got into the rhythm of it. There were enough crews to fly once out of every three missions per plane. For many weeks in the middle of it, it felt like it would go on forever. That it was all we were ever meant to do. We felt like we were saving India, and maybe saving the world. But it was India we were concerned with. No more deadly heat waves. So we hoped. It was a very emotional time.

Now, if I go anywhere in the world, and if someone speaks against what we did, I challenge them. You don't know anything, I tell them. It wasn't your people, so you don't care. But we know and we care. And there hasn't been a heat wave like that since. One may come again, no doubt of that, but we did what we could. We did the right thing. I must admit, I sometimes shout at people if they deny that. I damn them to hell. Which is a place we in India have already seen. So I have no patience for people who object to what we did. They don't know what they're talking about. They haven't seen it, and we have.

11

Ideology, n. An imaginary relationship to a real situation.

In common usage, what the other person has, especially when systematically distorting the facts.

But it seems to us that an ideology is a necessary feature of cognition, and if anyone were to lack one, which we doubt, they would be badly disabled. There is a real situation, that can't be denied, but it is too big for any individual to know in full, and so we must create our understanding by way of an act of the imagination. So we all have an ideology, and this is a good thing. So much information pours into the mind, ranging from sensory experience to discursive and mediated inputs of all kinds, that some kind of personal organizing system is necessary to make sense of things in ways that allow one to decide and to act. Worldview, philosophy, religion, these are all synonyms for ideology as defined above; and so is science, although it's the different one, the special one, by way of its perpetual cross-checking with reality tests of all kinds, and its continuous sharpening of focus. That surely makes science central to a most interesting project, which is to invent, improve, and put to use an ideology that explains in a coherent and useful way as much of the blooming buzzing inrush of the world as possible. What one would hope for in an ideology is clarity and explanatory breadth, and power. We leave the proof of this as an exercise for the reader.

12

Bypassing the imaginary relationship part for a moment, what about the real situation? Unknowable, of course, as per above. But consider this aspect of it:

Recent extinctions include the Saudi gazelle, the Japanese sea lion, the Caribbean monk seal, the Christmas Island pipstrelle, the Bramble Cay melomys, the vaquita porpoise, the Alagoas foliage-gleaner, the cryptic treehunter, Spix's macaw, the po'ouli, the northern white rhino, the mountain tapir, the Haitian solenodon, the giant otter, Attwater's prairie chicken, the Spanish lynx, the Persian fallow deer, the Japanese crested ibis, the Arabian oryx, the snub-nosed monkey, the Ceylon elephant, the indris, Zanzibar's red colobus, the mountain gorilla, the white-throated wallaby, the walia ibex, the aye-aye, the vicuna, the giant panda, the monkey-eating eagle, and an estimated two hundred more species of mammals, seven hundred species of birds, four hundred species of reptiles, six hundred species of amphibians, and four thousand species of plants.

The current rate of extinctions compared to the geological norm is now several thousandfold faster, making this the sixth great mass extinction event in Earth's history, and thus the start of the Anthropocene in its clearest demarcation, which is to say, we are in a biosphere catastrophe that will be obvious in the fossil record for as long as the Earth lasts. Also the mass extinction is one of the most obvious examples of things done by humans that cannot be undone, despite all the experimental de-extinction efforts, and the general robustness of life on Earth. Ocean acidification and deoxygenation are other examples of things done by humans that we can't undo, and the relation between this ocean acidification/deoxygenation and the extinction event may soon become profound, in that the former may stupendously accelerate the latter.

Evolution itself will of course eventually refill all these emptied ecological niches with new species. The pre-existing plenitude of speciation will be restored in less than twenty million years.

13

Anytime he broke a sweat his heart would start racing, and soon enough he would be in the throes of a full-on panic attack. Pulse at 150 beats a minute or more. It didn't matter that he knew he was safe, and that this panic reaction was to something that had happened long before. It didn't matter that he lived outside Glasgow now, and had a job in a meat-processing plant that gave him access to refrigerated rooms where the temperature was kept just a few degrees above freezing. By the time an attack started it was too late; his body and mind would be plunged instantly into another terrible tornado of biochemicals, pounding through his arteries like crystal meth at its paranoid worst.

This was what people called post-traumatic stress disorder. He knew that, he had been told it many times. PTSD, the great affect of our time. As one of his therapists had once explained to him, one of the identifying characteristics of the disorder was that even when you knew it was happening to you, that didn't stop it from happening. In that sense, the therapist admitted, the naming of it was useless. Diagnosis was necessary but not sufficient; and what might be sufficient wasn't at all clear. There were differing opinions, differing outcomes. No treatment had been shown to be fully effective, and most were still largely experimental procedures.

Exposure to events like the event: no.

He had tried this on a visit to Kenya, out every day in temperatures that crept closer and closer to the unlivable. That had resulted in daily panic attacks, and he had curtailed the trip and gone back to Glasgow.

Virtual environments in which to explore aspects of the event: no. He had played video games that made him relive parts of the experience in ways he could control, but these games were as ugly as any M-rated travesty. Panic attacks were frequent, during the games or not.

Rehearsal therapy: he had written accounts of the event while on beta-blockers, repeating this exercise over and over. Sleepy all the time because of the beta-blockers, memoir as automatic writing: *I tried to get people inside. I had a water tank in the closet. I hid what I had. The pistol barrel was a little black circle. Everybody was dead.*

No good. Just more blurry sick gasping panic attacks, more nightmares.

About half the time he fell asleep, he had nightmares that woke him in a cold sweat. Sometimes the images were sadistically cruel. After waking from one of these dreams he had to try to warm back up, so he would wiggle his cold toes, toss and turn, try to forget the dream, try to get back to sleep; but it took hours, and sometimes it didn't work. The next day he would operate like a slow zombie, get through the day by working mindlessly, or by playing video games, mostly games in which he bounced from point to point in a low-g environment. Asteroid hopping.

His therapists talked about trigger events. About avoiding triggers. What they were glossing over with this too-convenient metaphor was that life itself was just a long series of trigger events. That consciousness was the trigger. He woke up, he remembered who he was, he had a panic attack. He got over it and got on with his day as best he could. The command not to think about certain things was precisely a mode of thinking about that thing. Repression, forgetting; he had to learn to forget. Perpetual distraction was impossible. He wanted to get better but he couldn't.

Cognitive behavioral therapy was accepted by many as the best way to deal with PTSD. But CBT was hard. He pursued it like a religious calling, like a sidewalk over the abyss. One therapist said to him, when he used that phrase, everyone walks that sidewalk over the abyss, that's life. Mine's a tightrope, he replied. Focusing on balance was necessary at all times. In that sense, distractions were actually contra-indicated. If you were distracted enough, then a single misstep could send you plunging into the abyss. Constant vigilance—but this too was bad, as just another way of thinking about it, of paying attention to it. No. Hypervigilance was part of the disorder. So there was no way out. No way out but dreamless sleep. Or death.

Or certain drugs. Anti-anxiety drugs were not the same as antidepressants. They were meant to foil the brain's uptake of fight-or-flight stimulants. To give consciousness a bit of time to calm the system down by realizing there was no real danger. These drugs had unwanted side effects, sure. Flatness of affect, yes. It was even part of what you wanted. If you killed all feelings, the bad feelings would necessarily be among them and therefore less likely to appear, even if they were the first ones in line, ready to pop. But if you did accomplish that flattening of all feeling, then what? March through life like an automaton, that's what. Eat like fueling an old car. Exercise hoping to get so tired you could fall asleep and make it through the night. Try not to think. Try not to feel.

So, after many months of that, after years of that, he went back to India.

He needed to try it to see if it would help. Until he tried it, he wouldn't know. It would be something like aversion therapy, or rather immersion therapy. Go right back to the scene of the crime. Plus he had an idea that had begun to obsess him. He had a plan.

He landed at Delhi, took the train to Lucknow, got off at the station and got on a crowded bus out to his town. The sights and smells, the heat and humidity—they were all triggers, yes. But since consciousness was the real trigger, he steeled himself and looked out the bus's dusty window and felt the sweat pouring out of his skin, felt the air pulsing in and out of his lungs, felt his heart pound inside him like a child trying to escape. Take it! Live on!

He got off at the bus stop in the town's central square. He stood there looking around. People were everywhere, all ages, Hindu and Muslim as before, the differences subtle and sometimes not there at all, but his eye had been trained to note the signs—a tikka, a particular rounded cap. The usual mix that this town had always had, back to Akbar and before. All appeared to him as it had been four years earlier. There was no sign that what had happened had ever happened.

Surely there must be a memorial at least. He walked toward the lake, feeling his heart hammer, his skin burn. His clothes were soaked with his sweat, he drank from the water bottle he had in his daypack, just a sip each time, and yet soon it was empty. Everything pulsed, his eyes stung with sweat, behind his wrap-around sunglasses he was weeping furiously. The polarization of the glasses was not enough to keep bursts of light from shattering in his retinas. Sights incoming like needles in the eyeball, everywhere he looked.

The lake was the same. How could it be the same, how could they not have drained it, built over the site with some mausoleum or temple or just an apartment block or a bazaar?

Then again, who was there to remember what had happened here, or what it had been like that week? There were no survivors to be haunted by this place. As for those who had come in and cleaned it up, disposed of the bodies, well, it had been just one town of many, all of them the same. There was no reason to fixate on this one. No—he was the sole survivor. No one else had seen what he had seen and survived to remember it. For all the people walking on this crowded narrow sidewalk, this sad little pastiche of a corniche, it was only what it was today. No sign of a plaque, much less a memorial in the old martial style.

He walked back to the bus station. After some thought, and a trip to an open-walled store to buy more bottled water, he walked to his old offices. The building was still there. The offices were occupied by lawyers and accountants, and a dentist. Next door was a Nepali restaurant that hadn't been there before. It was just a building. What had happened up there in those rooms—

He sat on the curb, suddenly too weak to stand. He was quivering, he put his head in his hands. It was all there in his head—every hour of it, every minute of it. The water tank in that closet.

He got up and walked back to the bus station and took the next bus back to Lucknow. There he called a number he had been given. A man answered in Hindi. In his bad Hindi he asked if he could speak in English, then switched over when the man said, "Yes, what?"

"I was there," Frank said. "I was here, during the heat wave. I'm an American, I was with an aid group, here doing development work. I saw what happened. Now I'm back."

"Why?"

"I have a friend who told me about your group."

"What group?"

"I was told it's called Never Again. Devoted to various kinds of direct action?"

Silence from the other end.

"I want to help," Frank said. "I need to do something."

More silence. Finally the man said, "Tell me where you are."

★   ★   ★

He sat outside the train station for an hour, miserably hot. When it seemed he would wilt and fall, a car drove up to the curb and two young men jumped out and stood before him. "You are the firangi who called?"

"Yes."

One of them waved a wand over him while the one who had spoken patted him down.

"All right, get in."

When he was in the front seat passenger side, the driver took off with a squeal. The men behind him blindfolded him. "We don't want you to know where we are taking you. We won't harm you, not if you are what you say you are."

"I am," Frank said, accepting the blindfold. "I wish I weren't, but I am."

No replies. The car made several turns, fast enough to throw Frank against the door or the restraint of his seatbelt. A little electric car, quietly humming, and quick to speed up or slow down.

Then the car stopped and he was led out and up some steps. Into a building. His blindfold was removed; he was standing in a room filled with young men. There was a woman there too, he saw, alone among a dozen men. All of them regarded him curiously.

He told them his story. They nodded grimly from time to time, their gleaming eyes fixed on him. Never had he been looked at so intently.

When he was done they glanced at each other. Finally the woman spoke: "What do you want now?"

"I want to join you. I want to do something."

They spoke in Hindi among themselves, more quickly than he could follow. Possibly it was another language, like Bengali or Marathi. He didn't recognize a single word.

"You can't join us," the woman told him after they were done conferring. "We don't want you. And if you knew about everything that we did, you might not want us. We are the Children of Kali, and you can't be one of us, even if you were here during the catastrophe. But you can do something. You can carry a message from us to the world. Maybe that can even help, we don't know. But you can try. You can tell them that they must change their ways. If they don't, we will kill them. That's what they need to know. You can figure out ways to tell them that."

"I'll do that," Frank said. "But I want to do more."

"Do more then. Just not with us."

Frank nodded, looked at the floor. He would never be able to explain. Not to these people, not to anyone.

"All right then. I'll do what I can."

14

We had to leave. It was too dangerous to stay.

I was a doctor, I ran a small clinic with an assistant and three nurses and a couple who ran the office. My wife taught piano and my children went to school. Then rebels from our area began fighting the government and troops moved into town, and people were being killed right on the street. Even some kids from the school my children went to. And one day our clinic was blown up. When I went to it and saw the wreckage, looked from the street right into my examination room, I knew we had to leave. Somehow we were on the wrong side.

I went to a friend who had been a journalist before the war and asked him if he could put me in touch with a smuggler who would get us on the way to someplace safe. I did not have any particular idea about where that might be. Anyplace was going to be safer than where we were. When my friend understood what I was asking him, he rounded the table and gave me a hug. I'm sorry it's come to this, he said. I will miss you. This stuck me like a knife in the heart. He knew what it meant, this move. I didn't know, but he did. And when I saw this, saw what he knew in his face, I sat down on my chair as if shot. My knees buckled. People say this as a figure of speech, but really it is a very accurate account of what happens when you get a big shock. It's something in the body, a physiological thing, although I can't explain the mechanism.

The smuggler was expensive, so much so that only those who had some considerable savings would be able to leave by using one. Most of my townspeople were stuck. But we could afford it. So I met my friend one night at our usual café, and he had a man with him. The man was polite but distant. Professional. He asked to see my money, asked about my family, when I could be ready, that sort of thing. He said he could get us into

Turkey and then Bulgaria, and after that Switzerland or Germany. I went to the bank and withdrew the money, then went home and told my wife, and we told the children to pack one suitcase each, that we were going on a trip. That night at midnight a car pulled up to our apartment curb and we went downstairs and put our suitcases in the trunk and piled in the back of the car. As we drove off I looked out the car window at our apartment and realized I would never see it again. All that was over. I had had my routines, I liked to go down to the café after work or late at night when it had cooled, drink coffee and play backgammon and talk to friends. My wife and I got together with a few couples, made meals and watched their kids. We knew the people who ran the grocery and the local stores. We had all that, just like anyone. I remember what it was like. But just barely.

15

Taking notes for Badim on regular Monday meeting of ministry executive group. I'll clean these up later to give to him.

Mary Murphy, convening her leadership team in seminar room next to her office, Hochstrasse. I should have gone to the bathroom.

Badim on Mary's left, then thirteen division heads seated around the table, the rest of us behind them against the walls. George is going to fall asleep.

Tatiana V., legal. Just heard this morning World Court declined to take up her Indian case. Not happy.

Imbeni Halle. Infrastructure. Poached from Namcor.

Jurgen Atzgen. Zurcher. Gets to commute from his house down lake. Insurance and re-insurance. Swiss Re vet.

Bob Wharton, nat cat. American ecologist. Mitigation and adaptation.

Climate lead Adele Elia. French, coordinating our climate science. Started as a glaciologist, hates meetings like this. Once said so right in meeting. Lived eight years on glaciers, she said. Wants back there. As for world cryosphere, it's still melting.

Huo Kaming, ecologist, Hong Kong. Biosphere studies, habitat restoration, refugia creation, animal protection, rewilding, biologically based carbon drawdown, watershed governance, groundwater recharge, the commons, the Half Earth campaign. She can do it all.

Estevan Escobar. Chilean. Oceans. Prone to despair.

Elena Quintero, agriculture. Buenos Aires. She and Estevan joke about Argentina–Chile rivalry. She cheers him up very skillfully.

Indra Dalit, Jakarta. Geoengineering. Works with Bob and Jurgen.

Dick Bosworth, Australian, economist. A card. Taxes and political economy. Our reality check.

Janus Athena, AI, internet, all things digital. Very digital themself.

Esmeri Zayed. Third of the E gals. Jordanian Palestinian. Refugees, liaison to UNHCR.

Rebecca Tallhorse, Canada. Indigenous peoples' rep and outreach.

Mary starts meeting by asking for new developments.

Imbeni: Looking into plans to redirect fossil fuel companies to do decarbonization projects. Capabilities strangely appropriate. Extraction and injection both use same tech, just reversed. People, capital, facilities, capacities, all these can be used to "collect and inject," either by way of cooperation or legal coercion. Keeps oil companies in business but doing good things.

Tatiana looks interested. Rest of group looking skeptical. Carbon capture and reinsertion into empty oil wells are both dubious as a reality.

Mary: Look into it more. We've got to have it, from what the calculations say about how much the natural methods can grab.

Jurgen: Insurance companies in a panic at last year's reports. Pay-outs at about one hundred billion USD a year now, going higher fast, as in hockey stick graph. Insurance companies insured by re-insurance. These now holding short end of stick (tall end of stick?). Can't charge premiums high enough to cover pay-outs, nor could anyone afford to pay that much. Lack of predictability means re-insurance companies simply refusing to cover environmental catastrophes, the way they don't insure war or political unrest etc. So, end of insurance, basically. Everyone hanging out there uninsured. Governments therefore payer of last resort, but most governments already deep in debt to finance, meaning also re-insurance companies. Nothing left to give without endangering belief in money. Entire system therefore on brink of collapse.

Mary: What mean collapse?

Jurgen: Mean, money no longer working as money.

Silence in room. Jurgen adds, So you can see why re-insurance hoping for some climate mitigation! We can't afford for world to end! No one laughs.

Bob Wharton: Some things we can mitigate, some we can't. Some things we can adapt to, others we can't. Also, we can't adapt to some things we are now failing to mitigate. Need to clarify which is which. Mainly need to tell adaptation advocates they're full of shit. Bunch of economists, humanities professors, they have no idea what talking about. Adaptation just a fantasy.

Mary halts Bob rant very skillfully. Sympathetic squint as she chops air with hand. Preaching to choir, she suggests. Moves along to Adele and the rest.

Adele: You think that's bad! Joke gets laugh. The big Antarctic glacial basins, mainly Victoria and Totten, hold ice sliding downhill faster and faster. Will soon be depositing many thousands of cubic kilometers of ice into sea. Now looking like could happen in a few decades. Sea level rise two meters for sure, maybe more (six meters!) but two meters enough. Doom for all coastal cities, beaches, marshes, coral reefs, many fisheries. Would displace ten percent of the world's population, disrupt twenty percent food supply. Like a knock-out punch to dazed fighter. Civilization kaputt.

Jurgen throws up hands. Cost of this cannot be calculated!

Calculate it, Mary orders him.

J. frowns, pondering big picture in his head. A quadrillion. Yes, really. A thousand trillion is not too high. Maybe five quadrillion.

Dick: So just call it infinity.

Adele: Number of species threatened with extinction now at Permian levels. (Piling on here?) Permian the worst extinction ever. Now on course to match it.

Kaming: Ninety-nine percent of all meat alive is made of humans and their domestic beasts. Cattle, pigs, sheep, goats. Wild creatures one percent of meat alive. And suffering. Many species gone soon.

M: Soon?

K: Like thirty years.

Estevan: Only twenty percent of the fish now in oceans are wild fish.

Mary ends discussion, chop chop. Regards team. Speaks slowly.

MftF has budget 60B USD/year. Big. But world GDP 100 trillion/year. Half that GWP is so-called consumer spending by prosperous people, means non-essential buying of things that degrade biosphere. Ship going down. Parasite killing host. Even the productive half of GWP, food and health and housing, burning up world. In short: fucked.

Team watches her.

So. Have to find ways to spend our sixty billion that strike at leverage points.

Dick: Our money not enough to matter. Have to change laws—that's our leverage point. Spend our money on changing laws.

Tatiana likes this.

Imbeni: Critical infrastructure needs funding.

Elena: Ag improvements.

Mary chops discussion. Chop chop chop! Stop. We need to lever change, and fast. However we can. By whatever means necessary.

Badim surprised by this last statement, I'm not sure why. Looks at Mary, surprised.

16

Possibly some of the richest two percent of the world's population have decided to give up on the pretense that "progress" or "development" or "prosperity" can be achieved for all eight billion of the world's people. For quite a long time, a century or two, this "prosperity for all" goal had been the line taken; that although there was inequality now, if everyone just stuck to the program and did not rock the boat, the rising tide would eventually float even the most high-and-dry among them. But early in the twenty-first century it became clear that the planet was incapable of sustaining everyone alive at Western levels, and at that point the richest pulled away into their fortress mansions, bought the governments or disabled them from action against them, and bolted their doors to wait it out until some poorly theorized better time, which really came down to just the remainder of their lives, and perhaps the lives of their children if they were feeling optimistic—beyond that, *après moi le déluge*.

A rational response to an intractable problem. But not really. There was scientifically supported evidence to show that if the Earth's available resources were divided up equally among all eight billion humans, everyone would be fine. They would all be at adequacy, and the scientific evidence very robustly supported the contention that people living at adequacy, and confident they would stay there (a crucial point), were healthier and thus happier than rich people. So the upshot of that equal division would be an improvement for all.

Rich people would often snort at this last study, then go off and lose sleep over their bodyguards, tax lawyers, legal risks—children crazy with arrogance, love not at all fungible—over-eating and over-indulgence generally, resulting health problems, ennui and existential angst—in short, an insomniac faceplant into the realization that science was once again right, that money couldn't buy health or love or happiness. Although it has to be added that a reliable sufficiency of money is indeed necessary to scaffold the possibility of those good things. The happy medium, the Goldilocks zone in terms of personal income, according to sociological analyses, seemed to rest at around 100,000 US dollars a year, or about the same amount of money that most working scientists made, which was a little suspicious in several senses, but there it stood: data.

And one can run the math. The 2,000 Watt Society, started in 1998 in Switzerland, calculated that if all the energy consumed by households were divided by the total number of humans alive, each would have the use of about 2,000 watts of power, meaning about 48 kilowatt-hours per day. The society's members then tried living on that amount of electricity to see what it was like: they found it was fine. It took paying attention to energy use, but the resulting life was by no means a form of suffering; it was even reported to feel more stylish and meaningful to those who undertook the experiment.

So, is there energy enough for all? Yes. Is there food enough for all? Yes. Is there housing enough for all? There could be, there is no real problem there. Same for clothing. Is there health care enough for all? Not yet, but there could be; it's a matter of training people and making small technological objects, there is no planetary constraint on that one. Same with education. So all the necessities for a good life are abundant enough that everyone alive could have them. Food, water, shelter, clothing, health care, education.

Is there enough security for all? Security is the feeling that results from being confident that you will have all the things listed above, and your children will have them too. So it is a derivative effect. There can be enough security for all; but only if all have security.

If one percent of the humans alive controlled everyone's work, and took far more than their share of the benefits of that work, while also blocking the project of equality and sustainability however they could, that project would become more difficult. This would go without saying, except that it needs saying.

To be clear, concluding in brief: there is enough for all. So there should be no more people living in poverty. And there should be no more billionaires. Enough should be a human right, a floor below which no one can fall; also a ceiling above which no one can rise. Enough is a good as a feast—or better.

Arranging this situation is left as an exercise for the reader.

17

Today we're here to inquire who actually enacts the world's economy—who are the ones who make it all go, so to speak. Possibly these people constitute a minority, as it is often said that most people alive today would actively welcome a change in the system.

Only a stupid person would say that.

Well, and yet I've just said it.

Yes.

But to get back to the question in hand, who do we think actually enacts the market as such? By which I mean to say, who theorizes it, who implements it, who administers it, who defends it?

The police. It being the law.

So but can we then assume that those people who make the laws are deeply implicated?

Yes.

But lawmakers are often lawyers themselves, notoriously bereft of ideas. Can we assume they get their ideas about law from others?

Yes.

And who are some of those others?

Think tanks. Academics.

Meaning MBA professors.

All kinds of academics. And very quickly their students.

Economics departments, you mean.

The World Trade Organization. Stock markets. All the laws, and the politicians and bureaucrats administering the laws. And the police and army enforcing them.

And I suppose the CEOs of all the companies.

Banks. Shareholder associations, pension funds, individual shareholders, hedge funds, financial firms.

Might the central banks indeed be central to all this?

Yes.

Anyone else?

Insurance companies, re-insurance companies. Big investors.

And their algorithms, right? So, mathematicians?

The math is primitive.

And yet even primitive math still takes mathematicians, the rest of us being so clueless.

Yes.

Also I suppose simply prices themselves, and interest rates and the like. Which is to say simply the system itself.

You were asking about the people doing it.

Yes, but it's an actor network. Some of the actors in an actor network aren't human.

Balderdash.

What, you don't believe in actor networks?

There are actor networks, but it's the actors with agency who can choose to do things differently. That's what you were trying to talk about.

All right, but what about money?

What about it?

To my mind, money acts as if it worked as gravity does—the more of it you gather together, the more gathering power it exerts, as with mass and its gravitational attraction.

Cute.

Ultimately this is a very big and articulated system!

Insightful.

All right then, back to the ones who administer our economic system as such, and teach others how to work it, and by a not-so-coincidental coincidence, benefit from it the most. I wonder how many people that would turn out to be?

About eight million.

You're sure?

No.

So this would be about one in every thousand persons alive today.

Well done.

Thank you! And the programs they've written.

Stick to the people.

But if the non-human elements of the system were to break?

Stick to the people. You were almost getting interesting.

Who matters the most in that group of eight million?

Government legislators.

That's a bad thought.

No it isn't. Why would you say that?

Corruption, stupidity—

Rule of law.

But—

But me no buts. Rule of law.

What a weak reed to stand on!

Yes.

What can we do about that?

Just make it stick.

18

The PTSD model uses the word "trigger" as both noun and verb, to suggest the speed of a PTSD reaction, and the way it can be switched on by some incident that should be the equivalent of a small curved piece of metal, innocuous except when placed in a gun. One must learn not to pull these things.

Cognitive behavioral therapy is a hard thing to learn. One of the main strategies involved asks you to label the type of thought you are having, identify it as unhelpful or painful, and then switch tracks to a more positive train of thought. Often this strategy fails. You know what's happening, you know it's inappropriate—on it goes anyway. Your palms sweat, your heart pounds in your chest like a child trying to escape, and over that throbbing animal reality, you can be thinking to yourself, Wait, no danger here now—this isn't a situation to be frightened in, you're just sitting at a café table, midday, light wind, low clouds, all well, please don't do this, don't start crying, don't leap up and run away—just still your shaking hands, just pick up your coffee cup—

But the trigger is pulled, and you are looking right down the barrel of the gun.

Enough times like that, and looking down the actual barrel of an actual gun, its trigger under your actual thumb—not your forefinger, because you have the gun pointed at yourself, resting against your sternum, and it's the thumb that can best pull (or in this case push) the trigger—this can be seen as a huge relief, as a promise that the fear will finally stop. This happens all the time. It happens so often that one form of PTSD therapy goes like this—you don't have to worry so much, because if it stays this bad you can always kill yourself. And for some sufferers this thought is a real comfort, sometimes even the anchor point of a way back to sanity.

You can always end this misery by killing yourself; so give it another day and see how it goes.

It's not easy to stay unafraid. It can't always be done. Try as you might, want it ever so much, things are out of your control, even when they are in your mind, or especially because they are in your mind. The mind is a funny animal. If it were just conscious thought; or if conscious thought was something we could control; or if unconscious thoughts were conscious; or if moods were amenable to our desires...then maybe things could work. Things like cognitive behavioral therapy, or the project of sanity itself. Just make it happen!

But no. You're swimming in a river. You can get carried out to sea on riptides not of your making, or at least not under your control. You can find yourself swimming against a current much stronger than you. You can drown.

Frank was drowning. He had that same shortness of breath. Therapy had mostly made it crystal clear to him that he would never be cured.

In a sense, maybe that was progress. Abandon all hope, ye who enter here. What did that mean? Could you live without hope? There was a Japanese saying he read in a book: live as if you were already dead. But what did that mean, why would that be encouraging? Was it encouraging? It was enigmatic at best, a kind of double bind—first an injunction to live; but second, "as if you were already dead." How would one do that? Was it part of the samurai code, were you to be careless of your own life in defense of whomever you were charged with protecting? So, a kind of servant's stoicism? To let yourself be used like a shield, to become a human tool? Maybe so. In which case it was a matter of crushing your hopes into the proper channel.

Thus, in his case, no more hoping that he would become normal again. That he would live a normal life. That what had happened would not have happened. Forget all that. Therapy taught him to give up those hopes. Hope would have to reside in something like this: hope to do some good, no matter how fucked up you are.

This was worth writing down on a piece of paper, in shaky block letters, and then pinning the paper onto the mirror in his bathroom, along with various other encouragements in the form of phrases or images;

probably it looked like a madman's mirror, but he wanted to put things up there.

HOPE TO DO SOME GOOD, NO MATTER HOW FUCKED UP YOU ARE

Every time he remembered to brush his teeth or shave, which was getting less and less often, he would see that sign and ponder what he might do. This mainly made him feel confused. But it did seem like the urge to do something was there in him, sometimes so strong it was like a bad case of heartburn. When he was exhausted by sleeplessness, or groggy with too much sleep, that burn still sometimes struck him, radiating outward from his middle. He *had* to do something. Maybe he wasn't going to get to be a Child of Kali, clearly not, but something like that. A fellow traveler. A warrior for the cause. A lone assailant.

Over breakfast, soothing his stomach with a little plastic tub of yogurt, he pondered what he might do. One person had one-eight-billionth of the power that humanity had. This assumed everyone had an equal amount of power, which wasn't true, but it was serviceable for this kind of thinking. One-eight-billionth wasn't a very big fraction, but then again there were poisons that worked in the parts-per-billion range, so it wasn't entirely unprecedented for such a small agent to change things.

He wandered the streets of Glasgow, thinking it over. Up and down the hills to the east and north, enjoying the sidewalks so steep they had staircases incised into them. You could work off a lot of stress walking the streets of Glasgow, and the views kept changing their perspectives under the changing weather, reflecting the storms within, the fear, the sudden bursts of exhilaration, the black depths of ocean-floor grief. Or beautiful dreams, the world gone right. How to share that? How give? Saint Francis of Assisi: give yourself away, give up on yourself and all you thought you had. Feed the birds, help people. The positive of that was so obvious. Do like Saint Francis. Help people.

But he wanted more. He could feel it burning him up: he wanted to kill. Well, he wanted to punish. People had caused the heat wave, and not all people—the prosperous nations, sure, the old empires, sure; they all deserved to be punished. But then also there were particular people, many still alive, who had worked all their lives to deny climate change, to keep burning carbon, to keep wrecking biomes, to keep driving other species extinct. That evil work had been their lives' project, and while pursuing that project they had prospered and lived in luxury. They wrecked the world happily, thinking they were supermen, laughing at the weak, crushing them underfoot.

He wanted to kill all those people. In the absence of that, some of them would do. He felt the urge burning him from the inside out. He wouldn't live long with that kind of internal stress, he could feel that as surely as he could feel his triphammer heart pumping over-pressured blood through his carotids. Oh yes, high blood pressure. He could feel it trying to burst him from inside. Something in there would break. But first, some kind of action. Vengeance, yes; but also, preemption. A preemptive strike. This might stop some bigger bad from happening.

Twice a week he visited his therapist. A nice middle-aged woman, intelligent and experienced, calm and attentive. Sympathetic. She was interested in him, he could see that. Probably she was interested in all her clients. But for sure she was interested in him.

She asked him what he was doing, how he was feeling. He didn't tell her about his dreams of vengeance, but what he did tell her was honest enough. Earlier that week, he told her, a hot waft of steam from a giant espresso machine in a coffee emporium had caused him to freak out. Panic attack; he had had to sit down and try to calm his beating heart.

She nodded. "Did you try the eye movements we talked about?"

"No." He was pretty sure this was a bullshit therapy, but the truth was that in the heat of the moment, so to speak, he had forgotten about it. "I forgot. I'll try it next time."

"It might help," she said. "It might not. But nothing lost in trying it."

He nodded.

"Do you want to try it now?"

"Just move my eyes?"

"Well, no. You need to do it when you're dealing with what happened. I don't want you to re-experience anything in a way that feels too bad, but you know we've tried having you tell me what happened from various perspectives, and maybe, if you're up for it, we could try that again, and while you tell me about it you could try the eye movements. It would help build the association."

He shrugged. "If you think it will help."

"I don't know what will help, but it can't hurt to try this. If it's too upsetting just stop. Anytime you want to stop, be sure to stop."

"All right."

So he began to tell the story of how he had first come to his town, and how the heat wave had at first seemed like all the other hot weather they had had. As he spoke he moved his eyes, in tandem of course, as that was the only way he could do it, back and forth, looking as far to the left as he could, vague view of her bookshelves, then in a quick sweep to as far right as he could, catching a vague view of flowers in a vase in front of a window looking out onto a courtyard. This was a voluntary effort that stopped the moment he stopped thinking to do it, so he had to devote some of his attention to it, while at the same time continuing with his story, which as a result was halting and disjointed, unrehearsed and different from what he would have said if he was just telling her the same thing again as before. This he presumed was one benefit of the exercise.

"I got there in the winter so it wasn't that hot to begin with...but it wasn't cold, no. In the Himalayas it was cold, you could even see the snow peaks to the north on clear days, but most days...most days weren't clear. The air was dirty almost all the time. Not that different from anywhere else. So I got settled in and was taking classes in Hindi and working... working at the clinic. Then the heat wave came. It got way hotter than it had been up till then, but everyone...everyone said it was normal, that the time right before the monsoon was the hottest of all. But then it got hotter still. Then it all happened fast, one day it was so hot even the people were scared...and that night some of the older people and the littlest kids died. That sent everyone into shock, but I think they were thinking it was as bad as it could get. Then it got worse, and the power went out, and after that there was no air conditioning...and not much water. People freaked out, and rightfully so. The heat was beyond what the human body can stand. Hyperthermia, that's just a word. The reality is different. You can't breathe. Sweating doesn't work. You're being roasted, like meat in an oven, and you can feel that. Eventually a lot of them went down to the local lake, but its water was like bath temperature, and not...safe to drink. So that's where a lot of them died."

He stopped talking and let his eyes rest. He could feel muscles behind his eyes, pulsing at the unaccustomed efforts. Like any other muscles, they welcomed a rest. That felt odd.

The therapist said, "I noticed that this time you didn't really put yourself in the story."

"No? I thought I did."

"You always talked about them. They did things, things happened to them."

"Well, I was one of them."

"At the time, did you think of yourself as one of them?"

"...No. I mean, they were them, I was me. I watched them, I talked with some of them. The usual stuff."

"Of course. So, could you tell me your part of the story, moving your eyes like that?"

"I don't know."

"Do you want to try?"

"No."

"All right. Maybe some other time. And maybe next time we can try to create the bilateral action by having you hold those little buzzers in your hands. Remember I showed you those? They'll pulse left-right-left-right as you talk it through. It's easier than moving your eyes."

"I don't want to do that now."

"Next time, maybe."

"I don't know when."

"You don't want to?"

"No. Why should I?"

"Well, the theory is that if you tell the story, you're shaping the memory of it to some extent, by putting it into words. And if you do that while making the eye movements, or feeling the hand buzzers, that seems to create a kind of internal distance in you between your memory of the story as you told it, and the, what you might call the reliving of it, the spontaneous reliving of it by way of some trigger setting you off. So that if that were to happen and you wanted some relief from it, you could move your eyes and start maybe thinking of your spoken version of what happened, and it would relieve you from reliving it. If you see what I mean."

"Yes," Frank said. "I understand. I'm not sure I believe it, but I understand."

"That makes sense. But maybe worth a try?"

"Maybe."

One fall he took a Scottish friend's offer to work on a project in Antarctica. She was principal investigator of a small scientific team going to the Dry Valleys, to study the stream that ran there briefly every summer, the Onyx River. And she had room on the team for a field assistant, and wanted to help him out. Since he was having trouble handling the heat, she said, Antarctica ought to be a great place for him.

Sounds good, he said. He was running out of money from a small inheritance left to him by his grandmother, and he still didn't want to contact his parents or his organization, so it would help with that too. And so that fall he flew to Denver and went through the interviews, and altered his résumé to omit his time in India, and then he was hired and off to Auckland, then Christchurch, and from Christchurch south to McMurdo Station on Ross Island, just across McMurdo Sound from the Dry Valleys, which lay between the Royal Society Range and the frozen sea. Even the plane flight to McMurdo was cold, its interior a long open room like a warehouse floor. Same with all the old junky buildings of McMurdo, and the newer buildings too—like warehouses, institutional buildings, and never heated to much more than 60 degrees. Even the line that ran through the buffet in the kitchen was a cool experience. All very congenial.

Then, out in the Dry Valleys, the hut they ate their meals in was kept warm, but not exceptionally so; really it was only warm relative to the outside. The dorm huts were a little hot and stuffy, but it was possible to sleep out in a tent of his own. That was really cold, so cold that the sleeping bag he slept in weighed about ten pounds; it took that much goose down to hold in enough of his own heat to keep him warm. He stuck his nose out of this bag to breathe, and that repeating injection of frigid air reminded him that it was really cold out, even though it was sunny all the time. The continuous light was strange but he soon got used to it.

The problem was that extreme cold somehow led to thoughts of temperature itself, and to warm up their freezing hands after a session of field work, they would heat the dining hut to quite a high temperature, which would make for a stuffy steamy room, and Frank found himself slipping down the slippery slope. Out at this remove from any possibility of relief, freak-outs would be at best inconvenient, at worst a disaster. Medevacs by helo were rare and expensive, he had heard them say. So he had to stay cool. But sometimes he could only hide a freak-out and hope it would go away soon and not come back. Sometimes he torqued his eyes like he was watching a Ping-Pong match.

And they had a sauna hut there. He stayed away from it, of course, but one night, going out to his tent in the bright daylight, he passed it just as a group of scientists burst out of it half naked in bathing suits, shrieking in delighted agony at the instantaneous extreme shift of temperature, evaporative steam bursting off their bodies like they were big pink firecrackers. That sight, which ought to have been beautiful, and their shrieking, which sounded like pain though it was ecstasy, set him off instantly. His heart pounded so fast and hard that he went light-headed, then suddenly fell to his knees and pitched face first onto the snow. No warning, just the sight of the pink firecracker people, a racing heart, then he found himself laid out on the hard cold snow. He had fainted right in front of them. The sauna-goers naturally helped him up, and someone took his pulse as they lifted him and cried out in a panic, Hey feel this tachycardia, my God! Feel it! They said it was 240 beats a minute. Within two hours a helo was thwacking down to medevac him out of there. And once medevacked to McMurdo, and his condition and past experiences made fully known to the NSF brass on site, he was accused of lying on his application form and shipped back home.

19

We had been at sea for something like eight years. They said they would pay us when we landed but everyone knew they wouldn't. Wouldn't pay us, wouldn't land us. We were slaves. If we didn't work they locked us in our cabins and didn't feed us. We went back to work.

The food was trash, including fishheads and guts from the take, but it was that or starve, so we ate it. And we worked, we had to. Set the lines, ran the reels, tried to keep our fingers and arms out of the way. That didn't always happen. The southern Atlantic is rough, the Antarctic Ocean even worse. Accidents were common. Often guys just stepped over the rail into the water. One guy waved goodbye to us before the whitecaps rolled him under. We knew why he did it. It was probably the best option, but it took courage. You could always imagine something would happen to change things.

Then one day it did. A ship came over the horizon, this wasn't unusual, it happened all the time. Not only fishing boats like ours, either slave ships or not, there was no way to tell, but the transport ships that came out to transfer our catch into their holds and resupply us so we didn't have to land. That was the way they did it. We didn't even know what countries they came from.

So it looked like a transport ship, and it approached us, and it was clear the captain and his mates thought the same. The people on this ship must have known the signs and fooled them. Then after it came beside us and we had grappled it, men jumped over the side holding guns pointed at us. We put our hands in the air just like in the movies, but it would have been a funny movie, because most of us were grinning and it was all I could do to keep from cheering.

We were herded into the cabins and locked in. When the newcomers came into our cabin and asked us questions we answered eagerly. Maybe they were just pirates who would put us back to work for someone else, or even kill us, but even so we told them our stories, and who the captain and every single one of his men were. They left us in there and came back later. Get on our ship, they told us. We did what they said, not knowing what it would mean. All the slaves climbed a ladder onto the bigger ship, that was eight of us. All the captain's men and the captain were left on board our boat. That was five of them. They said some stuff but the men with the guns ignored them.

When we were about a hundred meters away I saw that some of the men on this new boat were filming our old one. Then the bow of our old boat blew up, just above the waterline. The boom wasn't very loud, but the bow shattered. There was a bit of flame but water poured in and doused it. In about fifteen minutes the boat tilted and started going down. Then another explosion in the stern finished the deal. It went down fast. The captain and his men climbed on the roof of the cabin and yelled at us. No one on our savior ship said anything. Everyone just watched it happen.

You're killing them? we asked the sailor nearest us.

He said, They've got life rafts, right?

We don't know, we said. Inflatables, you mean?

Yeah.

I guess so.

So, they'll either get those inflated and over the side or they won't. If they don't, they'll get what's coming to them. We'll post film of it on sites that other fishermen will see. If they get off in a life raft, they can try to make it to land. If they manage that, they can tell the story of what happened to whoever will listen. Either way, the point will be made.

So that meant these people were probably not police. That was not a good thing, but it wasn't as if we could choose who saved us.

What's the point? we asked.

No more fishing.

Good, we said.

20

The Gini coefficient, devised by the Italian sociologist Corrado Gini in 1912, is a measure of income or wealth disparity in a population. It is usually expressed as a fraction between 0 and 1, and it seems easy to understand, because 0 is the coefficient if everyone owned an equal amount, while 1 would obtain if one person owned everything and everyone else nothing. In our real world of the mid-twenty-first century, countries with a low Gini coefficient, like the social democracies, are generally a bit below 0.3, while highly unequal countries are a bit above 0.6. The US, China, and many other countries have seen their Gini coefficients shoot up in the neoliberal era, from 0.3 or 0.4 up to 0.5 or 0.6, this with barely a squeak from the people losing the most in this increase in inequality, and indeed many of those harmed often vote for politicians who will increase their relative impoverishment. Thus the power of hegemony: we may be poor but at least we're patriots! At least we're self-reliant and we can take care of ourselves, and so on, right into an early grave, as the average lifetimes of the poorer citizens in these countries are much shorter than those of the wealthy citizens. And average lifetimes overall are therefore decreasing for the first time since the eighteenth century.

Don't think that the Gini coefficient alone will describe the situation, however; this would be succumbing to *monocausotaxophilia*, the love of single ideas that explain everything, one of humanity's most common cognitive errors. The Gini figures for Bangladesh and for Holland are nearly the same, for instance, at 0.31; but the average annual income in Bangladesh is about $2,000, while in Holland it's $50,000. The spread between the richest and the poorest is an important consideration, but when everyone in that spread is pretty well off, this is a different situation than when everyone across the spread is poor.

Thus other rubrics to think about inequality have been devised. One of the best is the "inequality-adjusted Human Development Index," which is no surprise, because the Human Development Index is already a powerful tool. But it doesn't by itself reveal the internal spread of good and bad in the country studied, thus the inequality adjustment, which gives a more nuanced portrait of how well the total population is doing.

While discussing inequality, it should be noted that the Gini coefficient for the whole world's population is higher than for any individual country's, basically because there are so many more poor people in the world than there are rich ones, so that cumulatively, globally, the number rises to around 0.7.

Also, there are various ways of indicating inequality more anecdotally (perhaps we could say in more human terms) than such indexes. The three richest people in the world possess more financial assets than all the people in the forty-eight poorest countries added together. The wealthiest one percent of the human population owns more than the bottom seventy percent. And so on.

Also, note that these disparities in wealth have been increasing since 1980 to the present, and are one of the defining characteristics of neoliberalism. Inequality has now reached levels not seen since the so-called Gilded Age of the 1890s. Some angles of evidence now suggest this is the most wealth-inequal moment in human history, surpassing the feudal era for instance, and the early warrior/priest/peasant states. Also, the two billion poorest people on the planet still lack access to basics like toilets, housing, food, health care, education, and so on. This means that fully one-quarter of humanity, enough to equal the entire human population of the year 1960, is immiserated in ways that the poorest people of the feudal era or the Upper Paleolithic were not.

Thus inequality in our time. Is it a political stability problem? Perhaps in a controlocracy backed by big militaries, no. Is it a moral problem? But morality is a question of ideology, one's imaginary relationship to the real situation, and many find it easy to imagine that you get what you deserve, and so on. So morality is a slippery business.

So it is that one often sees inequality as a problem judged economically; growth and innovation, it is said, are slowed when inequality is high. This is what our thinking has been reduced to: essentially a neoliberal analysis and judgment of the neoliberal situation. It's the structure of feeling in our time; we can't think in anything but economic terms, our ethics must be quantified and rated for the effects that our actions have on GDP. This is said to be the only thing people can agree on. Although those who say this are often economists.

But that's the world we're in. And so people invent other indexes to try to come to grips with this issue. In fact we have seen a real proliferation of them.

Recall that GDP, gross domestic product, the dominant metric in economics for the last century, consists of a combination of consumption, plus private investments, plus government spending, plus exports-minus-imports. Criticisms of GDP are many, as it includes destructive activities as positive economic numbers, and excludes many kinds of negative externalities, as well as issues of health, social reproduction, citizen satisfaction, and so on.

Alternative measures that compensate for these deficiencies include:

the Genuine Progress Indicator, which uses twenty-six different variables to determine its single index number;

the UN's Human Development Index, developed by Pakistani economist Mahbub ul Haq in 1990, which combines life expectancy, education levels, and gross national income per capita (later the UN introduced the inequality-adjusted HDI);

the UN's Inclusive Wealth Report, which combines manufactured capital, human capital, natural capital, adjusted by factors including carbon emissions;

the Happy Planet Index, created by the New Economic Forum, which combines well-being as reported by citizens, life expectancy, and inequality of outcomes, divided by ecological footprint (by this rubric the US scores 20.1 out of 100, and comes in 108th out of 140 countries rated);

the Food Sustainability Index, formulated by Barilla Center for Food and Nutrition, which uses fifty-eight metrics to measure food security, welfare, and ecological sustainability;

the Ecological Footprint, as developed by the Global Footprint Network, which estimates how much land it would take to sustainably support the lifestyle of a town or country, an amount always larger by considerable margins than the political entities being evaluated, except for Cuba and a few other countries;

and Bhutan's famous Gross National Happiness, which uses thirty-three metrics to measure the titular quality in quantitative terms.

All these indexes are attempts to portray civilization in our time using the terms of the hegemonic discourse, which is to say economics, often in the attempt to make a judo-like transformation of the discipline of economics itself, altering it to make it more human, more adjusted to the biosphere, and so on. Not a bad impulse!

But it's important also to take this whole question back out of the realm of quantification, sometimes, to the realm of the human and the social. To ask what it all means, what it's all for. To consider the axioms we are agreeing to live by. To acknowledge the reality of other people, and of the planet itself. To see other people's faces. To walk outdoors and look around.

21

We were on the lakefront in Brissago, on the Swiss side of Lake Maggiore, partying on the lawn of Cinzia's place, just above the narrow park between her property and the lake. She had a celebrity chef there who cooked with a welder's torch he used to fire at the bottom of big fry-pans he held in the air, and a band with a brass section, and a light show and all that. Altogether a righteous party, and lots of happy people there, skewing young because that's the way Cinzia likes it.

But the narrow stretch of grass between her lawn and the lake was a public park, and as we partied we saw a guy down there on the shore, just standing there staring up at us. Some kind of beachcomber dude, holding a piece of driftwood. Nothing Cinzia's security could do about him, they told us. Actually they could have if they wanted to, but they didn't. The local police might make trouble if someone were objected to for just standing on a public beach. This is what one of them told us when we told him to make the guy go away. The guy was skinny and bedraggled and he just kept staring, it was offensive. Like some kind of Bible guy laying his morality on us.

So finally a few of us went down there to do what the security team ought to have done, and send this guy packing. Edmund led the way as usual, he was the one most annoyed, and we followed along because when he was annoyed Edmund could be really funny.

The guy watched us come up to him and didn't move an inch, didn't say a word. It was a little weird, I didn't like it.

Edmund got in the guy's face and told him to leave.

The guy said to Edmund something like, You fuckers are burning up the world with your stupid games.

Edmund laughed and said, "Dost thou think because thou art virtuous, there shall be no more cakes and ale?"

We laughed at that, but then this guy hit Edmund with the chunk of driftwood he was holding, so fast we had no time to react. Edmund went down like a tree, didn't get his hands up or anything, just boom. He had been cold-cocked.

The guy held his piece of wood out at us and we froze. Then he tossed it at us and took off right into the lake, swimming straight out into the night. We didn't know what to do—no one wanted to swim off after a nut like that, not in the dark, and besides we were concerned about Edmund. It just looked bad, the way he went down. Like a tree. Cinzia's security finally joined us, but they only wanted to hold the perimeter, they didn't chase the guy either. They took over checking out Edmund, and when they did that they quickly got on their phones. An ambulance showed up in about five minutes and took him away. After that it was a couple of hours before we got word. We couldn't believe it. Edmund was dead.

22

Adele Elia and Bob Wharton were at a meeting of the Scientific Committee for Antarctic Research, an international scientific organization formed to coordinate Antarctic research after the 1956 International Geophysical Year and the 1959 signing of the Antarctic Treaty. Over the years SCAR had become one of the main de facto governments of Antarctica, along with the US National Science Foundation and the British and other national Antarctic research programs, especially Argentina's and Chile's. The SCAR meeting was in Geneva this year, so it had been a morning's train ride for Adele and Bob, who during their work in the ministry had become friends.

Now they sat looking out at the lake from a bar on the second story of the meeting's hotel. Out the window beside their table they could see the famous fountain launch its spire of water into the air. To the south a stupendous set of thunderheads, as solid as the marble tabletops in their bar, lofted high over the Mont Blanc massif. They were enjoying this view and their drinks when they were approached by an American glaciologist they knew named Pete Griffen. Griffen was pulling by the arm another man they didn't know, whom Pete introduced as Slawek, another glaciologist. Adele and Slawek had read some of each other's papers, and even attended some of the same meetings of the AGU, but until now had never met.

A waiter appeared with a tray of drinks that Griffen had ordered: four snifters of Drambuie, and a carafe of water with water glasses. "Ah, Drambers," Adele said with a little Gallic smile. This liqueur was what Kiwis always drank in the Dry Valleys, they informed Bob as they took their first sips. When Bob made a face as he tasted it, Griffen explained that long ago a ship filled with cases of the weird sweet stuff had been stranded in Lyttelton when its shipping company went bankrupt, and the cases had been warehoused there and over the years sent south for cheap, year after year. So they drank a toast to the Dry Valleys and settled into their chairs.

When asked, Slawek said he had spent five years all told in the Dry Valleys, and Adele countered that she had spent eight years living on glaciers; Pete grinned and topped them both with twelve years total on the Ice. They quickly pointed out that he was older and so had an advantage, which he agreed to immediately. Slawek said he had become a glaciologist to indulge his introverted personality while still holding down a job, and Adele laughed and nodded.

"A lot of us are that way," she said.

"Not me!" Pete declared. "I like to party, but really the best parties are on the Ice." He rotated his hand at Slawek as if coaxing him. "Come on, Slawek, tell these guys your idea. I think they need to hear it."

Slawek frowned uncomfortably, but said, "You all heard the new data in there today."

They agreed they had.

"Sea level will rise so fast, the world is fucked."

It couldn't be denied, the others agreed. The data were clear.

"So," Pete prompted Slawek, "I've heard some people suggesting we just pump all the melted ice back up onto the polar plateau, right?"

Bob shook his head at hearing this. It was an old idea, he said, studied by the Potsdam Institute at one point, and the conclusions of their study had been bleak; the amount of electrical power needed to pump that much water up onto the east Antarctic ice cap came to about seven percent of all the electricity generated by all of global civilization. "It's too energy intensive," Bob concluded.

Slawek snorted. "Energy is the least of it. Since one percent of all electricity created is burned to make bitcoins, seven percent for saving sea level could be seen as a deal. But the physical problems are the stoppers. Have you run the numbers?"

"No?"

"Say sea level goes up one centimeter. That's three thousand six hundred cubic kilometers of water."

Adele and Bob glanced at each other, startled. Griffen was just smiling.

Slawek saw their look and nodded. "Right. It's six hundred times as much as all the oil pumped every year. Building the infrastructure to do that would not be feasible. And it would have to be clean energy pumping it, or you'd be emitting more carbon. That much clean energy would take ten million windmills, Potsdam said. And the water would have to be moved in pipes, and that's more pipe than has ever been made. And last but worst, the water has to freeze when it gets up there. Say a meter deep per year, I don't think you could go any deeper without problems—that means about half of eastern Antarctica."

"So it's too much in every way," Adele noted.

They drank more Drambers while they pondered it. Griffen said, "Come on, Slawek, get to your idea. Tell them."

Slawek nodded. "Reality of problem is that glaciers are sliding into the sea ten times faster than before."

"Yes."

"So, the reason for that is there's more meltwater created on the ice surface every summer, because of global warming. That water runs down moulins until it reaches the undersides of the glaciers, and there it has nowhere else to go. So it lifts up the ice a bit. It lubricates the ice flow over the rock beds. The ice used to be in contact with the rock bed, at least in some places, and usually in most places. The ice is so heavy it used to crush out everything under it. It bottomed out. Kilometer thick, that's a big weight. So the glacier scraped down its bed right on the rock, bottomed out, ice to rock. Even sometimes frozen to rock. Stuck. A good percentage of glacial movement at that point was viscous deformation of the ice downhill, not sliding at all."

Adele and Pete were nodding at this. Adele was beginning to look thoughtful, Griffen was grinning outright. "And so?" Bob said.

Slawek hesitated and Griffen said, "Come on!"

"Okay. You pump that water out from under the glaciers. Melt drill-holes like we already do there when we check out subglacial lakes, or to get through the ice shelves. Technology is well known, and pretty easy. Pump up the water from under the glaciers, and actually, the weight of the ice on it will cause that water to come up a well hole ninety percent of the way, just from pressure of all that weight. Then you pump it up the rest of the way, pipe it away from the glacier onto some stable ice nearby."

"How much water would that be?" Bob asked.

"All the glaciers together, maybe sixty cubic kilometers. It's still a lot, but it's not three thousand six hundred."

"Or three hundred and sixty thousand!" Adele added. "Which is what a single meter rise in sea level would be."

"Right. Also, the meltwater at the bottom of the glaciers is really from three sources. Surface water draining down moulins is the new stuff. Then geothermal energy melts a little bit of the glacier's bottom from below, as always. It never melted much before, except over certain hot spots, but a little. Third source is the shear heat created by ice moving downstream, the friction of that movement. So. Geothermal in most places raises the temperature at the bottom of the glacier to about zero degrees, while up on the surface it can be as cold as forty below. So normally the heat from geothermal mostly diffuses up through the ice, it dissipates like that and so the ice on bottom stays frozen. Just barely, but normally it does. But now, the moulin water drains down there and lubricates a little, then as glacier speeds up going down its bed, the shear heat down there increases, so more heat, more melting, more speed. But if you suck the bottom water out and slow the glacier back down, it won't shear as much, and you won't get that friction melt. My modeling suggests that if you pump out about a third to a half of the water underneath the glaciers, you get them to slow down enough to reduce their shear heat also, and that water doesn't appear in the first place. The glaciers cool down, bottom out, refreeze to the rock, go back to their old speed. So you only need to pump out something like thirty cubic kilometers, from under the biggest glaciers in Antarctica and Greenland."

"How many glaciers?" Pete asked.

"Say the hundred biggest. It's not so bad."

"How many pumps per glacier would you need?" Bob asked.

"Who knows? It would be different for each, I'm sure. Would be an experiment you'd have to keep trying."

"Expensive," Bob noted.

"Compared to what?" Pete exclaimed.

Adele laughed. "Jurgen said a quadrillion dollars."

Slawek nodded, mouth pursed solemnly. "This would cost less."

They all laughed.

Adele said to him, "So, Slawek, why didn't you bring this up at the session today? It was about this acceleration of glaciers."

Slawek quickly shook his head. "Not my thing. A scientist gets into geoengineering, they're not a scientist anymore, they're a politician. Get hate mail, rocks through window, no one takes their real work seriously, all that. I'm not ready for that kind of career change. I just want to get back on the Ice while I can still get PQ'ed."

"But the fate of civilization," Bob suggested.

Slawek shrugged. "That's your job, right? So I thought I'd mention it. Or really, Pete thought I'd mention it."

"Thanks, Slawek," Pete said. "You are a true glaciologist."

"I am."

"I think we should drink another round of Drambers to celebrate that."

"Me too."

23

It took a while to get a gun. Not easy. Finally he managed it by stealing a rifle and its ammunition from a Swiss man's closet, where Swiss men so often kept their service rifles. It turned out to be absurdly easy; the country was so safe that quite a few people never locked their doors. Of course Swiss men were ordered to keep their service rifles under lock and key, and they all did that, but some were more careless with their hunting rifles, and he found one of those.

He was ready to act.

He researched his subjects of interest. One of them would be attending a conference in Dübendorf in a month.

So he carried the stolen rifle in a backpack over the Zuriberg, to a parking garage near the conference center in Dübendorf. He took the stairs up to its top floor, then climbed up onto its roof. When he got to the roof he was looking down on the main entrance to the conference center. He assembled the rifle and set it on a wooden stand he had knocked together, and looked through its sight at the entrance area.

His target walked up broad stairs to the entry and turned to say something to an assistant. Blue suit, white shirt, red tie. He looked like he was making a joke.

Frank regarded the man in the crosshairs. He swallowed, aimed again. He could feel his face heating up, also his hands and feet. Finally the man went inside.

Frank put the rifle back in the backpack and carried it down the stairs. Up in the forest, on the hill separating the city from its suburban town, he cast the rifle onto the ground under a tree. Then down the hill, back to his shed.

That had been a criminal, something like a war criminal, there in the crosshairs of that gun. A climate criminal. Few war criminals would kill as many people as that man would. And yet Frank hadn't pulled the trigger. Hadn't been able to. A target whose life's work would drive thousands of species to extinction, and cause millions of people to die. And yet he hadn't done it.

Maybe he wasn't as crazy as he thought. Or maybe he had completely lost it. Or simply just lost his nerve. He wasn't sure anymore. He felt sick, shaky. Like he had just missed getting hit by a bus. That made him angrier than ever. He still wanted to kill one of the people killing the world. A Child of Kali, here in the West. A fellow traveler.

He had seen people die, had been broken by that very sight. So making more death was perhaps not the right response. Hurt people hurt people, one of his therapists had said. This was no doubt true. But it was too sweeping, too easy. It didn't have to play out like that every time. Though he did want to kill something. People like that oil executive, or that jerk he had lashed out at and knocked down, thoughtless assholes or great people as they might be, some of both, no doubt. They might get killed for their crimes someday, and deserve it too. But not by him. He didn't care about that guy he had hit, but that was an accident. Shooting someone was not like punching an asshole. He would have to find a different way. Though he still wanted to kill something. It was hard to think of anything else.

24

Just as we are all subject to some perceptual errors due to the nature of our senses and physical reality, we are also subject to some cognitive errors, built into our brains through the period of human evolution, and unavoidable even when known to us.

The perceptual illusions are easy. There are certain patterns of black and white that when printed on a paper circle, and the circle then spun on a stick, the paper will flash with colors in the human eye. Slow the spinning down and the printed pattern is obviously just black and white; speed it up and the colors reappear in our sight. Just the way it is.

Foreshortening is another perceptual distortion we can't correct for. When standing under a cliff in the mountains and looking up at it, the cliff always appears to be about the same height—say a thousand feet or so. Even if we are under the north wall of the Eiger and know the wall is six thousand feet tall, it still appears to be about a thousand feet tall. Only when you get miles away, like on the lakefront in Thun looking south, can you actually see the immense height of the Nordwand of the Eiger. Right under it, it's impossible.

Everyone can accept these optical illusions when they are demonstrated; they are undeniably there. But for the cognitive errors, it takes some testing. Cognitive scientists, logicians, and behavioral economists have only recently begun to sort out and name these cognitive errors, and disputes about them are common. But unavoidable mistakes have been demonstrated in test after test, and given names like anchor bias (you want to stick to your first estimate, or to what you have been told), ease of representation (you think an explanation you can understand is more likely to be true than one you can't). On and on it goes—online there is an excellent circular graphic display of cognitive errors—a wheel of mistakes that both lists them and organizes them into categories, including the law of small numbers, neglect of base rates, the availability heuristic, asymmetrical similarity, probability illusions, choice framing, context segregation, gain/loss asymmetry, conjunction effects, the law of typicality, misplaced causality, cause/effect asymmetry, the certainty effect, irrational prudence, the tyranny of sunk costs, illusory correlations, and unwarranted overconfidence—the graphic itself being a funny example of this last phenomenon, in that it pretends to know how we think and what would be normal.

As with perceptual illusions, knowing that cognitive errors exist doesn't help us to avoid them when presented with a new problem. On the contrary, these errors stay very consistent, in that they are committed by everyone tested, tend toward the same direction of error, are independent of personal factors of the test takers, and are incorrigible, in that knowing about them doesn't help us avoid them, nor to distrust our reasoning in other situations. We are always more confident of our reasoning than we should be. Indeed overconfidence, not just expert overconfidence but general overconfidence, is one of the most common illusions we experience. No doubt this analysis is yet another example: do we really know any of this?

Oh dear! What do these recent discoveries in cognitive science mean? Some say they just mean humans are poor at statistics. Others assert they are as important a discovery as the discovery of the subconscious.

Consider again the nature of ideology, that necessary thing, which allows us to sort out the massive influx of information we experience. Could ideology also be a cognitive illusion, a kind of necessary fiction?

Yes. Of course. We have to create and employ an ideology to be able to function; and we do that work by way of thinking that is prone to any number of systemic and one might even say factual errors. We have never been rational. Maybe science itself is the attempt to be rational. Maybe philosophy too. And of course philosophy is very often proving we can't think to the bottom of things, can't get logic to work as a closed system, and so on.

And remember also that in all of this discussion so far, we are referring to the normal mind, the sane mind. What happens when, starting as we do from such a shaky original position, sanity is lost, we defer to another discussion. Enough now to say just this: it can get very bad.

25

Zurich in winter is often smothered in fog and low clouds for months on end. Prevailing winds from the north ram the clouds coming in from the Atlantic against the wall of the Alps, and there they stick. Gray day after gray day, in a gray city by a gray lake, split by a gray river. On many of these days a train ride or drive of less than a hundred kilometers will lift one out of the fog into alpine sunlight. Then again, it's the time of year to get down to work.

So Mary worked. She read reports, she took meetings, she talked to people all over the world about projects, she wrote legislative proposals for stronger national laws governing legal standing for those future people and creatures and things without standing. Every day was full. In the evenings she often went out with Badim and others of the team. They would usually walk down to the Niederdorf, and either cross a bridge and eat at the Zeughauskeller, or stay on their side in the dark alleys around the Grossmünster, and gather around the long table at the back of the Casa Bar.

On this day the sun had broken through the clouds, so they took the tram down to Bürkliplatz and then another one toward Tiefenbrunnen, and ate at Tres Kilos, which they usually did only for birthday parties or other big occasions. But this was a day to celebrate. First sun of the year, February 19; by no means a Zurich record, as Jurgen noted lugubriously. The Swiss like to keep records for these kinds of weather phenomena.

A short day, so that by the time they got to the restaurant, night had long since fallen. The string of chili-shaped lights draping the entryway glowed like spots of fire. Inside someone else was celebrating too, and just as they walked in all the lights went off and a waitress carrying a cake with sparklers sparking off it walked in from the kitchen to the sounds of Stevie Wonder's "Happy Birthday." They cheered and sang with the rest, and sat back near the kitchen where they always did if they could. Mary sat next to Estevan and Imeni, and listened to them flirt by bickering. This was getting to be an old routine, but they seemed stuck in it.

"We are the Ministry for the Future," Estevan insisted to her, "not the ministry to solve all possible problems that can be solved now. We have to pick our battles or else it becomes just an everything."

"But everything is going to be a problem in the future," Imeni said. "I don't see how you can deny that. So if you pick and choose, you're just dodging our brief. And end up in a shitty future too by the way."

"Still we have to prioritize. We've only got so much time."

"This is a priority! Besides we're a group, we have time."

"Maybe you do."

Imeni elbowed him, then poured them both another margarita from the pitcher, which had just arrived. They shifted to a discussion of the day's news from Mauna Kea: carbon dioxide had registered at 447 parts per million, the highest ever recorded for wintertime. This despite all the reports from individual countries showing significant drops in emissions, even the US, even China, even India. Even Brazil and Russia. No: all the big emitters were reporting reductions, and yet the global total still grew. There had to be some unreported sources; or people were lying. Opinions as to which it was were divided around the table. Probably it was a bit of both.

"If people are lying, it means they know they are in the wrong. But if there are secondary emissions no one knows about, maybe stimulated by the heat already baked in, that would be worse. So we have to hope people are lying."

"Easy to hope for that, you always get it!"

"Come on, don't be cynical."

"Just realistic. When have people ever told the truth about this particular question?"

"People? Do you mean scientists or politicians?"

"Politicians of course! Scientists aren't *people*."

"I thought it was the reverse!"

"Neither scientists nor politicians are people."

"Careful now. Mary here is a politician, and I'm a scientist."

"No. You are both technocrats."

"So, that means we are scientific politicians?"

"Or political scientists. Which is to say, politicized scientists. Given that political science is a different thing entirely."

"Political science is a fake thing, if you ask me. Or at least it has a fake name. I mean, where's the science in it?"

"Statistics, maybe?"

"No. They just want to sound solid. They're history at best, economics at worst."

"I sense a poli sci major here, still living the trauma."

"It's true!"

Laughter around the table. Another round of margaritas. The bill was going to be stupendous—Tres Kilos, like all Zurich restaurants, maybe all restaurants, made most of its money by way of outrageous liquor prices. But probably her team thought the ministry would be paying for it tonight. Which was true. Mary sighed and let them refill her glass.

She looked at the table, listened to Estevan and Imeni flirt. Feeling each other out, but subtly, as they were with the group. In-house romances were never a good idea, and yet they always happened. No one was going to be doing this job for life; and that was true of all jobs. So why not? Where else were you going to meet people? So it happened. It had happened to her, long ago. She could recall this very kind of banter between her and Martin, long ago in London. Mary and Marty! Two Irish in London, a Prot and a Catholic, trying to find some way to get their claws into the system. Now he had been dead for over twenty years.

Quickly and firmly she refocused on the present. Estevan and Imeni were very much alive. And Mary could see why they might take to each other, despite their obvious differences. Well, who knew. Their banter was a little brittle, a little forced. And what drew people to each other was fundamentally unknowable. For all she knew they had already become lovers and then broken up, and were now negotiating a settlement. No way to be sure, not from the boss's angle.

At the end of the long meal she hauled herself to her feet and calculated her level of inebriation—mild as always these days, she was careful, she was with her colleagues, her employees, and it wouldn't do to be unseemly; and her youth had taught her some hard lessons, as well as given her a pretty high tolerance for alcohol. All was well there, she could glide along in her crowd to the nearest tram stop and get on one of the blue trams, transfer at Bürkliplatz and catch one going up the hill, saying goodbye to Bob and Badim and traveling on with Estevan and Imeni. Then she got off at Kirche Fluntern, waving goodbye to her two young companions as they headed onward, curious about them, but already thinking about other things: tea, bed, whether she would be able to sleep.

She was walking down Hochstrasse when a man walking the other way turned abruptly and began walking by her side. She looked at him, startled; he was staring at her with wild eyes.

"Keep going," he commanded her in a low choked voice. "I'm taking you into custody here."

"What?" she exclaimed, and stopped in her tracks.

He reached out and snapped some kind of clasp around her wrist. In his other hand he showed her a small snub-nosed pistol. The clasp now locked around her wrist was one half of some kind of clear plastic handcuffs, it appeared, with the other clasp locked around his wrist. "Come on," he said, starting to walk and pulling her along with him. "I want to talk with you. Come with me and I won't hurt you. If you don't come with me now, I'll shoot you."

"You won't," she said faintly, but she found herself walking by his side, tugged along by the wrist.

"I will," he said with a blazing glare at her. "I don't care about anything."

She gulped and kept her mouth shut. Her heart was racing. The drink she thought hadn't affected her was rushing through her body like fire, such that she almost staggered.

He walked her right to her building's door, surprising her.

"In we go," he said. "Come on, do it."

She punched in her code and opened the outer door. Up the flights of stairs to her apartment. Unlocked and opened the door. Inside. Her place looked strange now that she was a prisoner in it.

"Phone," he said. "Turn it off and put it down here." He indicated the table by the door where she indeed often left her stuff. She took it from her purse, turned it off, dropped it on the table.

"Do you have any other GPS on you?" he asked.

"What?"

"Are you chipped? Do you have any other GPS stuck to you?"

"No," she said. "I don't think so," she added.

He gave her a wondering look, incredulous and disapproving. From the same pocket that held the gun, he took out a small black box, flicked a tab on it, moved it around her body. He checked its screen, nodded. "Okay, let's go."

"But I don't—"

"Let's go! I have a place right around the corner."

"I thought you said you just wanted to talk to me!"

"That's right."

"Then do it here! Unhook me and I'll talk to you. If you drag me out of here, I'll fall down as soon as someone is looking at us and scream for help. If you only want to talk, it will go better here. I'll feel safer. I'll listen more."

He glared at her for a while. "All right," he said at last. "Why not."

He shook his head, looking baffled and confused. She saw that and thought to herself, This man is sick in the head. That was even more frightening. He reached down with his free hand, unlocked the clasp around her wrist. She kneaded that wrist with her other hand, stared at him, thinking furiously. "I need to go to the bathroom."

He glared. "I want to check it first."

They went in the bathroom, and he looked in the cabinet, and then behind the shower screen. She supposed she might have had some kind of lifeline system in there, or another phone or an alarm system. No such luck. When he was satisfied he walked out the door, leaving it open. She went in and stood there for a second. Then the door swung almost shut— he had given it a push. Propriety. A polite madman. Well, it was something. She sat on the toilet and peed, trying to think the situation through. Nothing came to her. She stood, flushed, went back out to him.

"Do you want some tea?" she asked.

"No."

"Well I do."

"Fine. We'll start there."

She boiled water in the teapot and brewed a cup of tea. Something to make her feel calmer. He refused a cup again, watching her work. Then they were sitting across from each other, her little kitchen table between them.

He was young. Late twenties or early thirties; hard to tell at that age. Thin drawn face, dark circles under his eyes. A lean and hungry look, oh yes. Spooked by his own action here, she thought; that would make sense; but that wildness she had seen from the moment he had accosted her was there in him too, some kind of carelessness or desperation. To do this thing, whatever it was, he had to be deranged. Something had driven him to this.

"What do you want?" she said.

"I want to talk to you."

"Why do it this way?"

His lip curled. "I want you to listen to me."

"I listen to people all the time."

He shook his head decisively, back and forth. "Not people like me."

"What do you mean, why not?"

"I'm nobody," he said. "I'm dead. I've been killed."

She felt a chill. Finally, not knowing if it was smart, she said, "How so?"

He didn't appear to hear her. "Now I'm supposed to have come back, but I didn't. Really I'm dead. You're here, you're the head of a big UN agency, you have important meetings all over the world, every hour of every day. You don't have time for a dead man."

"How do you mean?" she asked again, trying not to become more alarmed than she already was. "How did you die?"

"I was in the heat wave."

Ah.

He stared at her tea cup. She picked it up and sipped from it; his eyes stayed locked on the table. His face was flushing—right before her eyes the skin of his cheeks and forehead blushed, from a blanched white to a vivid red. Beads of sweat popped out on his forehead and the backs of his hands, tense there on the table before him. Mary swallowed hard as she saw all this.

"I'm sorry," she said. "That must have been bad."

He nodded. Abruptly he stood. He wandered around her little kitchen, his back to her. Out the window the night was black, the lights on the rise of the Zuriberg glowing all fuzzy in the night's smirr. He was breathing hard, as if recovering from a sprint, or trying not to cry.

She listened to his deep rapid breathing. Could be he was charging himself up to do something to her.

After a while he sat across from her again. "Yes it was bad," he said. "Everyone died. I died. Then they brought me back."

"Are you all right now?" she asked.

"No!" he cried out angrily. "I'm not all right!"

"I meant physically."

"No! Not *physically*. Not any way!" He shook his head, as if shaking away certain thoughts.

"I'm sorry," she said again. She sipped her tea. "So. You want to talk to me. About that, I assume."

He shook his head. "Not that. That was just the start. That was what made me want to talk to you, maybe, but what I want to say isn't about that. What I want to tell you is this"—and he looked her in the eye. "It's going to happen again."

She swallowed involuntarily. "Why do you think so?"

"Because nothing's changed!" he exclaimed. "Why do you ask me that!"

He stood up again, agitated. Now his flushed face turned an even darker red. His brows were bunched together. He leaned over her and said fiercely, in a low choked voice, "Why do you pretend not to know!"

She took a breath. "I don't pretend. I really don't know."

He shook his head, glaring down at her.

"That's why I'm doing this," he said, voice low and furious. "You do know. You only pretend not to know. You all pretend. You're head of the United Nations Ministry for the Future, and yet you pretend not to know what the future is bringing down on us."

"No one can know that," she said, meeting his eye. "And I have to say, the ministry is organized under the Paris Agreement. The UN isn't directly involved."

"You're the Ministry for the Future."

"I lead it, yes."

He looked at her silently for a long time. At some point in this inspection, still looking at her, he sat back down across from her. He leaned over the table toward her.

"So," he said, "what do you and your ministry know about the future, then."

"We can only model scenarios," she said. "We track what has happened, and graph trajectories in things we can measure, and then we postulate that the things we can measure will either stay the same, or grow, or shrink."

"Things like temperatures, or birth rates, or like that."

"Yes."

"So you know! I mean, in your exercises, is there any scenario whatsoever in which there won't be more heat waves that kill millions of people?"

"Yes," she said.

But she was troubled. This possibility that he was bringing up to her now was exactly what kept her awake at night, night after night. Scenarios with good results, in which they managed to avoid more incidents of mass deaths, were in fact extremely rare. People would have to do things they were not doing. His presence in her kitchen was all too much like one of her insomniac whirlpools of thought, as if she had stumbled into one of her nightmares while still awake, so that she couldn't get out of it.

"Ha!" he cried, reading this off her face.

She grimaced, trying to erase the look.

"Come on," he said. "You know. *You know the future.*" It sounded like he might hurt his voice, he was so intent to speak without shouting. He coughed, shook his head. "And yet you're not doing anything about it. Even with your job."

He stood again, went to her sink, took a glass out of the drainer, filled it from the tap, took a drink. He brought it with him back to the table, sat again.

"We're doing what we can," she said.

"No you're not. You're not doing everything you can, and what you are doing isn't going to be enough." He leaned toward her again and captured her gaze, his eyes bloodshot and bugging out, pale tortured blue eyes scarcely held in by his sweating red face—transfixing her—"Admit it!" he exclaimed, still strangling his voice to less than a shout.

She sighed. She tried to think what to say to him. The look in his eye scared her; maybe he was thinking that if he killed her now, someone more effective would replace her. It looked like that was what he was thinking. And here they were, after all. She had been kidnapped and taken to her own apartment. When this happened to women they often died.

Finally she shrugged, heart racing. "We're trying."

For a long time they sat there looking at each other. She got the impression he was letting her ponder her statement for a while. Letting her stew in the juices of her own futility.

Finally he said, "But it isn't working. You're trying, but it isn't enough. You're failing. You and your organization are failing in your appointed task, and so millions will die. You're letting them down. Every day you let them down. You set them up for death."

She sighed. "We're doing all we can with what we've got."

"No you're not."

His face flushed again, he stood up again. He circled in her kitchen like a trapped animal. He was breathing heavily. Here it comes, she thought despite herself. Her heart was really racing.

Finally he stopped over her. He leaned down at her yet again. He spoke again in the low choked voice that seemed all he could manage.

"This is why I'm here. You have to stop thinking that you're doing all you can. Because you're not. There's more you could be doing."

"Like what?"

He stared at her. He sat down again across from her, put his face in his hands. Finally he released his face, sat back in his chair. He looked her in the eye. There she saw something: a real person. A very troubled real person, a young man, sick and scared.

"I went back to India," he said. "I tried to join a group of people I had heard about. Children of Kali, you've heard of them?"

"Yes. But they're a terrorist group."

He shook his head, staring at her all the while. "No. You have to stop thinking with your old bourgeois values. That time has passed. The stakes are too high for you to hide behind them anymore. They're killing the world. People, animals, everything. We're in a mass extinction event, and there are people trying to do something about it. You call them terrorists, but it's the people you work for who are the terrorists. How can you not see that?"

"I'm trying to avoid violence," Mary said. "That's my job."

"I thought you said your job was to avoid a mass extinction event!"

"Did I say that?"

"I don't know, what did you say? What do you say now! Don't split hairs with me, I'm not here for hair-splitting! You're killing the world and you want me to remember what words you used to cover your ass? You tell me now! What is your job as head of the Ministry for the fucking Future?"

She swallowed hard. Took a sip of tea. It was cool. She tried to think what to say. Was it wise to try to talk things over with this distraught young man, who was getting angrier by the moment? Did she have any other choice than to do so?

She said, "The ministry was set up after the Paris meeting of 2024. They thought it would be a good idea to create an agency tasked with representing the interests of the generations to come. And the interests of those entities that can never speak for themselves, like animals and watersheds."

The young man gestured dismissively. This was boilerplate, known to him already. "And so? How do you do that, how do you defend those interests?"

"We've made divisions that focus on various aspects of the problem. Legal, financial, physical, and so forth. We prioritize what we do to portion out the budget we're given, and we do what we can."

He stared at her. "What if that's not enough?"

"What do you mean, not enough?"

"It's *not enough*. Your efforts aren't slowing the damage fast enough. They aren't creating fixes fast enough. You can see that, because everyone can see it. Things don't change, we're still on track for a mass extinction event, we're in the extinctions already. That's what I mean by not enough. So why don't you do something more?"

"We're doing everything we can think of."

"But that either means you can't think of obvious things, or you have thought of them and you won't do them."

"Like what?"

"Like identifying the worst criminals in the extinction event and going after them."

"We do that."

"With lawsuits?"

"Yes, with lawsuits, and sanctions, and publicity campaigns, and—"

"What about targeted assassinations?"

"Of course not."

"Why of course? Some of these people are committing crimes that will end up killing millions! They spend their entire lives working hard to perpetuate a system that will end in mass death."

"Violence begets violence," Mary said. "It cycles forever. So here we are."

"Having lost the battle. But look, the violence of carbon burning kills many more people than any punishment for capital crimes ever would. So really your morality is just a kind of surrender."

She shrugged. "I believe in the rule of law."

"Which would be fine, if the laws were just. But in fact they're allowing the very violence you're so opposed to!"

"Then we have to change the laws."

"What about violence against the carbon burning itself? Would bombing a coal plant be too violent for you?"

"We work within the law. I think that gives us a better chance of changing things."

"*But it isn't working fast enough.*" He tried to compose himself. "If you took your job seriously, you'd be looking into how to make change happen faster. Some things might be against the law, but in that case *the law is wrong.* I think the principle was set at Nuremberg—you're wrong to obey orders that are wrong."

Mary sighed. "A lot of our work these days goes to trying to point out the problems created by the currently existing legal regime, and recommending corrections."

"But it isn't working."

She shrugged unhappily, looking away. "It's a process."

He shook his head. "If you were serious, you'd have a black wing, doing things outside the law to accelerate the changes."

"If it was a black wing, then I wouldn't tell you about it."

He stared at her. Finally he shook his head. "I don't think you have one. And if you do, it isn't doing its job. There are about a hundred people walking this Earth, who if you judge from the angle of the future like you're supposed to do, they are mass murderers. If they started to die, if a number of them were killed, then the others might get nervous and change their ways."

She shook her head. "Murder breeds murder."

"Exile, then. Prisons that you contrive, on your own recognizance. What if they woke up one day with no assets? Their ability to murder the future would be much reduced."

"I don't know."

"If you don't do it, others will."

"Maybe they should. They do their part, we do ours."

"But yours isn't working. And if they do it, they get killed for it. Whereas you would just be doing your job."

"That wouldn't justify it."

"So you keep it in the black zone! A lot of the world's history is now happening in the dark. You must know that. If you don't go there yourself, you've got no chance."

She sipped her tea. "I don't know," she said.

"But you're not *trying* to know! You're trying *not* to know!"

She sipped her tea.

Abruptly he stood again. He could barely hold himself still, now; he twitched, he turned this way and that, took a step and stopped. He looked around as if he had forgotten where they were.

"What's the story here?" he said, gesturing at her tiny kitchen. "Are you married?"

"No."

"Divorced?"

"No. My husband died."

"What did he die of?"

"I'm not telling you my life," she said, suddenly repulsed. "Leave me alone."

He regarded her with a bitter gaze. "The privilege of a private life."

She shook her head, focused on her tea cup.

"Look," he said. "If you really were *from* the future, so that you knew for sure there were people walking the Earth today fighting change, so that they were killing your children and all their children, you'd defend your people. In defense of your home, your life, your people, you would kill an intruder."

"An intruder like you."

"Exactly. So, if your organization represents the people who will be born after us, well, that's a heavy burden! It's a real responsibility! You have to think like them! You have to do what they would do if they were here."

"I don't think they would countenance murder."

"Of course they would!" he shouted, causing her to flinch. He shook hard, he quivered where he stood; he reached up and held his head as if to keep it from exploding. His eyes looked like they might pop out of his head. He turned his back to her, kicked the front of her refrigerator, staggered and then went to the window and glanced out, hissing, fuming. "I'm a fucking dead person already," he muttered to himself. He put his hands to the windowsill, leaned his forehead on the glass. After a while his breathing steadied, and he turned and faced her again. He sat down across from her again. "Look," he said, visibly pulling himself together. "People kill in self-defense *all the time*. Not to do that would be a kind of suicide. So people do it. And now your people are under assault. These supposed future people."

She heaved a sigh. She kept her eyes on her tea cup.

He said, "You just want someone else to do it. Someone with less cover than you have, someone who will suffer more for doing it. That way you can keep your good life and your nice kitchen, and let the desperate people take the hit for trying. The very people you're tasked with defending."

"I don't know," she said.

"*I* know. I met with some of them in India. I wanted to join them, but they wouldn't have me. They're going to do what you should be doing. They'll kill and they'll be killed, they'll commit some petty act of sabotage and be put in prison for the rest of their lives, and all for doing the work you ought to be doing."

"You didn't join them?"

"They wouldn't have me."

A spasm contorted his face. He wrestled with the memory. "I was just another firangi, as far as they were concerned. An imperial administrator, like in the old days. An outsider telling them what to do. Probably they were right. I thought I was doing the best I could. Just like you. And I could have died. Just by helping a little health clinic. I *did* die, but for some reason my body lived on after my death. And here I am, still trying to do things. I'm a fool. But they didn't want my help. Probably they were right, I don't know. They'll do what they need to without me, they don't need me. They're doing what your agency should be doing. That'll be harder once they go outside India. They'll get killed for it. So I've been trying to do what I think they might want, here, where I can move around better than they can."

"You've been killing people?"

"Yes." He swallowed hard, thinking about it. "I'll get caught eventually."

"Why do you do it?"

"I want justice!"

"Vigilante justice is usually just revenge."

He waved her away. "Revenge would be okay. But more importantly, I want to help to stop it happening again. The heat wave, and things like it."

"We all want that."

His face went red again. Choked voice again: "Then you *need* to do *more*."

Her doorbell rang.

It was well after midnight. There was no one who would call on her now.

He saw that on her face and lunged toward her. They both had stood up instinctively. "You gave me away!" His terrified face inches from hers.

"I didn't!"

And because she hadn't, she could meet his wild-eyed glare with one of her own. For a second they stood there locked in a gaze beyond telling, both of them panicked.

"There are cameras everywhere," she said. "We must have been seen out on the street."

"Go tell them you're okay." He put his hand in the pocket with the gun.

"All right."

Heart pounding harder than ever, she went to her door, out onto the landing, down the flights of internal stairs to the building's outside door on the ground floor. She opened it, keeping the chain on.

Two police officers, or perhaps private security. "Minister Murphy?"

"Yes, what is it?"

"We received a report that you were seen entering your apartment with a man."

"Yes," she said, thinking hard. "He's a friend, there's no problem."

"He's not listed among your known friends."

"I don't like the implications of that," she said sharply, "but for now, just know that he's the son of an old schoolmate of mine from Ireland. I tell you it's all right. Thanks for checking on me."

She closed the door on them and went back upstairs.

Her apartment was empty.

She wandered around. No one there. Finally she checked the door that led out onto the little balcony hanging over the back lot of the place. It had been left ajar. All dark down there. Overhead, the bare branches of the giant linden that covered the yard blocked the stars with a black pattern. She leaned over the metal railing, looked down. Probably one could downclimb one of the big square posts at the outer corners of the balcony. She wouldn't have wanted to try it herself, but the young man had looked like someone who wouldn't be stopped by having to downclimb a single story.

"I told them you were a friend," she said angrily to the darkness.

She was angry at him and at herself. Her mind raced. She felt sick. It took about a minute for her to run through her options and realize what she had to do. She ran back inside and down the stairs to the outside door, ran out into the street calling, "Police! Police! Come back here! Come back!"

And then they were back, hustling to her fast and staring at her curiously. She told them that she had had to lie to them to keep all three of them from being shot by the man who had been with her. He had seized her in the street, held her at gunpoint; he was gone now, having left while she was talking to them. All true, though ever so incomplete. She realized as she spoke to them that she would never be able to tell anyone what had really happened. Things like that hour were not tellable.

They were on their radios, alerting their colleagues. They led her upstairs, guns drawn. She sat back down on her kitchen chair, took a deep breath. Her hands were shaking. It was going to be a long night.

26

You can hide but you can't run.

There were cameras everywhere, of course. On the public transport systems, in the stores, on the streets. If he went out to get food, he would pass by surveillance cameras several times no matter where he went. If he tried to leave the country, there would be passport control. Not so much within Europe, but still, a check could happen; if he traveled, he would eventually get checked. He had a fake passport, but he didn't trust it would work in situations like that. He was stuck here, in one of the most surveiled countries on Earth.

But he had already learned where and how to hide. He had a place to live; high on the side of the Zuriberg, overlooking the city, were several blocks of community gardens. These plots of terraced and cultivated land were studded by little wooden garden sheds, storage containers for tools and fertilizers and pesticides and the like. One of them had a side wall panel he had pried off in a way he could glue back into position, and once inside the shed he could replace the panel and lie on the floor in his down sleeping bag, and be gone by dawn, leaving no sign of his incursion. This shed he could use as his base. He could hide in Zurich itself.

His fake passport had been the passport of a man back in the States who had died and never had a death certificate filed. Frank's photo had been inserted into it, and to that extent he was this other man, Jacob Salzman. It would stay good for another three years. And he had created a credit card in that name, with some money in the account; and he had converted most of his remaining money to fifty-euro bills. He had a visa for his fake ID that allowed him to stay in Switzerland for most of the coming year, and a supposed job back in America, and an apartment rented in that name down in the city. All that had gone well, over a year ago. There were cameras at the entrance to that apartment building, of course, but the back door that led to the trash compactor didn't have a camera, so he could come and go by that back way sometimes, and use the apartment's bathroom. As Salzman he was a member of one of the lake swim clubs, and could use those facilities too.

And all day he could walk. He could wander without revealing his presence in any obtrusive way. He ate from food stalls that didn't have cameras, and bought other food from Migros, and spent his days in parks where the cameras were few. Thus he lived a life that was not much registering in the system, but was not flagrantly off-grid either. Just low profile. No doubt after kidnapping a UN ministry head and forcibly entering her apartment and spending a night haranguing her, he was the subject of a big police search. Not to mention what had happened down at Lake Maggiore. Salzman might have to be abandoned. But for a good while, at least, he could hide just a kilometer or two from the minister's apartment. And so, for the time being, he did.

27

Now Mary had a new problem: twenty-four-hour police protection.

Of course there were worse problems to have, but it was surprising how upsetting it was to Mary, until she considered it: she had lost her life. Or at least her habits, her privacy. Sad to consider that these were much of what her life had come to.

After the police were done with their questions and investigations that night, she went to bed and tried to get some sleep, and failed. Police officers were still in her kitchen, and downstairs outside the front door. That was likely to remain true for the foreseeable future. She damned her kidnapper with a million damns; she was hating him more and more, the more she thought about it.

Even though what he had said nagged at her. Even though the memory of his face troubled her. His wild-eyed conviction that he was right. Usually she disliked and distrusted people like that, but he had been different, she had to admit. A terrible conviction had been forced into him. His brush with death had made him mad. Although in the end he had only shouted at her. Kidnapped her to argue with her—then also, her bathroom door swinging shut on her—in some ways he had fought hard to hold it together, to do nothing more than persuade her. Words like fists to the face. Paper bullets of the brain. It was enough to make her heart hammer all over again. Her face burned with the memory.

So she went into the office that morning in a very foul mood. They had promised her they would keep her apprised of their progress in the case, but she doubted she would learn anything important or timely. The young man had seemed confident of his ability to hide. That in itself was strange. No one should have that confidence, especially not in Switzerland. She wondered if he had some hideaway in Zurich, or near it, so that he could get to it quickly and go to ground, wouldn't have to go on the run.

She would find out later. Or not. Meanwhile she would be accompanied by bodyguards wherever she went, and some polite Swiss woman or man, or a trio of them, would be installed in her apartment with her. Damn damn damn. The damned fool—she could have killed him.

At her office people crowded around her and commiserated and such. She ordered them to get back to work, and contacted Badim. He was on a train back from Geneva and texted his condolences; he had just heard, he would come to her office when he got in. He hoped she could meet him for lunch. That was good, actually; get out of the office and talk frankly where no one could overhear. Require her new bodyguards to keep a distance.

So just after noon Badim entered her office, and he went to her and briefly held her hands, looked at her closely, gestured at a hug he didn't enact. They left the office and walked to the tram stop, bought sandwiches and a chocolate bar and coffees, and walked back toward the office until they came to the little park that overlooked the rounded green copper roofs of the ETH, and the city across the river to the west. They sat on one of the park benches. Her bodyguards were a couple benches down, the obvious bodyguards anyway.

Mary had had to think about what to say to Badim, and in this matter, as in everything else this morning, her exhaustion caused a flurry of contradictory thoughts to ricochet around in her. Something in her was resisting the idea of telling Badim the full story of her night. Not that she could avoid it entirely.

"I had quite a night last night," she said.

"I'm so sorry," he said. "Are you all right?"

"I'm all right. I guess I was being surveiled? And they saw me go into my apartment with the guy, and they came to check up on me. An hour later, I must say. Must have been camera stuff, and when they saw it they came over. He slipped away while they were inquiring."

"So I heard. I'm glad you're okay. Are you okay?"

"No!"

"I'm sorry."

They ate in silence for a while. The city was its usual forest of cranes over gray stone. No sight of the Limmat from this vantage; to the south, just a narrow arc of the lake, with the long hill that ran south from the Uetliberg backing it.

"I've been thinking about our situation," she said when she had taken the edge off her hunger. "Our dilemma."

"Which is?"

"That we're charged with representing the people and animals of the future, in effect to save the biosphere on their behalf, and we're not managing to do it. We're failing to do it, because the tools at our disposal are too weak. You said something like that the time we walked to the lake. The world is careening along toward disaster, and we can't get it to change course fast enough to avoid a smash."

Badim chewed on his sandwich for a while. "I know," he said.

"So what are we going to do about it?"

"I don't know."

She regarded him. A small dark man, very smart, very calm. He had seen a lot. He had worked for the Indian government and for the government of Nepal, which had begun as a Maoist revolutionary organization. He had worked for Interpol. She said, trying it on, "I think maybe we need a black wing."

That surprised him. He looked at her for a while, blinking, and then said, "What do you mean?"

"I think we need to set up a secret division of the ministry, working in secret to forward the cause."

He considered it. "To do what exactly?"

"I don't know." She chewed for a while, thinking it over. Her kidnapper's vivid glare. The fear she had felt.

"I don't like violence," she said after a while. "I mean, really. I'm Irish. I've seen the damage done. I know you have too. Secret wars, civil wars, the damage never goes away from those. So, I don't mean killing people. Or even hurting them physically. We're not the CIA here. But still, there are other things in the black, I'm thinking. Actions that are maybe illegal, or in some senses ill-advised. Undiplomatic. That would nevertheless forward the cause. We could consider them in secret, on a case-by-case basis, and see if any of them were worth pursuing. Things that we could defend doing if we got caught."

He had stifled a smile, and now he shook his head a little bit. "That's not sounding very black to me. One aspect of a black agency is that they must be uncatchable. Nothing can be written down, nothing can be hacked, no one can talk to outsiders. The people in charge aren't to know about them. If there is any break in the secrecy, you as head of the agency would have to be able to deny all involvement, even any knowledge of such a thing, without explanation or defense."

"You sound like you've had experience with such things."

"Yes." He was looking out at the city now.

"When was that?"

He regarded the gray city, thinking it over. He heaved a small sigh, took another bite of his sandwich, washed it down with a big sip of coffee.

"Now..." he said, as if starting a sentence and then not continuing it.

"Now what?" she said, after he had paused for a while.

"Now," he repeated more firmly, and then looked at her. "Always, in other words."

"What do you mean?"

"I'm sorry. I can't talk about it."

She found she was standing over him. Her paper coffee cup was quivering in her hand, half-squished. He was wincing as he regarded her. He glanced at her coffee as if she might sling it in his face.

"Tell me what you mean," she demanded in a grating voice. "Now."

He sighed again. "Imagine that there might already be a black wing of the Ministry for the Future. That even I myself might have started it after you hired me as your chief of staff."

"You started it?"

"No, I'm not saying that, please. I'm like everyone else. I have some friends in the office. Whatever they do as friends, it would not be quite right to tell you about, precisely so that if something got out and you were asked questions, you could honestly say that you didn't know about it."

"Plausible deniability?"

"Well, no. That I think refers to having a good lie in hand. This would be more like proper functioning of an agency, keeping things as they should be. If ever questions were asked, which hopefully would not happen, but if they did, you would say you didn't know about it. If there were improprieties, others would take the fall, and on you could go with the ministry undamaged."

"As if it would happen that way!"

"Well, it has before. Pretty often in fact. It's a common form of organization. Most unorthodox operations take place without their political heads knowing they exist. Certainly never the details, even if they're aware in a general sense that such things exist. As I thought you might be."

The previous night's events all of a sudden crashed into her and shook her to the core, and she shouted, "I won't be lied to!"

The bodyguards across the little park looked over at them.

"I know," Badim said, looking pinched and unhappy. "I'm sorry. I haven't felt good about it. I'll tender my resignation this afternoon, if you like. Happy to fall on my sword right here and now. But I will remind you that you were just talking about the possible need for such a thing."

"The fuck I was," she said. She waited for her heart to slow down, thinking that over. Too many thoughts were jamming her head at once, creating something like a roar. "I didn't really mean it. And even if I did, I'll be damned if I'll have secrets being kept from me in my own fucking agency!"

"I know," he said, looking down. He took a deep breath, steeling himself. "Tell me, did this not happen to you when you were a minister in Ireland's government?"

"What do you mean?" she cried.

"I mean, you were head of foreign affairs, wasn't it? Did you think you knew everything your agency was doing?"

"Yes, I did."

He shook his head. "Surely not. Of course I don't know for sure. But Ireland was in a civil war for a long time, and that always has aspects that are out of the country. Right? So, your security forces almost certainly kept things from you, and they probably understood you to be knowing about that, and wanting to keep it that way." He shrugged. "It's certainly been true in Nepal and India, and at Interpol too for that matter." Interpol was the agency Mary had hired him out of.

She sat down beside him, hard. Her coffee cup was smashed; carefully she sipped the remainder of the coffee out of one of the folded spots in the rim, put the cup on the ground.

"So I'm naïve, is it? An innocent stateswoman in the world of Realpolitik? Which is no doubt the reason they gave me this job?"

"I'm not saying that. I think they assumed you knew what you were doing."

"So what has this black wing of yours done so far? And who does it consist of?"

"Well, but this is just what you should not be asking. No no, please"—he held up a hand as if to ward off the blow she was about to give him—"people might resign if they knew that their actions were known to higher-ups. Anyway you might not even know these people, I'm not sure how acquainted you are with our whole staff."

He watched her calm down. She picked up her crushed coffee cup and sipped from it again. "I'm head of whatever happens operationally," he went in a low voice, as if sharing a secret to conciliate her. "That's what a chief of staff does. People who help me when doing the needful gets tricky, they would naturally come from various divisions. Cyber security of course, it's in the nature of their work to be preemptive sometimes. Nat cat guys are often helpful, they're used to getting dirty—I mean just physically dirty, you know, working with machines and such. They're at the coal face. They see the damage being done, and they get impatient to do something about it."

That struck a nerve, and her sleepless night came back all in a rush, not that it had ever left. Her stomach was a knot that the sandwich didn't quite fit into.

"I met someone like that last night," she said. She reached out and grasped the back of Badim's hand. "That's the kind of person kidnapped me last night! He was sick of doing nothing, of nothing happening."

"I'm sorry," he said again. He turned his hand up so that it held hers. "Tell me what happened?"

She took her hand back and told him briefly, leaving parts out.

"We probably should have been guarding you more closely," he said when she was done.

"It was close enough, as it turned out. Besides, I think you offered it once, but I said no. I hate that kind of thing."

"Even so. It might be just for a while. They'll probably pick this guy up soon."

"I'm not so sure."

"It's hard to stay off the grid for long."

"But not impossible."

"No," he admitted. "Not impossible."

She shifted back into the bench, shuddered. Really things were fucked. She needed to sleep. But here they were.

"Look," she said, thinking it over. "You grew up in Nepal, right? And I grew up in Ireland. In both places there was a lot of political violence. Which really means murder, right? Murder and all that follows murder. Fear, grief, anger, revenge, all that. The damage never goes away, I can tell you that, and you probably already know it. And then the better murderers in these murder contests tend to take over in the end. It's not at all clear it has ever done any good in the world at all."

Badim wagged his head side to side, not agreeing with her.

"What?" she cried. "You know it's true! The damage has been tremendous!"

He sighed. "And yet," he said. "What was the damage before? And did doing some of these things lessen the damage overall? This is what is never clear."

"So what has this black wing of yours done!" she exclaimed, suddenly frightened.

"I really can't say." He saw her face, temporized. "Possibly some coal plants have experienced problems. They've had to go offline, and the investment crowd has seen that and understood that they won't ever be good investments again. In a sense you could say that worked."

"How do you mean worked?"

"The plants stayed shut, and solar's gotten another investment boost. And new coal plant construction worldwide is down eighty percent since these things started happening."

"That could be because of the Indians shifting to solar."

"Yes."

"Has anyone been hurt or killed?"

"Not on purpose."

"Has anyone been terrorized?"

"You mean scared away from burning carbon?"

"Yes."

"I think it might be good if that had happened, don't you?"

"But how?"

"Well, you know. What really scares people is financial."

"What really scares people is being fucking kidnapped!"

"Granted. Threat of violence. Although access to their money, if people get cut off from that, they are definitely scared."

"Fuck. So you're playing the god game."

"Excuse me?"

"Playing god. Putting people through experiences they think are real, then seeing what they do."

"Maybe. But it wouldn't be just to see what they do. It would be to make them change."

"So you terrorize them!"

"Well, but terrorism means killing innocent people to scare other innocent people into doing what you want. That's what it means today, right? It isn't just boo in the dark."

"No, I suppose not. But you scare people. You use intimidation."

"If we had, it might be a good thing. It might be doing the needful. As you were pointing out yourself, I think."

She nodded. She remembered the young man from the night before. He had scared her, no doubt about it. On purpose. To get her attention. In fact he had had to calm her down a little, maybe, just to get started, so that she would better take in what he had wanted so desperately to convey to her. He had wanted her to pay close attention to what he was saying, so that she wouldn't forget it. A mammal never forgets a bad scare; and they were mammals. And indeed she would not be forgetting.

"I just had that done to me," she said. "That man wanted to scare me."

"It sounds like it."

"It worked," she said, looking at him.

He looked back, letting her think that over.

She considered it. She felt like her stomach was going to implode.

"All right," she said. "Look. I want to know what's going on. I'll lie if it comes to that, or take the hit if need be. But I want to know."

"Do you really?"

"I do. You have to promise me you'll tell me. Do you promise?"

For a long time he didn't speak. He looked out at the city, then at the ground. Finally: "All right," he said. "I'll tell you what you need to know."

"Everything!"

"No."

"Yes!"

"No." He looked her in the eye. "I can't tell you everything. Because look: there might be some people who deserve to be killed."

She stared at him. The sandwich inside her was getting trash-compacted, it felt like; it would be better if she could just throw it up.

Finally she said, "Maybe I can help you to focus this program better than it is now. I know some things now that I didn't yesterday."

"I'm truly sorry about that."

"Don't be. They're things I should have known all along."

"Maybe."

She thought it over. "Fuck!"

"I know."

"But... Well, we have to do something. Something more than we've been doing."

"I think maybe so."

"Because right now we're losing."

"It's a fight. That's for sure."

28

The Hebrew tradition speaks of those hidden good people who keep the world from falling apart, the *Tzadikim Nistarim*, the hidden righteous ones. In some versions they are thirty-six in number, and thus are called the *Lamed-Vav Tzadikim*, the thirty-six righteous ones. Sometimes this belief is connected with the story of Sodom and Gomorrah, and God's promise that if he could be shown even fifty good men in these cities (and then ten, and then one) he would spare them from destruction. Other accounts refer the idea to the Talmud and its frequent references to hidden anonymous good actors. The hidden quality of the *nistarim* is important; they are ordinary people, who emerge and act when needed to save their people, then sink back into anonymity as soon as their task is accomplished. When the stories emphasize that they are thirty-six in number, it is always included in the story that they have been scattered across the Earth by the Jewish diaspora, and have no idea who the others are. Indeed they usually don't know that they themselves are one of the thirty-six, as they are always exemplars of humility, *anavah*. So if anyone were to proclaim himself to be one of the *Lamed-Vav*, this would be proof that actually he was not. The *Lamed-Vav* are generally too modest to believe they could be one of these special actors. And yet this doesn't keep them from being effective when the moment comes. They live their lives like everyone else, and then, when the crucial moment comes, they act.

If there are other secret actors influencing human history, as maybe there are, we don't know about them. We very seldom get glimpses of them. If they exist. They may be just stories we tell ourselves, hoping that things might make sense, have an explanation, and so on. But no. Things don't make sense like that. The stories of secret actors are the secret action.

29

We set up camp on the Thwaites Glacier, about a hundred kilometers inland from the coast of Antarctica. Thwaites was chosen because it was one of the fastest-moving big glaciers, combined with a fairly narrow gateway to the sea relative to other glaciers in its class. There were about fifty glaciers in Antarctica and Greenland that were going to dump ninety percent of the ice that was going to end up in the sea in the next few decades, and in that group, this one looked like one of the best test subjects. So there we were.

The camp was a typical Antarctic field camp, of the larger variety. A runway long enough for C-130s was secured; this meant landing first in Twin Otters, then checking a two-mile stretch of ice to make sure there weren't any unseen crevasses. Blow up and bulldoze any crevasses you find. Eventually you get a full landing strip. After the C-130s could land, a few Jamesways were flown in and assembled to serve as galley and commons. These Jamesways are basically insulated Quonset huts of World War Two vintage: they are floored half cylinders, very simple, easy to assemble, and pretty energy efficient. These that we set up were powered and heated mostly by solar panels, as this was a summer camp and the sun would be up all the time. A few tents for people who liked to sleep away from the huts—I'm like that myself—and a couple of yurts like the Russians use. When we were done we had a little nomad village, colorful against the white background: yellow, orange, khaki, red.

The drilling equipment was one of the ice coring systems that have been operated for a long time in Antarctica to get core samples, or drill down to subglacial lakes, or get through an ice shelf to the ocean below. They shoot hot water at the ice through a thing like a giant shower head, and melt it. As the ice melts the drill head goes lower, and down it all goes.

The meltwater gets pumped out, and the hole can be sleeved with a heated sleeve if you want to keep it open, which we did. Some of the meltwater gets recycled into the drill head's feeder tank to be reheated. The rest gets piped away and dumped where it can spill out and freeze without messing anything up. Progress down the hole is slow by some standards, fast by others. A typical speed for a two-meter-diameter hole is about ten meters per hour. Fast, right? It's a lot easier than drilling in earth or anything else hard, although you do need a lot of power to heat the water. That used to mean burning a lot of diesel fuel, but solar will work too if you've got enough of it.

This time, when we got to the bottom of Thwaites, about nine hundred meters down from the surface, water came up the hole, but not all the way to the surface. It was under stupendous pressure from the weight of the ice on it, but no matter how thick the ice, the water under it gets shoved up the hole only about ninety percent of the way. The physics of hydrology dictates this, although we ran a pool anyway to see who could guess the actual height of the rise the closest, because there are always variants in play that mean the actual level in the hole will range a few meters one way or the other. In any case about ninety percent of the way to the top, so the energy needed to pump water the rest of the way to the surface was not that great. So we did that, but no matter how much we pumped out, it was replenished from below. This was the crucial question; could we empty out the water down there? Would we be able to pump up so much that there was nothing left to pump?

There turned out to be a lot of water under the Thwaites, as predicted. Bigger and bigger summer pools of meltwater on the surface had run down moulins, which are like vertical rivers that run down cracks in the ice. That water bottoms out on the bedrock and then lubricates the slide of the ice over it, until the ice is like riding down a water slide. The glaciers are therefore becoming more like rivers than ice fields, flowing almost as fast as some flat-country rivers, but with a hundred times more water in them than the Amazon, or even more. And the water in the Amazon was rain the week before, but the ice in Antarctica has been perched up there for the last five million years at least. So we're going to see sea level rise, big time.

So if we could pump that subglacial water out from under the glacier, the ice would thump back down onto bedrock and slow down to the grind-it-out speed that used to be normal. After that we would keep pumping subglacial water out, and the ice would stay grounded on the bedrock, and it would stay at its old speed, deform viscously, shatter in crevasse fields, all the usual behaviors, and at the old speeds. Thus the plan.

We were here to test the method. Some said the water at the bottom would get mixed with glacial silt until it was the consistency of toothpaste, and hard or impossible to pump up. Others said the silt was long since gone, the bottom clean as a whistle after millions of years of scraping, and the new water down there would be pure, and therefore gush into the sky the moment we punctured the ice, wrecking everything and maybe drowning or freezing us, or both. People good at hydrology scoffed at that last prediction, but you had to try it to be sure. We could only be sure by testing it.

So we got that first well melted, and the water came up about eighty-seven percent of the hole, so that I won the pool. Come on, Pete, they said, you can't win a pool you set up yourself! Sure I can! I replied. And we pumped water from the top of the column and it kept refilling to the same level. This went on for four days. All good!

But then the water from below cut off abruptly. Some shift in the ice down there had presumably cut off our hole. Like capping an oil well, some said, but not really. A shift in the ice, I reckoned. There was a crevasse field about thirty kilometers upstream that made me wonder how far downstream the ice was broken up.

Anyway, not good. Clearly, if the method was going to work, the wells were going to have to be kept open. So it was a question of whether our cut-off was a typical thing or an unusual accident. We also had to find out if we could fix it. Re-open it, in other words, and then prevent whatever had happened to it from happening again. Failing that, we would have to abandon it and presumably start a new one. But if we couldn't figure out the problem, and if this was what was going to happen all the time on these quickening glaciers, maybe the whole idea wasn't going to work.

A team was dispatched out to us to run some seismic tests, also to bring in more cameras and other monitors to drop down the hole to see what we could see. We had to wait out a windstorm for them to be able to fly in, so for a couple of days there was nothing to do. I thought we were going to be able to figure it out, but the mood in the dining hall got more and more what you might call apprehensive.

Pete, this might turn out to be another fantasy solution, one of my postdocs said to me. One of those geoengineering dreams of redemption. Silver bullet fix that just shoots us in the head kind of thing.

I sure hope not, I said. I like the beach.

Hey, someone else said, geoengineering isn't always just a fantasy. The Indians did that sulfur dioxide thing and that worked. Temperatures dropped for years after that.

Big deal, someone else replied.

It was a big deal!

But it didn't do anything to solve the bigger problem.

Of course not, but it wasn't meant to do that! It was a fix!

That's why we're here, I pointed out. This is a different fix.

Right. But look, Dr. G, even if we could get this to work, the glacier would still move downstream, so eventually this whole pumping system would get swept out to sea. It would have to be rebuilt up here again.

Of course! I said. It's like painting the Golden Gate Bridge. All kinds of things have to be done like that. Maintenance stuff.

Besides, what's the alternative? someone pointed out.

It'll cost a ton!

What's *cost*? I said. Postdocs can be so stovepiped, it would be funny if it weren't so alarming. I clarified reality for them: Look, if you have to do something, you have to do it. Don't keep talking about *cost* as if that's a real thing. Money isn't real. Work is real.

Money is real, Dr. G. You'll see.

This method is the only way that will work.

But it didn't work! It cut out on us!

Yes, but this was just the start. If at first you don't succeed—

You'll never get funded again.

30

When thinking about the suspended years before the Great Turn, what some have called the Trembling Twenties, historians have speculated whether it was part of the Great Turn itself, or the last exhausted moments of the modern period, or some sort of poorly theorized interregnum between the two. Comparisons have been made to the period 1900–1914, when clearly the twentieth century had not yet properly begun and people were unaware of the stupendous catastrophe approaching. The calm before the storm. But there is nothing like consensus here.

Of course attempts are always made to divide the past into periods. This is always an act of imagination, which fixes on matters geological (ice ages and extinction events, etc.), technological (the stone age, the bronze age, the agricultural revolution, the industrial revolution), dynastic (the imperial sequences in China and India, the various rulers in Europe and elsewhere), hegemonic (the Roman empire, the Arab expansion, European colonialism, the post-colonial, the neo-colonial), economic (feudalism, capitalism), ideational (the Renaissance, the Enlightenment, Modernism), and so on. These are only a few of the periodizing schemes applied to the flux of recorded events. They are dubiously illuminative, perhaps, but as someone once wrote, "we cannot not periodize," and as this appears to be true, the hunt is on to find out how we can best put this urge to use. Perhaps periodization makes it easier to remember that no matter how massively entrenched the order of things seems in your time, there is no chance at all that they are going to be the same as they are now after a century has passed, or even ten years. And if on the other hand things feel chaotic to the point of dissolution, it is also impossible that some kind of new order will not emerge eventually, and probably sooner rather than later.

"If things feel" like this or that: these feelings too are linked to periodization, because our feelings are not just biological, but also social and cultural and therefore historical. Raymond Williams called this cultural shaping a "structure of feeling," and this is a very useful concept for trying to comprehend differences in cultures through time. Of course as mammals we feel emotions that are basic and constant: fear, anger, hope, love. But we comprehend these biological emotions by way of language, thereby organizing them into systems of emotions that are different in different cultures and over time. Thus for instance, famously, romantic love means different things in different cultures at different times; consider ancient Greece, China, medieval Europe, anywhere.

So how you feel about your time is partly or even largely a result of that time's structure of feeling. When time passes and that structure changes, how you feel will also change—both in your body and in how you understand it as a meaning. Say the order of your time feels unjust and unsustainable and yet massively entrenched, but also falling apart before your eyes. The obvious contradictions in this list might yet still describe the feeling of your time quite accurately, if we are not mistaken. Or put it this way; it feels that way to us. But a little contemplation of history will reveal that this feeling too will not last for long. Unless of course the feeling of things falling apart is itself massively entrenched, to the point of being the eternal or eternally recurrent individual human's reaction to history. Which may just mean the reinscription of the biological onto the historical, for we are all definitely always falling apart, and not massively entrenched in anything at all.

31

India is now leading the way on so many issues! We remain horrified by the memory of the heat wave, galvanized, and if not unified then nearly so, in a broad coalition determined to re-examine everything, to change whatever needs changing. You see it all around you!

This situation is partly an accident of history, admittedly, in that BJP was in power when the heat wave hit and so they were associated with it, with whatever justice who can say, and yet as a result they were thrown out of power and discredited forever, such that hopefully they will never come back, and likewise good riddance to their RSS fake-traditional Hinduistic ethnic-nationalist triumphalism. It was not the true Indian way, as all see now, or most. The heat wave made the revulsion against that pernicious nonsense very great. The true Indian way has always been syncretic, right from the very beginning of our civilization.

Meanwhile the Congress party was clapped out, a thing of the past, its heroic era buried by decades of corruption. Possibly a cleaning-up can someday happen for the party that did so much for India, but that's for another time, a work in progress; in any case, the upshot post–heat wave was a complete loss of faith in both Congress and BJP, such that the world's biggest democracy was left with no nationally dominant party. This was an opportunity, as many realized, and the work being done now has been accomplished by the creation of a broad coalition of forces, many of them representing very large Indian populations which never had much political power before, or even very good political representation. The energy from this new coalition is palpable everywhere. Things are changing.

And there are good models to study and use, Indian models. Kerala has been big in this regard as a high-functioning state for almost a century now, devolving power to the local, and the state government alternating in a scheduled way between Left and Congress leadership. Much has been taken from Kerala and applied nationally. Then also Sikkim and Bengal have been developing an organic regenerative agriculture that, at the same time it provides more food than before, also sequesters more carbon in the soil, and this too has been taken up across the country. Indian agriculture moving into its post–green revolution is also a giant step toward independent subtropical knowledge production, achieved in collaboration with Indonesian and African and South American permaculturists, and its importance going forward cannot be over-emphasized.

Land reform is part of that, because with land reform comes a return to local knowledge and local ownership and thus political power. The new agriculture is also labor intensive, as to a certain extent people must replace the power of fossil fuels and pay close attention to small biomes, and of course we have that labor power and that close attention. Especially as the caste system is once again acknowledged to be a bad remnant of our past, a remnant also very associated with BJP, who along with selling our country to global financial predators also demonized so many ethnicities, and told so many Indians that they weren't really Indian. Now it's time to wipe that slate clean, and you see all the castes, including Dalits, joining society fully—also all the languages put on the same footing legally, as well as all the ethnic groups and religions across the country. Now we are all Indian together. It is truly a coalition. Which makes the name of the party very apt. May the Coalition hold!

Certainly its work so far has been admirable. Since sweeping the elections, the national Coalition government has completed the nationalization of all the country's energy companies, and set to work decommissioning all coal-fired plants. Completing the clean electrification of the country is being accomplished by construction of massive solar power arrays, and then electricity-storing facilities, and a refurbished national grid. This again has been labor intensive, but India has lots of people. And lots of sunlight. And lots of land.

The Coalition was very willing to take ideas from elsewhere, even from the Chinese, who have much to teach us about nationalization. It's a very strong feeling now, that India belongs to Indians—that Indians are not to be sold by other Indians as cheap labor to fill the economic needs of global capital—that the colonial and even now the post-colonial days are over. It is the New India now, and everything is to be reconsidered. Many a time we now say to each other, when the arguments get intense, as they always do, Look, my friend—never again. Never again. This reminds us of the heat wave and the stakes involved, but it is also a more general rejection of the bad parts of the past. Never again, we remind each other, and then go on to consider, So what must we do here, to get to an agreement and act on it?

And so India is coming into its own. We are the new force. People around the world have begun to take notice. This too is new—no one elsewhere has been used to thinking of India as anything but a place of poverty, a victim of history and geography. But now they are looking at us with a little bit of confusion and wonder. What is this? A sixth of humanity on one big triangular patch of land, caught under the blazing sun, cut off by a mighty range of mountains: who are these people? A democracy, a polyglot coalition—wait, can it be? And what can it be? Do we make the Chinese, who so decisively stepped onto the world stage at the start of this century, look dictatorial, monolithic, brittle, afraid? Is India now the bold new leader of the world?

We think maybe so.

32

Mary: Dick, what are you and your team doing to make current economics more helpful to the people of the future?

Dick: We've been looking at discount rates. We're studying what India is doing to their discount rate, it's very interesting.

Mary: How does that relate to future people?

Dick: It's very central. We discount the future generations. It works by analogy to how we treat money. With money, a euro you own now is worth somewhat more than the promise of a euro that will come to you a year from now.

Mary: How come?

Dick: If you have it now, you can spend it now. Or you can bank it and earn interest on it. Like that.

Mary: So how much is the discount? How does it work?

Dick: The rate varies. It works like this: if you would take ninety euros now rather than the promise of a hundred euros a year from now, that discount rate is point nine (0.9) a year. Applying that rate, a hundred euros coming to you in twenty years is worth the same as about twelve euros today. If you go out fifty years, that hundred euros you would get then is worth about half a euro today.

Mary: That seems like a steep rate!

Dick: It is, I'm just using it to make it clear to you. But steep rates are pretty common. Someone once won the pseudo-Nobel in economics for suggesting a four percent discount rate on the future. That's still quite high. All the different rates and time intervals get traded, of course. People bet on whether the value will go up or down relative to what got predicted. The time value of money, it's called.

Mary: But this gets applied to other things?

Dick: Oh yes. That's economics. Since everything can be converted to its money value, when you need to rate the future value of an action, to decide whether to pay to do it now or not, you speak of that value using a discount rate.

Mary: But those future people will be just as real as you and I. Why discount them in the same way you do money?

Dick: It's partly to help decide what to do. See, if you rate all future humans as having equal value to us alive now, they become a kind of infinity, whereas we're a finite. If we don't go extinct, there will eventually have been quite a lot of humans—I've read eight hundred billion, or even several quadrillion—it depends on how long you think we'll go on before going extinct or evolving into something else. Whether we can outlast the death of the sun and so on. Even if you take a lower estimate, you get so many future people that we don't rate against them. If we were working for them as well as ourselves, then really we should be doing everything for them. Every good project we can think of would be rated as infinitely good, thus equal to all the other good projects. And every bad thing we do to them is infinitely bad and to be avoided. But since we're in the present, and trying to decide which projects to fund, with limited resources, you have to have a finer instrument than infinity when calculating costs and benefits. Assuming you're going to only be able to afford a few things, and you want to know which of them get you the most benefits for the least cost.

Mary: Which is what economics is for.

Dick: Exactly. Best distribution of scarce resources and so on.

Mary: So given that, how do you pick a discount rate?

Dick: Out of a hat.

Mary: What?

Dick: There's nothing scientific about it. You just pick one. It might be a function of the current interest rate, but that shifts all the time. So really you just choose.

Mary: So the higher the discount rate, the less we spend on future people?

Dick: That's right.

Mary: And right now everyone chooses a high rate.

Dick: Yes.

Mary: How does that get justified?

Dick: The assumption is that future people will be richer and more powerful than we are, so they'll deal with any problems we create for them.

Mary: But now that's not true.

Dick: Not even close to true. But if we don't discount the future, we can't quantify costs and benefits.

Mary: But if the numbers lie?

Dick: They do lie. Which allows us to ignore any costs or benefits that will occur more than a few decades down the line. Say someone asks for ten million to enact a policy that will save a billion people in two hundred years. A billion people are worth a huge number of dollars, if you take a rough average of the insurance companies' monetary valuations for a human life. But using the point nine discount rate, that huge number might equal only five million dollars today. So do we spend ten million now to save what is calculated as being worth five million after the discount rate is applied? No, of course not.

Mary: Because of the discount rate!

Dick: Right. Happens all the time. Regulators go to government budget office to get a mitigation project approved. Budget office uses the discount rate and says, absolutely no. Doesn't pencil out.

Mary: All because of the discount rate.

Dick: Yes. It's a number put on an ethical decision.

Mary: A number which can't be justified on its merits.

Dick: Right. This often gets admitted. No one denies future people are going to be just as real as us. So there isn't any moral justification for the discounting, it's just for our own convenience. Plenty of economists acknowledged this. Robert Solow said we ought to act as if the discount rate were zero. Roy Harrod said the discount rate was a polite expression for rapacity. Frank Ramsey called it ethically indefensible. He said it came about because of a weakness of the imagination.

Mary: But we do it anyway.

Dick: We kick their ass.

Mary: Easy to do, when they're not here to defend themselves!

Dick: True. I like to think of it as a rugby match, with present-day people as the New Zealand All Blacks, playing against a team of three-year-olds, who represent the people of the future. We kick their ass. It's one of the few games we're good at winning.

Mary: I can't believe it.

Dick: Yes you can.

Mary: But what do we do?

Dick: We're the Ministry for the Future. So we step in and play for the three-year-olds. We substitute for them.

Mary: We play rugby against the All Blacks!

Dick: Yes. They're pretty good.

Mary: So they'll kick our asses too.

Dick: Unless we get as good as they are.

Mary: But can we do that?

Dick: We may be sticking with this analogy a little too far here, but let's do that for the fun of it. So now I'm thinking about that movie about the South African football team, when the World Cup was held in South Africa. They were a beginner team, but they ended up winning it all.

Mary: How did they do that?

Dick: You should watch the movie. Basically, they were playing for more than the game. The other teams were playing because that's what they did. It was their profession. But those South Africans, they were playing for Mandela. They were playing for their lives.

Mary: So . . . is there a way we can make the calculations better?

Dick: This is where India comes into it. Since the heat wave, they've been leading the way in terms of re-examining everything. So regarding this issue, you could just set a low discount rate, of course. But Badim tells me that in India it was traditional to talk about the seven generations before and after you as being your equals. You work for the seven generations. Now they're using that idea to alter their economics. Their idea is to shape the discount rate like a bell curve, with the present always at the top of the bell. So from that position, the discount rate is nearly nothing for the next seven generations, then it shifts higher at a steepening rate. Although they're also modeling the reverse of that, in which you have a high discount rate but only for a few generations, after which it goes to zero. Either way you remove the infinities from the calculation, and give a higher value to future generations.

Mary: Good idea.

Dick: We've been running modeling exercises to see how various curves play out in the creation of new cost-benefit equations. It's pretty interesting.

Mary: I want to see that. Run with that.

Dick: The All Blacks will be trying to tackle us, I warn you. They tackle to hurt. They'll be trying to get us to cough up the pill.

Mary: When you get hit, pass the ball to me. I'll be on the inside to receive I will.

Dick: Good on ya mate.

33

They killed us so we killed them.

Everyone in our cell had helped to clean up after the heat wave. You don't forget a thing like that. I myself didn't speak for three years. When I did I could only say a few things. It was like I was two years old. I was killed that week, and had to start over again. Lots of the Children of Kali had gone through similar experiences. Or worse. Not all of my comrades were human.

It was a question of identifying the guilty and then finding them and getting to them. The research and detective work was done by another wing. A lot of the guilty were in hiding, or on fortress islands or otherwise protected. Even when identified it wasn't easy to get near them. They knew the danger.

Methods were worked up over many iterations. We took a lot of losses at first. Of course suicide bombing is often effective, but this is a crude and ugly way to go about it, and uncertain. Most of us didn't want to do it. We weren't that crazy, and we wanted to be more effective than that. Much better to kill and disappear. Then you can do it again.

For that, drones are best. Much of the job becomes intelligence; finding the guilty, finding their moments of exposure. Not easy, but once accomplished, boom. The drones keep getting faster and faster. The guilty often have defenses, but these can often be overwhelmed by numbers. A swarm of incoming drones the size of sparrows, moving at hundreds or even thousands of meters per second—these are hard to stop. The guilty died by the dozens in those years.

Eventually they stayed indoors for the most part. At that point, a decade into the campaign, they knew they were in trouble. Security redoubled. It became a question, or several questions. Were there still people left so guilty they deserved to die? Yes to that one. Could we get to them? Harder.

We sometimes joined domestic staffs or landscaping crews, and worked for years. Other times it was a matter of breaking and entering. Sometimes they could be caught during transit, while less protected. Sometimes their bodyguards would have to be killed too. They shouldn't have taken work like that. Protecting mass murderers makes you complicit in mass murder. So we didn't worry about them.

The only thing we worried about was what the guilty ones always call "collateral damage." In other words, the accidental killing of innocents to kill your target. The guilty do it all the time, it's one sign of their guilt, but we don't. It's a principle. Kali is very fair and very meticulous. If to kill a hundred guilty you had to kill one innocent, no. It's against the law.

So it was often very tricky. One time I had to crawl in through air ducts. One intake had been left unmonitored, a mistake. All in the dark, but the building's plan was clear in my mind. I got to the intake not monitored. Broke into it and crawled and crawled. Left right, up down. I had with me a plastic knife, pliers, and screwdriver. So. Unscrew the screws holding the master bedroom's ceiling vent screen in place, from above using the pliers, in silence and moving very slowly. Took two hours. Then get to feet, making sure feet have not gone to sleep, confirm location of guilty one by night vision goggles and micro-periscope looking through vent slots. This one was a weapons manufacturer. There are a lot of them, but the ones at the top, who own majority shares, they aren't so many. Several hundred identified at this point. All death dealers. Mass murderers for cash. You may know some.

Leap in the air, come down on the vent, crash down into room right on bed, trailing rope ladder. Stab the guilty one in the torso quickly four times, then the neck, several times. Night vision goggles make blood look black. Guilty one dead for sure.

Back up rope ladder, ignoring the other person also in bed, now on floor, shocked into immobility, or perhaps trying to avoid attention. Good idea. Back through air ducts, crawling fast. Out onto compound wall, up to roof, drone waiting to carry me up and away like packaged goods.

Now to spread the headset photos, spread the story. The guilty need to know: even in their locked compounds, in their beds asleep at night, the

Children of Kali will descend on you and kill you. There is no hiding, there is no escape.

One down, several hundred to go. Although the list might get extended. Because Kali sees all. And the Children of Kali are not going away until all the guilty are gone. Be advised.

34

Notes for Badim again, on trip with B and Mary to India.

Fly into Delhi, met by Chandra, no longer in government, but asked by B to meet us and introduce us to new minister and staff. C takes us into government house and introduces us to her replacement and his staff. Meet and greet, then updates. Discussion of solar radiation management applied post–heat wave. They claim to have depressed temperatures in India two degrees and globally one degree, for three years, with decreasing effect, until six years later back to pre-operation levels. No discernible effect on monsoon during that time.

M questions this last assertion and C testy in response. Monsoon variability increasing for last thirty years, somewhat like California weather in that the average is seldom hit, most years much higher or lower than average, which is an artifact only. M objects, says thought monsoon was regular as rain in Ireland, crucial to crops and life generally, July through September daily rain, how variable could it be? Very variable, C replies. Not happy to be challenged on this. Daily rain a myth. Weeks can pass in August, etc. M looking skeptical, as is B.

B intervenes. What are graphs showing? Total rainfall year to year, monsoonwise. Why no graph?

Staff peck around a bit and bring up graph. Monsoon rain indeed fluctuating more through last two decades, and after their geoengineering maybe a bit more so. Second year after application particularly low, semi-drought, especially in the west. Another problem, C points out. Monsoon not the same east to west, always that way.

M asks what plan is going forward. Are they going to do it again? Because global average temperatures rising again. A few wet-bulb 34s in the previous few years, lots of deaths. Wet-bulb 35s very likely to happen again somewhere, and soon.

Exactly, C says. And wet-bulb 35 is deadly to all, but even wet-bulb 33 is also bad enough to kill lots of people.

Of course, M says. So does that mean you're going to do it?

C defers to new minister, Vikram. V says, We are certainly ready to do it. It will be a more orderly procedure this time (not looking at C as he says this), involving democratic processes and expert consultation. But we are ready.

B asks, A double Pinatubo this time, I hear?

V: Probably yes. That is what first intervention was understood to be.

M and B not looking at each other. Finally M says, There are questions of sovereignty here, I know. But India signed the Paris Agreement along with all other nations, and the Agreement has protocols for this kind of thing that all signatories have agreed to adhere to.

We may break the treaty, V says. Again. That's what we have to decide.

B points out penalties for this may be high. Won't be like after heat wave.

V: We are aware of that. Part of the deliberative process. Is benefit worth cost?

C adds sharply, We won't allow another heat wave just so we can be in compliance with a treaty written up by developed nations outside of trop-ics and their dangers.

Mary: Understood.

Meeting ends. No one looking happy.

Mary asks if she can be taken to see the site of the heat wave.

C says no. Nothing to see there. Not a tourist site.

Even less happiness.

B: Where to then? Anything you would like to show us?

V and C exchange look. Yes, V says. Must stay in Delhi himself, but C can take M and B out to see farms in Karnataka.

Farms?

New paradigm for farms. Come see.

Happy to, claim M and B unconvincingly. Not ag people really.

Next day in Karnataka after short flight. Green. Hills to east, also green. Terraces on slopes, but much land flat. Green of several shades, but also rect-angles of yellow, orange, red, purple, dark brown, even pale blue. Flowers of spice crops apparently. Everything grows here, new local host says, name Indrapramit. Best soil and climate on Earth. Now banking 7 parts per thousand of carbon per year, immense drawdown. Also food for millions. Local tenure rights for local farmers, no absent landlords anymore except for India herself, the Indian people, as represented by Karnataka state, also the district and village. Stewards of land. Keep room for wild animals in the hedgerows and habitat corridors. Tigers back, dangerous but beautiful. Gods among us. And all organic. No pesticides at all. Sikkim model now applied to ag all over India. Kerala model for governance, the same.

Communist organic farmers, B notes. He thinks it's funny, M doesn't think he should joke. Locals happy to agree to his characterization. Also these changes mean end of caste's worst impacts, they claim. Dalits now involved, women always half of every panchayat, an old Indian law now applied for real. Now there are farm tenure rights, full ownership of one's work, its surplus value. Women and all castes equal, Hindu and Muslim, Sikh and Jain and Christian, all together in New India. Communist organic farmers just the tip of the iceberg.

Riots we heard about? B asks. Again M annoyed with him.

But again hosts happy to explain. Indra: Some absentee landlords, on being eminent-domained, hired BJP thugs from city to come here and beat people up. It was one-on-one, so *satyagraha* not so effective. Can't lie down in front of a thug to much effect. Better tactic against armies, ironically. So had to fight it out, defend selves. Swarmed the invaders, beat them pretty thoroughly. Reported as riot but actually repelling of invaders. Sort of like organic ag; you have to use integrated pest management.

B: Is one tool of integrated pest management targeted assassinations? Global community not happy with murders of private citizens. A terror tactic, he says, like thugees of yore, these Children of Kali.

M again staring at him as if thinking he is deliberately trying to be offensive.

C also not pleased. Says sharply, Murder rate never really a concern when poor people are being killed by terrorists hired by rich people. Turnabout may not be fair play, but not everything that happens is government sponsored, as distinguished visitors well know. Lots of forces unleashed now. Kali just one of the gods in action now, remember that. So, integrated pest management—yes.

M seizes on this to change topic. How does it work if you get a crop infestation? Insects you don't want, no insecticides to use, what do you do?

Indra: There are insecticides, they just aren't poisonous chemicals. Other bugs, mainly. Biological warfare.

Does that work?

Not always, no. But when crops are lost to infestations we can't stop, we clear the fields and send all the waste plant material to the vats. They are part of our system too, and include bugs on our side you might call them, that will eat spoiled crops no problem, eating also the pests that spoiled them. All of that is food for the vats. Vat amoeba don't care what they eat, being omnivores. And from them we get a kind of flour, also the ethanol that powers any machine that still needs to run on liquid fuels. Then for some kinds of infestations we burn the land, let it lie fallow a season or two, then back in business. Try something different each time. We keep learning things, it's a work in progress.

So you have microbacterial Children of Kali, B jokes. Again M annoyed.

Everything under the sun cycles, Indra says.

This perhaps also acknowledges we are all getting cooked in midday sun. Mad dogs and Englishmen, and Irish women. Even B looks overheated.

Locals bear down on their enthusiasm by pointing up. So much sun! It's power, right? We can use solar power to pull water right out of the air, hydrogen out of the water, grow the plants that provide for bioplastics and biofuels for whatever still needs liquid fuel, use hydrogen to power turbines. Sun also helps grow forests that draw down carbon, and fuel the biochar burners, and provide the wood for building. We are a fully recycling solar powerhouse. A green power. Other countries don't have our advantages in sunlight, and minerals, and people, especially people. And ideas.

M and B nodding as politely as they can manage. Heard it before. Preaching to converted. Hotter than hell out here.

The New India: all agree. Our Indian hosts very pleased with how things are going here. But there's an edge to them too. M and B definitely aware of that edge, emanating strongly from C, also from locals. Aggressive pride. Don't tread on me. No outsider gets to tell India what to do, not anymore. Never again. Post-colonial anger? Post-geoengineering defensiveness? Tired of the Western world condescending to India? All of the above?

35

We came into Switzerland on a train from Austria. Austria was sending closed trains from Italy through to Switzerland, and in St. Gallen the Swiss stopped them and required passengers to get off and go through a registration process. That happened to us. Most of the people on our train were from Algeria or Tunisia, boat people who had landed in Italy hoping to get to France or Suisse Romande, where we would speak the language. Getting into Switzerland would be a big step on the way.

We were herded through rooms in a giant building, like immigration control buildings at airports, but older. We were interrogated in French, and then we were separated into men and women, which caused a lot of distress and anger. No one understood it until we were led into smaller examination rooms and given a cursory physical that included stripping to the waist and submitting to a chest X-ray. Apparently they were looking for signs of tuberculosis. That was offensive and disturbing enough that when we were dressed again and reunited with the women, and we found they had been forced to undergo the same process, which had been administered by women when it came to the X-rays, but run by men in the other parts of the process, we got mad. The whole thing was dehumanizing, and of course this was not the first time it had happened, refugees are by definition less than human, having lost their homes, but perhaps it was some kind of last straw. Something about being in Switzerland, which had a reputation as a clean orderly lawful place, and then being treated like animals, made us mad. Of course the irony was not lost on some of us that it was only because we were in Switzerland and expecting to be treated well that such behavior was offensive; in Egypt or Italy such demeaning treatment would have been expected and thus submitted to. Worse things had happened to us already. Howsoever that may be, we were mad, and when guards marched us back to the trains and we were led to the more southerly tracks, which seemed to be heading back into Austria, many cried out to object, and nothing the guards could say comforted us. We refused to get on trains set on those tracks, because it made no sense to us; we had seen other trains using the southern tracks to head east, and we were sure our train would head that direction too. And being herded onto trains has a very bad feel to it, of course.

Reinforcements appeared to help the guards herd us, and these had long rifles slung over their shoulders. We shouted at them too, and when they started to unsling their rifles and shift them into firing position, some of the young people charged them and the rest of us followed. It had been seven months since leaving Tunisia, and something had snapped in us.

None of the Swiss soldiers fired at us, but as we left the railroad sidings and rushed the buildings in a mass, many more soldiers appeared, and suddenly the air was filled with tear gas. Some of us fled, some charged the police line, and in front of the building the fighting got intense. It seemed clear the police had orders not to shoot us, so we went at them hard, and somehow a gang of young men got one policeman down and got his rifle from him, and one of them shot at the police and then everything changed. Then it was war, except our side had only that one rifle. We started going down left right and center, screaming. Then someone said They're rubber bullets, they're just rubber! And we charged again, and in all the confusion a group of us got inside the building. It was the safest place to be given what was happening outside. But by that time we had lost our minds, we had seen our people shot down, rubber bullets notwithstanding, so we thrashed everyone we caught in that building, and someone found something flammable and torched the big receiving room, and though it was a small blaze, by no means was the building on fire, still there was a lot of smoke, not as painful as the tear gas but probably worse for one's health over the long haul. We didn't know, we had no idea, we were just lashing out, trying to keep the police out of the building at the same time we were trying to set it on fire, even with us inside it. I suppose at that point we were on a kind of suicide mission. None of us cared at that point. I will never forget that feeling, of lashing out irregardless, of not caring whether I lived or died, of just wanting to maximize damage whatever way I could. If I got killed doing it that was fine, as long as there was damage. I wanted the world to suffer like we had.

Finally there was nothing to do but lie on the concrete floor and try to get under the smoke. That worked for most of us, but it meant they could come in and scoop us up and carry us off trussed like sheep for slaughter.

Later, after the inquiry, we were sent through to France. We were reunited with our families, and ultimately we found that only six people died in that riot, none of them Swiss. And damage to the facility was minimal. What remained was that feeling. Oh I will never forget it. When you lose all hope and all fear, then you become something not quite human. Whether better or worse than human I can't say. But for an hour I was not a human being.

36

The Arctic Ocean's ice cover melted entirely away in the late summer of 2032, and the winter sea ice that formed in the following winter was less than a meter thick, and broken up by winds and currents into jumbled islands of pancake ice, separated by skim ice or brash ice or even open leads and internal ice-bordered seas many kilometers wide, so that when spring came and the sun hit, that winter's skimpy ice quickly melted and the constant summer sunlight penetrated deep into the Arctic waters, which it warmed quite a bit, which thus thinned the next winter's ice even more.

This was a feedback loop with teeth. The Arctic ice cap, which at its first measurement in the 1950s was more than ten meters thick, had been a big part of the Earth's albedo; during northern summers it had reflected as much as two or three percent of the sun's incoming insolation back into space. Now that light was instead spearing into the ocean and heating it up. And for reasons not fully understood, the Arctic and the Antarctic were already the most rapidly warming places on Earth. This meant also that the permafrost ringing the Arctic in Siberia and Alaska and Canada and Greenland and Scandinavia was melting faster and faster; which meant the release of a great deal of permafrost carbon, and also methane, a greenhouse gas twenty times stronger than CO_2 in its ability to capture heat in the atmosphere. Arctic permafrost contained as much stored methane as all the Earth's cattle would create and emit over six centuries, and this giant burp, if released, would almost certainly push Earth over an irreversible tipping point into jungle planet mode, completely ice-free; at which point sea level would be 110 meters higher than at present, with global average temperatures at least 5 or 6 degrees Celsius higher and probably more, rendering great stretches of the Earth uninhabitable by humans. At that point civilization would be over. Some remainder of humanity might adapt to the new biosphere, but they would be a post-traumatic remnant, in a post-mass-extinction world.

That being the case, efforts were being made to thicken the Arctic sea ice in winter, which would allow it to hold on longer through the summers.

These efforts were awkward at best. Arctic winters were still sunless and cold, and a skim of ice covered most of the sea. And there was no good and obvious method to thicken that ice. So to begin with, a number of methods were tried and evaluated. One involved running autonomous amphibious craft over the outer edge of the winter sea ice as soon as it formed, each craft pumping up seawater from the still liquid ocean nearby, then spraying it into the air in a fine spray that would freeze before it came down, this flocking making thicker the outermost border of the sea ice, and hopefully thus slowing its breakup when the sun arrived in spring.

This worked, in a limited way, but it would take thousands of such vehicles to adequately thicken the sea ice, and each vehicle could only be created at the expense of a certain amount of carbon burned into the atmosphere. Nevertheless it was felt worth doing, despite the expense in money, materials, and carbon burn associated with construction of the amphibians. Now that the cost of losing the sea ice was clearly seen to be hugely more important than the financial cost of preventing the loss, the financial cost was no longer a stopper to this plan proceeding.

Then, as airship factories were proliferating all around the world, it was possible to make some that would fly over the Arctic sea ice every winter, powered by batteries, pumping up water from holes punched in the thinnest sea ice, to fill tanks and then spray that water onto the surface ice below, where it also froze and fell as flocking, and thus thickened the ice. As in the Antarctic pumping efforts, it felt at first like sucking the ocean through a drinking straw: miniscule efforts! Teensy effects! A sick little joke! But every good work has to begin somewhere. And really, if these projects didn't succeed, what would follow was so dire to contemplate that the efforts were judged worth making, miniscule or not.

The third method was to drop on the winter ice clouds of small plastic machines, each of which had a solar panel and a drill and a pump and a little sprayer, so that the machine could puncture the ice, draw up some water, and cast it into the air to freeze. Eventually these machines would get buried in snow by way of their own efforts, and then they would shut off and wait for spring and summer to come, when they would float to the surface, if enough snow melted, and wait to be refrozen into the top of the ice the following fall, where they could start all over again. Thousands and then even millions of these could be distributed. If their work went well enough, they would eventually stay trapped in the thickening sea ice for centuries to come—they were even small bits of carbon sequestration, as some pointed out in little attempts at a joke.

But still, it was awkward, very awkward. Awkward as hell. Nothing they tried in this effort worked very well. Which meant...

37

This is hard to write. I was born in Libya, I'm told, and after my father disappeared, no one knows how or why, my mother took my sister and me to Europe, on a boat that carried mostly Tunisians. They made it to Trieste and were transferred by train to St. Gallen, Switzerland, where we were caught up in the riot there. That's my first memory—lots of us running into a building and everyone screaming. And my eyes burning from the tear gas. My mom tried to shield us inside her sweater, so I didn't see very much, but my eyes still burned. A woman and her two little girls, all crying our eyes out.

I don't remember much about the days that followed. The Swiss took care of us better than the sailors on the boat had. We were fed and had beds in a big dormitory, with showers and toilets in a compound next to it. It felt good to be clean and dry and not hungry. Mother finally stopped crying.

Then we were taken to a room and introduced to a group of people who spoke in French to us. Eventually Mother was invited to a refugee shelter just outside of Winterthur, and she eagerly and gratefully agreed to go. My sister and I were scared to move again, but Mother assured us it was for the best, so we got on a train again and said goodbye to the shelter in St. Gallen, which in truth had been the nicest place we had ever lived. But off we went.

The shelter outside Winterthur was in a beautiful garden. On certain days we could see the Alps in Glarus, very far away. We hadn't known that the world was as big as that, and at first it made me scared. How could we get along in a world so big?

Jake was one of the regular visitors to our shelter. His French was slow but clear, and he had a look on his face that right from the start I knew was different. As if he was suffering even more than we were. I wanted to tell him that we were all right.

He taught English to both children and adults, in different classes. Mornings for children, afternoon and evenings for adults. He spent most of every day there, Sundays included. At lunch he sat with us and ate. Sometimes at meals he sat there looking at us with his eyes moving back and forth, side to side, as if he was tracking a bird or having a thought. He seemed fond of us, and like all the sponsors, he grew to spend more time with particular refugees, greeting us by name and asking how we were doing, in both French and later English.

That went on for a long time, later I learned it was almost a year, and then Mother told us that she was going to marry Jake, and we would all move in with him in a nearby village. My sister and I had had no inkling that this might happen, and at first we were surprised and uncertain; the shelter was again the nicest place we had ever known, and going off with a single one of our helpers into the unknown struck us as a bad idea. We didn't know what was going on between Mother and this man with the twitchy eyes, and we suspected the worst.

But in fact we moved nearby into a little two-story white house with a walled garden beside it, and we settled in quickly and went back often to the shelter to see our friends there. Jake and Mother were always warm and cordial to each other, although they were never openly affectionate in front of us. But we could see that Mother was fond of him, and grateful to him, and he was always very kind to us, and always spoke to us in a mix of English and French, so that it seemed like the two languages were one, and later it took some sorting out on our part to get the two into their separate places in our heads. In that effort, Arabic seemed to slip away.

So we were a little family for a few years, from when I was seven until I was eleven. We went to school in Winterthur, played with friends from school and from the shelter, and all was well. In those years my mother was happy.

Then, when I started going to the middle school, I began to see signs that things were not well between my mother and Jake. They would sit in our kitchen after dinner looking at their screens or out the window. Watching them together I saw something that struck me very strongly; even just sitting there doing nothing, they were very different people. My mother is a calm person. She pours herself into a chair and relaxes there like a cat. Her eyes will move, her hands will do some sewing or knitting, but her body is still as can be. This is somewhat her nature. We're lucky to have her.

Jake on the other hand would sit there and yet he wasn't even close to still. Not that he fidgeted, or tapped his foot or anything like that; it was just that you could see that he was spinning inside. It was like you could see all his atoms spinning the way they are said to do. If people could be rated for their spins, like atoms or car engines, then Mother would be almost motionless, while Jake was always spinning, at thousands or even millions of revolutions per minute. RPM ten million, he said once; this whole image I am giving you comes from one of his own ways of assessing people. He would say we are all like quarks, which are the smallest elementary particles, he told us—smaller even than atoms, such that atoms are all made up of quarks held together by gluons. He made us laugh with these stories. And like quarks, everyone had a certain amount of strangeness, spin, and charm. You could rate everyone by these three constants, and our mother was the most charming person on Earth, but not very strange, and with almost zero spin. Jake confessed to having a high spin rate, also strangeness; and we found him charming too. He didn't agree to that.

So sometimes he would sit in his chair at the end of a day at the shelter, obviously exhausted, and his eyes would be moving left–right–left–right, which I think takes a lot of effort, and somehow it was clear that he was spinning. There was something dark inside him. Mother said he had done development work in his youth and had seen some bad things. We believed it. Sometimes he would stare at us, and sit hard on the floor and give us a hug: How I love you, he would say, you are such wonderful girls. Other times he would stare at us with his body rigid and his face contorted, clutching the sides of his chair as if preparing to leap to his feet and dash from the room. It was frightening to see that.

Then times came when he would shout at Mother, and even at us. He would leap to his feet, he would dash from the room; but sometimes first he would shout at her, in English it seemed but we couldn't understand it, and besides we were too scared to listen, we ran from the room at those times. It was so shocking at first; then it became something that could happen, something we were watchful for, so that when he was friendly, or contrite and remorseful, we would take it with a grain of salt, not knowing if he might turn on us in a second. Volatile people, you can't trust them, that's the thing; and they know it. So that even if they feel remorse, it does no good, and they know that too. So they get lonely. And they feel the remorse less and less, maybe. They give up. In any case, he left. One day Mother woke us, she was crying as she told us that he wouldn't be coming back, that we would have to move again. We all sat on the stairs and cried.

38

Today we're here to discuss potential alternatives to the global neoliberal order, which seems to be in such imminent danger of collapse. Are there any already existing alternatives we can look to?

China. Obviously.

But China seems very deeply implicated in the global economy.

They have a command economy that overrules the free market to bolster Chinese interests.

That's so interesting! What could we call such a mysterious new amalgam?

Socialism.

Oh my. How very transgressive of you, not to say nostalgic. But I seem to recall that China is always careful to add the phrase "with Chinese characteristics" whenever they use the word socialism, and it seems to many that those Chinese characteristics make for a completely new thing.

Yes but no.

You don't agree?

No. It's socialism with Chinese characteristics.

These characteristics including a huge dollop of capitalism, it seems.

Yes.

So might we learn things from them?

No.

How come?

Because we don't like them.

Isn't that rather prejudiced of us?

They don't like us either.

So, no hope of change from that quarter. What about the poor? The four billion poorest people alive have less wealth than the richest ten people on the planet, so they're not very powerful, but no one can deny that there are a lot of them. Might they force change from below?

There are guns in their faces.

What about the so-called precariat, then? Those middle billions just scraping by, what Americans still call the middle class, speaking of nostalgia? Could they rise up and change things by way of some kind of mass action?

Guns in their faces too.

And yet we do sometimes see demonstrations, sometimes quite large ones.

Demonstrations are parties. People party and then go home. Nothing changes.

Well, but what about coordinated mass action? That sounds like more than partying to me. The so-called fiscal strike that we hear so much about, leading to a financial crash and the subsequent nationalization of the banks, for instance. National governments would then be back in control, coordinating a complete takeover of global finance. They could rewrite the WTO rules, and create some kind of quantitative easing, giving new fiat money to Green New Deal–type causes.

We call that legislation.

So again we come back to legislatures! These are usually thought to be features of representative democracies. To the extent that such democracies still exist, if they ever did, their legislatures would have to be voted in by voting majorities, by definition. Fifty-one percent at the least, or more if possible, in all the major countries where such systems obtain. They would all have to join the plan.

Yes.

So this seems quite practical! What keeps us from doing that?

People are stupid. Also the rich will fight it.

Again this presumption that the rich have more power than the poor!

Yes.

But might it also be the case that there would be some kind of systemic resistance to change also, in that all these laws that need to change are intertwined, and therefore can't be easily disentangled?

Yes.

You could even say that money itself would resist this change. Indeed it seems to be the case that there's simply a kind of inherent, inbuilt resistance to change!

Constipation is a bitch. Sometimes you just have to sit on the box and push harder.

Well put! I guess you could call that the story of our decade. Or the entire century for that matter.

Why stop there?

Such a trenchant image for history, I must say.

Tremendous relief when all that shit is out of you.

No doubt! Well, that about wraps it up for this week. Perhaps it's time to pull down our pants and have a seat. I invite everyone listening to join us, next week this same time.

It might take longer.

39

Davos is one of my favorite parties. The World Economic Forum, held every year at the end of January. It's touted as an international gathering of power-brokers, those "stateless elites" who come to congratulate themselves and talk about how their plans for the future will make everything all right, especially for the elites themselves, who sometimes get called "Davos Man," that newly emergent subspecies of *Homo sapiens*, eighty percent male and in the top ten percent of the top one percent when it comes to personal wealth among other attributes. All true! And thus of course a great party. Even though some people think the partying itself is kind of sedate, despite the great liquor. Once some years ago Mick Jagger was spotted dancing by himself to a juke box in the corner; he was bored. But most of the people in attendance are happy just to be there and get seen by all the others.

Davos meets for a week, though few stay for the whole thing. About 2,500 businessmen and political leaders, with a few entertainers added for entertainment purposes; thus Jagger. The days of the conference are devoted to panel discussions and long meals, and all the problems of the day get discussed, mainly variations on the theme of riding herd on an increasingly fractious world by helping those most in need. Charity Inc.! With immense effort the percentage of women there has gone from six percent to twenty-four percent, we were told, and the organizers congratulated themselves on this progress and promised to keep working on the problem, which was difficult to solve, as most wealthy people and most political leaders are just by coincidence male. This may be one reason Jagger was bored.

Security costs for the conference are shared by the organizers and the Swiss canton of Graubünden, plus the Swiss federal government. Some in Switzerland criticized the cost of this, but then again, if the annual meeting of the rulers of the world wants to be in Switzerland, this probably helps Switzerland hold on to its weird position as one of the wealthiest countries on Earth despite having nothing at all to base that on. Maybe the beauty of the Alps and the brains of its people, but I'm dubious about both. Call me Doubtful in Davos.

There used to be protests at Davos, but not now. For one thing, the town is hard to get to and easy to defend. For another, the conference is more and more regarded as irrelevant, just a bunch of rich guys partying; which is true, as I said. So protests had mostly gone away. This perhaps represented an opportunity, or so people said afterward.

At this particular meeting, we had just gathered and gotten down to the serious business of eating and drinking and talking, when the power went off and we were left in the dark. Generators! we shouted merrily. Turn on the fucking generators!

But not. And the security people were suddenly seen to be not the same security people, these new ones were in masks guarding us in a different sense than we had been guarded before. We all said what the fuck and they ignored us, we all tried to get outside and see what was happening; no luck. Doors all locked. The whole town was physically closed. After a couple of hours, word spread that the Swiss road stoppers, installed the century before to foil Nazi or Soviet tank invasions, had popped up out of the pavement like giant shark's teeth, all over the valley and up in the few road passes in and out of the valley. And the airport and heliports in the area were all dark and similarly studded with shark's teeth. Even the Alpine mountain trails into the valley were said to be foamed with some kind of instant concrete that made the trails temporarily impassable. And the security on hand was there to guard us in this new way. They would not respond to us. We could hear the airspace over the town humming with circling drones, and people said they had clustered on a few approaching helicopters and forced them away, including a couple of crashes.

This thing is finally getting interesting, someone said. But most of us thought it was getting too interesting.

Announcements over loudspeakers were made to the effect that we would not be harmed, and would be released to the world at the end of the week. Only the schedule of events was being hijacked, we were told, not we ourselves, although obviously this was not true, as we were all quick to point out. But to no one, as all the guards on hand were helmeted with visors down, and not responding to us in any way, unless someone assaulted one, in which case the response was decisive and unpleasant, in the usual fashion seen on news clips. Clubs, pepper spray, dragged off to small rooms to chill; people stopped trying that. And the loudspeakers were not replying to our objections.

Then the services starting breaking down. In particular the plumbing stopped working, and we had to improvise a system for relieving ourselves. Shit! Poor Davos! There was no recourse but to head out into the woods and do it. So a fair amount of shit was distributed around the town, but quickly we created a system of impromptu sort-of latrines, and made do as best we could.

Then the taps stopped running, which to tell the truth was kind of scary. You can always shit in the woods, but you can't live on whisky, much as some people try. Some were pleased to stay hydrated entirely on four-thousand-dollar bottles of wine, of which there were many on hand. But turned out there were also two Alpine streams crashing down through the town in stone-walled channels, sometimes tunnels under streets but often just deep stone-walled channels, so we made use of some buckets someone found, and drank from these streams, either boiling the water or not. It looked clean to me. Snow just hours before.

Food was provided in boxes, and we were allowed into the town's various kitchens to cook for ourselves. We coped with that and were proud of ourselves for doing so. It beat just sitting around. Some of us were excellent cooks.

On the third day we found the town square filled with pallets of chemical toilets, which we assembled and placed in the bathrooms, now re-opened for use, even though there was still no running water. That was a relief, so to speak, as we could go back to relieving ourselves in more or less the usual manner, although it was nasty. It was like being trapped at Woodstock but with no music.

Water came back on the fourth day, and the boxes of food were never deficient. When we weren't cooking or cleaning up for ourselves, we were asked to attend what we called the reeducation camp. We figured we must have been captured by Maoists, that only Maoists would have such a naïve faith in propaganda lectures. These bounced right off us, and in fact were a considerable source of mirth, as we were already educated and knew what was what. Still, it was either attend or get locked in rooms where nothing at all happened. So most of us were willing to listen to the propaganda of our captors rather than spend the day stuck in an empty room.

The educational materials we were exposed to got universally bad reviews. So many clichés! First films of hungry people in poor places. It wasn't quite like looking at concentration camp footage, but the resemblances were there, and these images were of living people, often children. It was like looking at the longest charity advertisement ever made. We booed and made critical comments, but really the 2,500 most successful people in the world did not get to that status by being stupidly offensive. Often some diplomatic skill had been required and acquired. Also we were pretty sure we were being filmed in order to be later packaged into some kind of reeducative reality TV. So most of us just sat and watched the show and muttered to each other like you do in movie theaters.

And it was despite all a sobering sight to see how the poorest people on Earth still lived. Time travel to the twelfth century, for sure. That we ourselves had no bathrooms and were a bit hungry no doubt added to the effect of this footage, even when it was completely obvious that this was why they were doing it to us, the intended point of their pointless exercise, some kind of sick aversion therapy.

Often statistics appeared on the big screen; yes, PowerPoint shows, a true punishment. That a tenth of one percent of the human population owned half humanity's wealth—that was us, yay! That half the human population alive at that moment had no assets except their own potential labor power, which was much weakened by poor health and education, that was definitely too bad. But blaming this on capitalism was wrong, we told these non-listening boring people; there would be *eight* billion poor people if it weren't for capitalism! But whatever. The figures kept coming, graph after graph, repeated in ways that were not even close to compelling. Bored, sleepy, hypnotized, we tried to figure out which ideological or ethnic group had assembled such a stew. And the soundtrack! Sad music, jaunty music; horribly unforgettable earworms destined to stick in your head forever; tragically depressing music, like dirges played at one-third their intended speed; and so on.

By now the stresses of living rough and watching PowerPoint were getting to a lot of us, we were about to go into full Lord of the Flies mode.

I encouraged people to suck up and deal, to enjoy it as a kind of vacation, it was still glamping I told them, who cares about this shit? But turns out quite a lot of them did care.

These people are communists, they declared loudly. So what, I said. We were in a fucking commie glamping reeducation camp, it was going to be a story we could tell in the bars for years to come.

The finale to all the propaganda was a long lecture telling us that the current world order was only working for the elites, and even for us it wouldn't work for long. We were simply *strip-mining the lifeworld*, as one Germanic voice from the screen put it, sounding like Werner Herzog to a lot of us, and I have no doubt he could have been involved, and that in German words like *lifeworld* would be real words already. Those of us with some German had fun making more examples of this Herzogian English, backtranslating so to speak, *Ich bin zu herzgerschrocken! Ich bin zu rechtsmüde! Ich habe grossen Flughafenverspätungsschmerz!* which last, for the unGermanic among us, was explained to mean "big airport delay sadness," a word every modern language should definitely have.

An hour of film was then devoted to young people with wealthy parents. Earnestly apologetic or brashly arrogant, these made a sad fucking crew. It had to be a skewed sample, they had selected for awfulness no doubt about it, and the crowd around me murmured objections, My kids aren't like that, no way. Although as it went on and on, all kinds of pathetic angry supercilious kids, the room got quieter, and actually it became clear somehow that some of the whining faces onscreen had an actual parent right there in the room. And the graphed statistics charting these rich kids were disheartening as well. If the various kinds of anti-depressants used by them were aggregated, it could be made to look like well over one hundred percent of them were on anti-depressants. Which though clearly artifactual was indeed depressing.

Another chart graphed individual happiness in relation to personal wealth, and as so often that week, and in life generally, it was another damn bell curve. This one showed that poverty made for unhappiness, duh, then people got quickly happier once adequacy was reached; then, at a high middle class income, the very income that scientists usually demanded for themselves—having studied their own graphs maybe, the guy next to me muttered darkly, as if to imply they were rigging either their study or the system of compensation, or both—you got the highest happiness; then, as wealth rose, happiness decreased—not to the level of the poorest people, but far below the happiness of the middle peak. That middle income was the Goldilocks zone, the happy median ha ha, or so the PowerPointers claimed, but we shook our heads knowingly. Statistics can "prove" anything, but they could never beat the obvious truth that more is better. The soundtrack here toggled between "All You Need Is Love" and "Can't Buy Me Love," to add ridiculous Beatles wisdom to this part of the show.

It was actually getting pretty annoying at this point. How could it go on for so long? Where were the fucking Swiss police? Were they in on this? Was this a Swiss plot, like the Red Cross or something?

You are one of the Davos Hostages, a voice said at the end of this income comparison film. You will have been a participant at the Captured Davos. What will you do with that? *We will be interested to watch you live the rest of your lives.*

With that the incarceration ended. The drones in the airspace overhead flew away, our helmeted guards were suddenly nowhere to be seen.

We cheered when we realized this, and told each other that our indoctrination had been a complete failure and a sad example of bankrupt leftist notions that couldn't get purchase in the great marketplace of ideas. It had been like getting flown in a time machine back to 1917 or 1848 or 1793, though few of us knew why these years in particular were named by those with history degrees. 1848?

The real Swiss police finally appeared, to loud boos. Interpol was ordered to find the perpetrators, and the Swiss government came in for huge amounts of criticism, not to mention lawsuits for personal damages and emotional distress. They did their best to defend themselves from all this, saying it had been an unprecedented new form of hostage taking, which could not be defended against in advance of knowing it could exist. A new thing! And none of us had died, nor even been hurt much, except in our feelings. So much criticism of our lifeway! But we all had a lot of practice ignoring that kind of yelling, the dogs bark the caravan moves on, and indeed we all caravanned away as fast as we could. Back home we found ourselves minor celebrities, and opportunities to tell our story would last forever. Some of us took that opportunity, others slipped back into comfortable anonymity. I myself decided to decompress in Tahiti.

So, effect of this event on the real world: zero! So fuck you!

40

Jevons Paradox proposes that increases in efficiency in the use of a resource lead to an overall increase in the use of that resource, not a decrease. William Stanley Jevons, writing in 1865, was referring to the history of the use of coal; once the Watt engine was introduced, which greatly increased the efficiency of coal burning as energy creation, the use of coal grew far beyond the initial reduction in the amount needed for the activity that existed before the time of the improvement.

The rebound effect of this paradox can be mitigated only by adding other factors to the uptake of the more efficient method, such as requirements for reinvestment, taxes, and regulations. So they say in economics texts.

The paradox is visible in the history of technological improvements of all kinds. Better car miles per gallon, more miles driven. Faster computer times, more time spent on computers. And so on ad infinitum. At this point it is naïve to expect that technological improvements alone will slow the impacts of growth and reduce the burden on the biosphere. And yet many still exhibit this naiveté.

Associated with this lacuna in current thought, perhaps a generalization of its particular focus, is the assumption that efficiency is always good. Of course efficiency as a measure has been constructed to describe outcomes considered in advance to be good, so it's almost a tautology, but the two can still be destranded, as they are not quite the same. Examination of the historical record, and simple exercises in *reductio ad absurdum* like Jonathan Swift's "A Modest Proposal," should make it obvious that efficiency can become a bad thing for humans. Jevons Paradox applies here too, but economics has normally not been flexible enough to take on this obvious truth, and it is very common to see writing in economics refer to efficiency as good by definition, and inefficient as simply a synonym for bad or poorly done. But the evidence shows that there is good efficiency and bad efficiency, good inefficiency and bad inefficiency. Examples of all four can easily be provided, though here we leave this as an exercise for the reader, with just these sample pointers to stimulate reflection: preventative health care saves enormous amounts in medical costs later, and is a good efficiency. Eating your extra children (this is Swift's character's "modest proposal") would be a bad efficiency. Any harm to people for profit is likewise bad, no matter how efficient. Using an over-sized vehicle to get from point A to point B is a bad inefficiency, and there are many more like it; but oxbows in a river, defining a large flood plain, is a good inefficiency. On and on it goes like this; all four categories need further consideration if the analysis of the larger situation is to be helpful.

The orienting principle that could guide all such thinking is often left out, but surely it should be included and made explicit: we should be doing everything needed to avoid a mass extinction event. This suggests a general operating principle similar to the Leopoldian land ethic, often summarized as "what's good is what's good for the land." In our current situation, the phrase can be usefully reworded as "what's good is what's good for the biosphere." In light of that principle, many efficiencies are quickly seen to be profoundly destructive, and many inefficiencies can now be understood as unintentionally salvational. Robustness and resilience are in general inefficient; but they are robust, they are resilient. And we need that by design.

The whole field and discipline of economics, by which we plan and justify what we do as a society, is simply riddled with absences, contradictions, logical flaws, and most important of all, false axioms and false goals. We must fix that if we can. It would require going deep and restructuring that entire field of thought. If economics is a method for optimizing various objective functions subject to constraints, then the focus of change would need to look again at those "objective functions." Not profit, but biosphere health, should be the function solved for; and this would change many things. It means moving the inquiry from economics to political economy, but that would be the necessary step to get the economics right. Why do we do things? What do we want? What would be fair? How can we best arrange our lives together on this planet?

Our current economics has not yet answered any of these questions. But why should it? Do you ask your calculator what to do with your life? No. You have to figure that out for yourself.

41

In the twelfth year of continuous drought our city ran out of water. Of course we had been warned it would happen, but even in droughts some rain occasionally falls, and with conservation and fallowing of agriculture and construction of new reservoirs, and pipelines to distant watersheds, and digging of deeper wells and all the rest of those efforts, we had always squeaked through. And that in itself made us think it would keep on going that way. But one September there was an earthquake, not a very strong one, but apparently enough to shift something in the aquifer below us, and very quickly all the wells went dry; and the reservoirs were already dry; and the neighboring watersheds connected to us by pipeline were dry. Nothing came out of the taps. September 11, 2034.

Our city is home to about a million people. About a third of those had moved to town in the last decade, and were living in cardboard shacks in the hilly districts on the west side; this was partly another result of the drought. They were already living without running water, buying it in hundred-liter barrels or just cans and jugs. The rest of us were in houses and used to water coming out of the tap, of course. So on this day there was no question that the housed residents were most distressed by the change; for the poor in the hills it wasn't a change, except in this sense: there was no longer any water to buy either.

In our part of the world you can only live a couple of days without water. I suppose that's true everywhere. And this cessation of water, despite all the warnings and preparations, was sudden. Taps running, perhaps weakly, but running; then on September 11, not.

Panic, of course. Hoarding, for sure—if one could do it! Many of us had already filled our bathtubs, but that wasn't going to last long. A rush to the city's river, but it was still dry, in fact drier than ever. And then, no other choice, none at all: a rush to the public buildings. To the football stadium, the government house, and places like that. We needed a solution.

In from the coast came trucks with water from the desalination plants. Also a convoy appeared from inland; these vehicles included water trucks, also mobile machines that could suck water from the air, even dry air like ours. Humidity of ten percent feels dry as bone on the skin, but that air still has a lot of water in it. Luckily for us.

It had to be done in an orderly way. That was clear to everyone. You could fight your way to the front of the line, but to what effect? There was nothing there to grab; whatever water there was in the city was guarded by the military. The army and police were out in force. They directed people to stadiums and indoor assembly places like gymnasiums and libraries, and assembly halls of all kinds, and ordered us to get into lines, and water was brought to these places in heavily guarded trucks, and there was nothing to do then but wait your turn with your containers, and receive your allotted amount and then start conserving it.

Everything relied on the whole system working. If it didn't we would die, one way or the other, from thirst or from fighting each other. This was all completely clear to everyone but the crazies among us. There are always such people, but in this case they were outnumbered a hundred to one at least, and subdued by the police if they made trouble. For the rest of us, it came down to this: we had to trust our society to work well enough to save us. As it had never worked very well before, this was a big leap of faith, for sure. No one was confident. But it was the system or death. So we congregated in the places announced over the radio and online and on the streets themselves, and waited our turn.

Water kept coming into the city on trucks, or was dragged out of the air by the machines. The initial allotments were ridiculously small. We had enough to drink, but not enough to cook with. Quickly we learned to drink any water we did use for cooking, treating it as a soup. Conservation was taken to ridiculous levels; many bought filters said to clarify urine back to potable water. Sure, why not; we bought some too. But other methods of conservation were a bit more practical. Those filters don't last very long.

It was sobering in all this to see how many we were. You congregate the whole population into just a few places, you see how many you are.

And all of them strangers. In a city of a million you know something like, what, a hundred people? And maybe five hundred faces, at most a thousand. So you know at most one person in a thousand. Walk into a full football stadium, where you all have come to file through the line with your water jugs in hand or on dollies, water being so amazingly heavy—and you see that everyone there is a stranger. You are alone in a city of strangers. That's every day of your life! We saw that undeniably, right there with our own eyes. Alone in the city. Just a few friends, a family of sorts, these friends make, but so few, and them all lost in the larger crowd, doing business elsewhere. Well, people hung together with those they knew, sure, to go get water. Possibly to protect each other from the crazies, if someone lost it or whatever. But that seldom happened. We were so afraid that we behaved well, that was how bad it was. Mostly we went with friends just for the company. Because it was very strange to see with your own eyes that you live among strangers. Even though every night of your life, when you go to a restaurant and don't see anyone you know, that's the same thing—those are strangers.

But they are your fellow citizens! This is what makes it less than desolating—your fellow citizens are a real thing, a real feeling.

Seeing this, my friend Charlotte pointed something out to me one day as we stood in line waiting for a refill, feeling unwashed and parched and apprehensive, Charlotte's usual cynical sardonic attitude now almost jaunty, almost amused—she gestured at the line in front of us and said, Remember what Margaret Thatcher said? There is no such thing as society!

We laughed out loud. For a while we couldn't stop laughing. Fuck Margaret Thatcher, I said when I could catch my breath. And I say it again now: fuck Margaret Thatcher, and fuck every idiot who thinks that way. I can take them all to a place where they will eat those words or die of thirst. Because when the taps run dry, society becomes very real. A smelly mass of unwashed anxious citizens, no doubt about it. But a society for sure. It's a life or death thing, society, and I think people mainly do recognize that, and the people who deny it are stupid fuckers, I say this unequivocally. Ignorant fools. That kind of stupidity should be put in jail.

Then on the twenty-third day of our crisis, on October 4, it rained. Not just the ambivalent sea mist we often get in the fall, but a real storm, out of nowhere. How we collected that rainwater! Individually and civicly, the rain fell on our heads and in our containers, and I don't think a single drop of it made it downstream in our little river out of the city limits. We caught it all. And we danced, yes, of course. It was carnival for sure. Even though we knew it wasn't the total solution, not even close, as the drought was forecast to continue, and we still didn't have a good plan— nevertheless, we danced in the rain.

42

Asked Mary for a meeting with her and Dick Bosworth, to go over some of the economic plans the software team was developing. She cleared an hour at the end of a Friday and meeting convened in her seminar room.

What's up, Janus Athena? she said, a bit brusquely. She's always visibly skeptical that AI could contribute anything substantive to her project.

Went to the whiteboard and tried to show her how AI could help. Always awkward to explain things to computer illiterates, a translation problem, a matter of deploying metaphors and finding gross generalizations that aren't too gross.

Started this time with rehearsal of Hayek's argument that markets deliver spontaneous value, and are therefore the best calculator and distributor of value, because central planning can't collect and correlate all the relevant information fast enough. So planning always got things wrong, and the market was just better as a calculator. The Austrian and Chicago schools had run with that opinion, and thus neoliberalism: the market rules because it's the best calculator. But now, with computers as strong as they've gotten, the Red Plenty argument has gotten stronger and stronger, asserting that people now have so much computing power that central planning could work better than the market. High-frequency trading has been put forth as an example of computers out–achieving the market proper, but instead of improving the system it's just been used to take rents on every exchange. This a sign of effective computational power, but used by people still stuck in the 1930s terminology of market versus planning, capitalism versus communism. And by people not trying to improve system, but merely to make more money in current system. Thus economists in our time.

In fact, entirely new organizational possibilities now emerging with power of AI. Big data analyzed for best results, all money tracked in its movement all the time, allocations made before price competition distorts real costs into lies and universal multi-generational Ponzi scheme, and so on. Particulars here got both pretty technical and pretty theoretical at the same time, but important to do one's best to sketch out some things Mary might both understand and consider worth ordering team to do. Dick already up to speed on most of this.

Mary sighed, trying to focus on computer talk without boredom. Tell me how, she said.

So often they don't even understand the nature of the need. Reminded her that Raftery modeling still showed the vast bulk of the most probable twenty-first centuries experiencing an average temperature rise of 3.2 degrees Celsius. Chances of keeping average temperatures below 2 degrees C were five percent. Keeping it under 1.5 degrees C were one percent.

Mary just stared. We know it's bad, she said acidly. Give us your ideas to help!

Told her about the Chen paper, useful for its clarity, and now getting discussed in several different discourse communities, it being one of the earlier of various proposals to create some kind of carbon coin. This to be a digital currency, disbursed on proof of carbon sequestration to provide carrot as well as stick, thus enticing loose global capital into virtuous actions on carbon burn reduction. Making an effective carrot of this sort would work best if the central banks backed it, or created it. A new influx of fiat money, paid into the world to reward biosphere-sustaining actions. Getting the central banks to do that would be a stretch, but them doing it would be the strongest version by far.

Mary nodded grimly at that. A stretch, she repeated.

Persisted with arguments for carbon coin. Noted that some environmental economists now discussing the Chen plan and its ramifications, as an aspect of commons theory and sustainability theory. Having debunked the tragedy of the commons, they now were trying to direct our attention to what they called the tragedy of the time horizon. Meaning we can't imagine the suffering of the people of the future, so nothing much gets done on their behalf. What we do now creates damage that hits decades later, so we don't charge ourselves for it, and the standard approach has been that future generations will be richer and stronger than us, and they'll find solutions to their problems. But by the time they get here, these problems will have become too big to solve. That's the tragedy of the time horizon, that we don't look more than a few years ahead, or even in many cases, as with high-speed trading, a few micro-seconds ahead. And the tragedy of the time horizon is a true tragedy, because many of the worst climate impacts will be irreversible. Extinctions and ocean warming can't be fixed no matter how much money future people have, so economics as practiced misses a fundamental aspect of reality.

Mary glanced at Dick, and he nodded. He said to her, It's another way to describe the damage of a high discount rate. The high discount rate is an index of this larger dismissal of the future that J-A is describing.

Agreed to that.

And this Chen line of thought solves that? Mary asked. It extends the time horizon farther out?

Replied, Yes, it tries to do that.

Explained how the proposal for a carbon coin was time-dependent, like a budget, with fixed amounts of time included in its contracts, as in bonds. New carbon coins backed by hundred-year bonds with guaranteed rates of return, underwritten by all the central banks working together. These investments would be safer than any other, and provide a way to go long on the biosphere, so to speak.

Mary shook her head. Why would people care about a pay-off a hundred years away?

Tried to explain money's multiple purposes. Exchange of goods, sure, but also storage of value. If central banks issue bonds, they're a sure thing, and if return set high enough, competitive with other investment. Can be sold before they mature, and so on. Bond market. Then also, since this is a case of central banks issuing new money, as in quantitative easing, investors will believe in it because it's backed by long-term bonds. And this money could be created and given to people only for doing good things.

Like what? Mary asked. Issued for what?

For not burning carbon.

Started writing on the whiteboard, feeling she was oriented enough to be ready for some figures. Not equations, which might just as well be Sanskrit to her, only some numbers.

For every ton of carbon not burned, or sequestered in a way that would be certified to be real for an agreed-upon time, one century being typical in these discussions so far, you are given one carbon coin. You can trade that coin immediately for any other currency on the currency exchanges, so one carbon coin would be worth a certain amount of other fiat currencies. The central banks would guarantee it at a certain minimum price, they would support a floor so it couldn't crash. But also, it could rise above that floor as people get a sense of its value, in the usual way of currencies in the currency exchange markets.

Mary said, So really this is just a form of quantitative easing.

Yes. But directed, targeted. Meaning the creation, the first spending of the new money, would have been specifically aimed at carbon reduction. That reduction is what makes the new money in the first place. The Chen papers sometimes call it CQE, carbon quantitative easing.

Mary said, So anyone could get issued one of these coins after sequestering a ton of carbon?

Yes. Or also a fraction of a coin. There would have to be a whole monitoring and certification industry, which could be public-private in nature, like the bond rating agencies are now. Probably see some cheating and gaming the system, but that could be controlled by the usual kinds of policing. And the carbon coins would all be registered, so everyone could see how many of them there were, and the banks would only issue as many coins as carbon was mitigated, year by year, so there would be less worry about devaluing money by flooding the supply. If a lot of carbon coins were being created, that would mean lots of carbon was getting sequestered, and that would be a sign of biosphere health that would increase confidence in the system. Quantitative easing thus directed to good work first, then free to join economy however.

Mary said, So if you combined this thing with carbon taxes, you would get taxed if you burn carbon, but paid if you sequester carbon.

Agreed, and added that any carbon tax should be set progressively, meaning larger use more pay, to keep it from being a regressive tax. Then it becomes a good thing, and feebates can be added that pass some of this tax income back to citizens, to make it even better. A carbon tax thus added to the carbon coin was said by Chen and others to be a crucial feature of the plan. When both taxes and carbon coins were applied together, the modeling and social experiments got much better results than when either strategy was applied by itself. Not just twice as good, but ten times as good.

Mary said, Why is that?

Confessed did not know. Synergy of carrot and stick, human psychology—waved hands. Why people did what they did—that was her bailiwick.

Dick pointed out that for economists, carrots and sticks are both just incentives, and thus the same, although they tend to assume sticks are more efficient than carrots.

Mary shook her head vigorously. No fucking way, she said. We're animals, not economists. For animals, negative and positive are generally regarded as quite distinct from each other. A kick versus a kiss. Jesus Christ. She looked back and forth at us, said, It's a question which of the two of you are the more inhuman, the computer geek or the economist.

Both referents nodded at this. Point of pride, in fact. Trying to out-do each other; attempt to attain Spocklike scientific objectivity a worthy goal, and so on. Dick quite hilarious on this matter.

Mary saw the nods and sighed again. All right, when you align both negative and positive reinforcements to press us toward a certain behavior, we then do that behavior. It's just Pavlov, right? Stimulus and response. So how could we get this started?

Said, If the dozen biggest central banks agreed to do it together, it would go.

But that's true of almost anything! Mary exclaimed. What's the minimum you think it would need to succeed?

Said, Any central bank could experiment with it. Best would be the US, China, and the EU. India might be the most motivated to go it alone, they're still very anxious to get carbon out of the air fast. But the more the merrier, as always.

She asked to be led through the time element again.

Explained how the central banks could simply publish the rate of return that they planned to pay out in the future, no matter what. Investors would therefore have a sure thing, which they would love. It would be a way to go long, and to securitize their more speculative bets. The stick, the carbon tax, also needed to rise over time. With that tax rate and its angle of increase published in advance, and a long-term rate of return guaranteed for investing in carbon coins, one could then calculate the cost of burning carbon, and the benefits of sequestering it. Normal currencies float against each other in the exchange markets, but if one currency is guaranteed to rise in value over time no matter what, then it becomes more valuable to investors. It will always stay strong in the currency market because it's got a time stamp guarantee of a rise in value. The carbon coin designed in that way would eventually probably replace the US dollar as the world's benchmark currency, which would strengthen it even more.

It's like compound interest again, Dick remarked to Mary.

Said, Yes, but this time guaranteed by being delinked from current interest rates, which often hit zero, or even go negative. With this coin, you're good to go no matter what happens.

Dick said, That could make for a liquidity trap, because investors would stash money there for safety rather than put it to use.

Shook head at that. Set the rate low enough that it's seen as more of a back-up.

Dick said, If the central banks announced they were upping the amount of carbon needed to earn a coin, they could then balance it with other safe asset classes like treasury bonds and infrastructure bonds. That would add liquidity and give traders something about this that they could short, which is something they like to do.

Agreed this might be good.

Mary said, Could we issue these carbon coins ourselves from the ministry?

Shook head. You have to be able to buy them all back at some floor rate, to make people believe in them. We might not have the reserves to do that.

We can barely pay our staff, Mary said.

We've noticed, Dick joked. Good to see he liked this plan.

Mary brought the meeting to a close. Work this up into a full proposal, she said. One I can take to the central banks and defend. I've got meetings with them already scheduled. We'll see where it takes us.

43

I am a secret so everyone can know me. First you must count every part of me, then translate those parts into signs that do not describe me. Together we are shackled, and with the sign that does not describe me you can open me up and read me as I am. People will give you their promises for me, and if wrongdoers try to take me away from you, you can find me and tell the world where I am hidden. I began as a silent speaking, a key to open every door; now that I have opened all the front doors, I am the key that locks the back doors by which wrongdoers try to escape the scene of the crime. I am the nothing that makes everything happen. You don't know me, you don't understand me; and yet still, if you want justice, I will help you to find it. I am blockchain. I am encryption. I am code. Now put me to use.

44

The part of Antarctica that holds its ice the longest is near the middle, between the Transantarctics and an ice-submerged mountain range called the Gamburtsev Mountains. The Gamburtsevs are almost as high as the Alps, yet still completely buried by ice; they were only discovered by overflights using ice-penetrating radar. Between this newly discovered range and the Transantarctics there's a flat plain, surrounded by mountains in such a way that scientists estimate the ice sitting on it won't reach the coast for at least five thousand years. In other areas of the continent ice will get to the sea in the next couple of decades. So it's another case of location, location, location.

Naturally this Point of Maximum Ice Sequestration is a long way from the sea, and the polar ice cap there is ten thousand feet thick, meaning about that high above sea level, as the bedrock under all that ice lies just slightly below sea level. So if you were thinking of pumping seawater up onto this part of the ice cap to keep sea level from rising, it was going to take a lot of energy. And a lot of pipeline too. Run the numbers and you can see: not going to happen.

Still there were people who wanted to try it. Non-quantitative people, it would seem, and yet rich despite that. Chief among these curious people was a Russian billionaire from Silicon Valley, who felt Antarctica as a place to dump seawater just had to be tested, so much so that he was willing to fund the test. And you take grant money where you can find it, when it comes to getting to Antarctica. At least that's been my working method.

So an austral spring came when a fleet of private planes flew south from Cape Town, South Africa, where there's a permanent gate at the airport that says ANTARCTICA (I love that) and we landed on the Ronne Ice

Shelf, overlooking the frozen Weddell Sea. There we unloaded and set up a village of yurts, Jamesways, and tents, which looked small in the vast expanse of ice, because it was. Even the tourist villages at Pioneer Hills and under the Queen Astrid Range were larger. But this one served as the drop site for an ever-growing collection of specialized equipment, some of it lent to the operation by Transneft, the Russian state-owned oil pipeline corporation. The biggest piece of equipment was brought to the edge of the Ronne Ice Shelf by a massive Russian icebreaker, and unloaded in a tricky operation: a giant pump. Intake pipes were punched through the sea ice, and a transport pipeline was attached to the pump and run inland, across the Ronne Ice Shelf and up to the polar cap, past the South Pole to Dome Argus, the highest point on the Eastern Antarctic Ice Sheet. Because it was higher, this was felt to be the energy equivalent of the even more distant Gamburtsevs.

The power for the pumping, also the heating of the pipeline to keep the water liquid in the pipes, was provided by a nuclear submarine reactor donated for the occasion by the Russian navy. If it turned out to be feasible, the billionaire had explained to people back in Russia, this operation might turn into one of the biggest industries in the world. And save St. Petersburg from drowning. The fact that this supposed industry would require the power of about ten thousand nuclear subs was apparently left out of the discussion. But okay, an experiment in method, sure. Why not.

All the ice melting around the world was now raising sea level at a rate of some 5 millimeters a year, which did not sound too bad until one remembered that it had been 3 millimeters a year just twenty years before, and this rapid rate of increase was also itself speeding up. If the current rate doubled every year, then very quickly the sea would be rising so quickly that the coastlines of the world would be inundated, and that catastrophe would greatly complicate an already tricky ecological situation.

Many had pointed out that if sea level rise did increase in speed in any significant way, it would overwhelm any possible attempt to pump that water back up onto Antarctica or anywhere else. If it got as bad as even a centimeter a year, which could easily happen if things went south, ha ha, the amount of water in that rise would equal a cube roughly the size of the District of Columbia at its base, thus twice as tall as Everest. And moving that would take far more pipe than ever made in all history.

But since the rate of future sea level rise was unknown, it was felt by many, or at least by the billionaire in question, to be worth looking into. It would provide some real costs to check against the modeling exercises, and also would test what happens to seawater when released high on the polar cap. How far would it spread, how it would affect the ice already up there, and so on.

When we got the first line attached to the pump and started running it south, we took helos inland to get to the front edge of the operation. A longer flight every week. Looking down we could see the pipeline below us, like a black thread on white cloth.

It's like sucking up the ocean in a drinking straw, I said, and spitting your mouthfuls onto shore.

It's true, someone replied. But if you had ten million straws...

No, I said. It's stupid, this notion. But it's gotten us down here this year, and we might learn something useful from it.

So keep quiet about how stupid it is!

I will. My lips are sealed. I never said a thing. And if I did, I didn't mean it.

Griffen, you are such a smartass.

Hey! I said. Another great day in Antarctica!

45

Mary flew to San Francisco, where the US Federal Reserve was host-ing a meeting of some of the other big central banks. There was an annual meeting in Basel of all the central banks, convened by the Bank for International Settlements, but those were pro forma things; the real dis-cussions usually happened elsewhere, and when the US Federal Reserve called for a consultation, the other central banks usually showed up. This meeting was one of those, and the head of the Fed had welcomed Mary and given her a slot on the program. So it was time to talk things over with them in person, make the pitch for a carbon coin.

Before that meeting began, she dropped in on the annual gathering of California Forward, having been invited to it by a young woman who had once interned for her. So a morning came when Mary walked with Esther over San Francisco's urban hills to the Moscone Convention Center. It was a brisk morning, the air almost as cool as in Zurich, but oceanic and windy. This and something intangible, maybe the hills, or the light, gave the city a wild, open feel, very unlike staid old Zurich. Mary was very fond of Zurich, but this city overlooking its bay struck her in a different way as rather superb, basking under the windy Pacific sun and giving her with each block they walked new views in all directions.

The California Forward meeting was an annual summit gathering for several score organizations. California, if it had been a nation, would now constitute the fifth biggest economy on Earth, and yet it also ran at carbon neutrality, having established strong policies early on. They were intent to continue that process, and obviously the people at the meeting felt what they were doing was a model other people could learn from. Mary was happy to be taught.

Esther introduced her to people from the State Water Board, the

California Native Plant Society, the University of California's clean energy group, also its water group; also the head of the department of fish and wildlife, the state's biodiversity leader, and so on. Together a group of them walked her to a cable car terminus, and they all got on an open-sided cable car and rode it north to Fisherman's Wharf. Mary was surprised, thinking these quaint cars, canting up and down steep hills like Swiss cable cars attached to slots in the streets, were for tourists only, but her hosts assured her that they were as fast as any other transport across the city, and the cleanest as well. Up and down, up and down, squealing and clanking in the open air, and again she got the sense of a place enjoying its own sublimity. In some ways it was the topological reverse of Zurich. The California Forward crowd, enthusiastic enough already, were now all bright-eyed and red-cheeked, as if on holiday.

From Fisherman's Wharf they took a small water taxi to Sausalito, where a van drove them to a big warehouse. Inside this building the US Army Corps of Engineers had created a giant model of the California bay area and delta, a 3-D map with active water flows sloshing around on it. Here they could walk over the model landscape on low catwalks to see features better, and as they did that, the Californians told her and showed her how the northern half of the state was now functioning.

The state's Mediterranean climate, they told her, meant warm dry summers and cool wet winters, nurturing an immense area of fertile farmland, both on the coastal plains and in the state's great central valley. This central valley was really big, bigger than Ireland, bigger than the Netherlands. One of the chief breadbaskets of the world: but dry. Water had always been the weak link, and now climate change was making it worse. The entire state was now plumbed for water, they moved it around as needed; but when droughts came, there was not much to move. And droughts were coming more and more frequently. Also occasional deluges. Either too little or too much was the new pattern, alternating without warning, with droughts predominating. The upshot would be more forest fires, then more flash floods, and always the threat of the entire state going as dry as the Mojave desert.

Hydrologists pointed at the model below as they explained to Mary the water situation. Typically, the Sierra snowpack held about fifteen million acre-feet of water every spring, releasing it to reservoirs in a slow melt through the long dry summers. The dammed reservoirs in the foothills could hold about forty million acre-feet when full. Then the groundwater basin underneath the central valley could hold around a thousand million acre-feet; and that immense capacity might prove their salvation. In droughts they could pump up groundwater and put it to use; then during flood years they needed to replenish that underground reservoir, by capturing water on the land and not allowing it all to spew out the Golden Gate.

To help accomplish all this they had passed a law, the Sustainable Groundwater Management Act, which they called "Sigma." In effect it had created a new commons, which was water itself, owned by all and managed together. Records were kept, prices were set, allotments were dispensed; parts of the state had been taken out of agricultural production. In drought years they pumped up groundwater, keeping close track, conserving all they could; in flood years they caught water in the valley and helped it to sink into the basin.

How they did this last part was a particular point of pride for them, as they had discovered that the central valley's floor was variably permeable. Much of it was as hard as a parquet floor, as one of them put it, but they had located several "incised canyons," created when powerful flows of melted ice had poured off the Sierra ice cap at the end of the last two or three ice ages. These canyons had subsequently filled with Sierra boulders and been slowly covered with dirt, so that they now looked just like the rest of the valley floor; but in fact, if water was trapped over them, they would serve as "gigantic French drains," allowing water to sink into and through them, thus recharging the groundwater basin much faster than other areas would allow. So California's state government had bought or otherwise claimed the land over these French drain areas, and built dams, dikes, levees, baffles, and channels to and fro, until now the entire valley was plumbed to direct heavy rainfall floods onto these old incised canyons, holding water there long enough for a lot of it to percolate down rather than run out to sea. Of course there were limits to how much they could retain, but now pretty good flood control was combined with a robust recharge capacity, so they could stock up in wet years and then pump again in the drought years that were sure to follow.

Good in itself; great, in fact. And not only that, this necessity to replumb the great valley for recharge had forced them to return a hefty percentage of the land to the kind of place it had been before Europeans arrived. The industrial agriculture of yesteryear had turned the valley into a giant factory floor, bereft of anything but products grown for sale; unsustainable, ugly, devastated, inhuman, and this in a place that had been called "the Serengeti of North America," alive with millions of animals, including megafauna like tule elk and grizzly bear and mountain lion and wolves. All those animals had been exterminated along with their habitat, in the first settlers' frenzied quest to use the valley purely for food production, a kind of secondary gold rush. Now the necessity of dealing with droughts and floods meant that big areas of the valley were restored, and the animals brought back, in a system of wilderness parks or habitat corridors, all running up into the foothills that ringed the central valley on all sides. These hills had always been wilder than the flat valley floor, and now they were being returned to native oak forests, which provided more shelter for wild creatures. Salmon runs had been reestablished, tule marshes filled the old dry lake beds; orchards were now grown that could live through periods of flooded land; rice terracing was also built to retain floodwater, and they had been planted with genetically engineered rice strains that could stay flooded longer than previous strains.

All these changes were part of an integrated system, including major urban and suburban retrofits. California's first infrastructure had been very shoddy and stupid, they told Mary; cars and suburbs, plywood and profit—another secondary gold rush, which had repeated the ugliness of the first one. Recovering from that crazy rush, redesigning, restoring, rebuilding—all that was going to take another century at least. But they were already a carbon-neutral society of forty million people, headed to carbon negative; this was a work in progress, of course, and they were still grappling with equity issues, being tied to the rest of the world. But those too were being worked on, until it would finally be the Golden State at last.

All this they told Mary while looking down on the pretty model of the landscape filling the warehouse, as if from a small airplane or satellite. Like a three-dimensional quilt, it seemed to her, the Lilliputian valley floor a vivid patchwork of greens, bordered and criss-crossed by what looked like hedgerows but were actually miles-wide habitat corridors, reserved for wild animals. The surrounding foothills were pale blond, dotted densely with dark green forest copses.

"It looks great," Mary said. "I hope we can do this everywhere."

"Models always look good," Esther said cheerfully. But she was proud of it—not just the model, but the state.

Back in San Francisco, she met the chair of the US Federal Reserve on the top floor of the Big Tower, still the tallest skyscraper in the city. This meeting room gave them a view that made it seem as if they were almost as high as the peak of Mount Tamalpais, looming there to the north of Sausalito. It reminded Mary of the day before, looking down on the model California, but this time it was real, and vast. Alcatraz was a gray button in the blue plate of the bay, the bay's far shores were green and hilly. Out there to the west, the Farallon Islands spiked out of the ocean like a sea serpent's black back. Little bridges crossed the bay here and there. The city itself was rumpled and urban, the white buildings and grid of streets interrupted by a few parks. It was a grand view, and as the meeting began Mary found herself distracted by it. But quickly enough, what was happening in the room focused her attention.

The Fed chair, Jane Yablonski, had held her job for nine years, with three more to go. She had the patient look of someone who had heard a lot of exciting proposals in her time and had come to distrust all of them; a little too patient for Mary's liking. About Mary's age, black-haired, attractive in a silvery gray suit. Surrounded by assistants and some other Federal Reserve board members. She had been appointed by the previous president, and it was said that the current one was looking forward to replacing her.

Joining the Americans were several other central bank representatives, including China, the European Union, the Bank of England, and the Bundesbank in Germany. Just the people Mary wanted to talk to.

All central banks were curious hybrids, and the US Federal Reserve was no exception. It was a federal agency, therefore a public bank, but it was funded by private banks, at the same time that it oversaw them. It created money, being along with the Treasury the seigniorage function of the United States, and between that and setting interest rates it was thus responsible for the state of the US dollar, the strongest currency in the

world, the currency every other currency was pegged to—the one everyone ran to whenever there was a currency scare of any kind. The money
of last resort, so to speak. In this sense, a very crucial sense, the American
empire was alive and well.

After some preliminary business, it was Mary's turn to make her pitch.
She explained to them Janus Athena's idea for a carbon coin, wishing that
she had J-A there, or Dick, to help her with the explanation. But she
needed to be able to describe it herself, if it was going to make any head-
way. And the general ideas were clear to her.

Yablonski and the others had already heard of carbon quantitative easing, it turned out, and they were even familiar with the Chen paper and
the associated discussions of it. Yablonski did not look impressed.

"I don't see how we can get into the business of backing a currency that
isn't the US dollar," she said when Mary was done. "The Federal Reserve
exists to protect and stabilize the dollar, nothing else. That means stabilizing
prices more generally, which means we pay attention to unemployment levels
too, and try to help there as we can. So this idea is not really in our purview,
and if we tried this new alternative currency and it somehow destabilized or
harmed the status of the dollar, we would be worse than derelict in our duty."

Mary nodded. "I understand. But you can't just quantitatively ease your
way to a rapid transition out of the carbon economy. It costs too much and
it isn't profitable, and if you tried to simply QE your way to paying for it,
that would undercut the dollar even more than our plan. And yet something's got to be done. It's the vital work of our time. If we don't fund a
rapid carbon drawdown, if we don't take the immense amount of capital
that flows around the world looking for the highest rate of return and
redirect it into decarbonizing work, civilization could crash. Then the
dollar will be weak indeed."

Yablonski nodded, grimly amused. "If the world ends, the dollar is in
trouble. But aside from that contingency, we're here to defend it in the ways
we've been given. That's what we're tasked with. Monetary policy, not fiscal policy. And the carbon tax proposals are gaining momentum, we feel."

Mary said, "But you need a carrot to go with the stick. The modeling
shows that, not to mention common sense."

"Not our purview," Yablonski said. The Europeans nodded in agreement; the Chinese official, an elderly man, looked on more sympathetically.

"But maybe it should be," Mary said.

Now Yablonski looked displeased. This was her meeting, after all. Mary was there as a guest, making a pitch. And even though the global situation was urgent, and the new tool promising, Yablonski wasn't going to expose herself and her institution to that kind of heat without being ordered to do it by Congress. This was expressed very clearly just by the look on her face.

It was the same with the Europeans. China might be different, but Mary didn't think China would lead on this issue. It needed everyone on board for it to work; all the central banks would have to agree to both the problem and the solution. If they decided not to back this plan, no one could coerce them to change their minds; they were de-linked from their legislatures precisely to be able to avoid political pressure of any kind.

Mary regarded them, thinking it through. Because money ruled the world, these people ruled the world. They were the world's rulers, in some very real sense. Bankers. Non-democratic, answerable to no one. The technocratic elite at its most elite: financiers. Mary thought of her group back in Zurich. It was composed of experts in the various fields involved in the matter, people with all kinds of expertise, many of them scientists, all with extensive field experience of one sort or another. Here, she was looking at a banker, a banker, a banker, a banker, and a banker. Even if they understood an idea, even if they liked an idea, they wouldn't necessarily act on it. One principle for bankers in perilous times was to avoid doing anything too radical and untried. And so they were all going to go down.

Looking out over San Francisco and its bay, at Mount Tamalpais and the broad stretch of the Pacific, Mary heaved a sigh. This was not just a meeting of two women and three men, but of five teams, five institutions, five nation-states at the heart of the global nation-state system. The Paris Agreement's Ministry for the Future was small and impoverished, these central banks were big and rich. Just because the need was urgent and her case was good, that didn't mean anything would change. You couldn't change things with just an idea, no matter how good it was. Was that right? It felt right. Power was entrenched—but that phrase caught just a hint of the situation—actually the trenches were foundations that went right to the center of the earth. They could not be changed.

The meeting dragged to a close. Time to get a drink. Nothing had happened.

46

I was born small, as so many things are. A marsupial perhaps. People came to me and reached inside me to pass things to each other. I helped them do that. When I was young I had no blood, and people moving things around inside me had to do it by feel. They had to decide by feel alone which things were equally useful to them. And so few things are equally useful. Indeed only two identical things are equally useful; two hearts, two livers, two drops of blood. So in people's attempts to pass things around, there was friction. It was time-consuming and unsatisfactory. People would sometimes say "all things being equal," but this was never the case, so I was judged to have a difficult and unsatisfactory body, until my blood finally came into me, and stomach acids. All the fluids of metamorphosis, of life. Then things dropped into me could be digested and moved elsewhere in my body to do something else.

My stomach made disparate things the same by way of digestion into blood. This made food of all the things brought into me, and I quickly grew. I am an omnivore. And as I grew I ate more and more.

Every thing fed to me made other things. I digested things and turned them to blood, which moved around in me and helped to reconstitute some other thing of use; bone, or muscle, or some vital organ. Helping in this process were my mouth, esophagus, intestine, arteries, and veins, all of which grew with me, making a whole body out of which new things also useful could grow, things people wanted. I grew and grew and grew.

In this process, as in any body, there were useless residues not taken up in the new process, which left me in the usual ways. Thus sweat, urine, shit, tears.

My body worked so well that eventually all things everywhere were swallowed and digested by me. I grew so large that I ate the world, and all the blood in the world is mine. What am I? You know, even though you are like everything else, and see me from the inside. I am the market.

47

He spent his days around Zurich. On the shores of the Zurichsee, mostly the eastern shore. Sometimes he would feel an urge to see women's bodies, a useless and hopeless urge, but since it could be indulged in this city, he would go to Tiefenbrunnen park on the lakeshore, pay at the gate and go in and sit down by the water and try not to be too obvious as the topless women walked by him going in and out of the lake, Swiss women so stolidly Swiss in their voices, so gorgeous in their pale skin wet in the sun. It was a bit much, a bit too clear what he was up to, and he saw also the other single men sitting around "just by coincidence," not looking at anyone and yet undoubtedly there to drink in the sight of so many women's bodies, no, it was too obvious, too much in several different ways, and after a while he would leave and wander the streets of the neighborhood behind the lake. There was a house there that had been turned into a small art museum, apparently the house owner's personal collection, now open to the public. Some of the greatest paintings he had ever seen were just hanging on the walls of an ordinary little house. Or walk up along the shore to the bridge that spanned the lake's outlet, where the Limmat left the lake, always interesting to look at, that first dip and pour in the black sheet of water. Swans floated under the stone wall just east of this bridge, the wall dropping direct from park grass to lake surface, the swans hoping for children to throw bread crumbs down to them. Unearthly, incandescent white birds, floating on black water. Or over to the little park where the statue of Ganymede held his arm up against the distant Alps.

Other days he would walk the trails around the top of the Zuriberg, over to the cemetery where James Joyce was buried—a life-sized bronze statue of the writer, always ready for a silent conversation, sitting there reading his bronze book through round bronze spectacles which nicely

emphasized his near-blindness, a bronze cigarette held on his bent bronze
knee. The tall trees on the Zuriberg were mostly clear of underbrush,
and one could leave the dusty trails through the trees and wander among
them, looking for nothing. Distant views of the Alps, better from up here
than down in the city. The peaks farthest south were snow-capped and
looked completely vertical, like cardboard cut-out mountains at the back
of a stage set. Then through the trees to the path circling the top of the
hill, and onto one of the steep residential streets dropping into the city
proper. Past his secret garden shed, down through houses with their tiny
yards, often with statuary in them; a big naked concrete woman held a
green looped hose over her extended arm, reminding him of the casual
women at Tiefenbrunnen.

Or of Syrine. Syrine and her little girls Emna and Hiba. How he regretted all that, his inability to hold it together, to be there for them. He
couldn't think about it. Miserable at the thought of it. But he still couldn't
avoid obsessing about what hurt him. Something was broken in his head.
He wanted to get better but it didn't happen. He wanted to go back to
working with refugees, though it would have to be at a different center,
not the one where he had met them. That wouldn't be a problem, there
were many of them in the region, but when he tried one, he realized it
was just reminding him of how badly he had fucked up with Syrine and
the girls. Just another trigger. It didn't matter that it wasn't the same place,
because they were all the same place really. Always the same place, always
the same day.

So, down to the Utoquai schwimmbad, where you could rent a locker
and suit up and step down into the lake itself, and go for a swim until you
were too cold to think, too cold to feel. Then get out and shower and
have a *kafi fertig* afterward, a drink that would indeed finish you, but in a
good way.

And yet always he was hiding. Always he felt sick and broken. There
was no way to ignore the surveillance cameras mounted almost everywhere. Not that Jacob Salzman was being looked for, apparently. Someone had posted a map of all the cameras in the city, supposedly, but there
were newer cameras that were much smaller than the previous ones. The
bigger ones were there to remind you that you were under surveillance,
whereas the smaller ones were there to surveil. So it didn't make much

sense to avoid the bigger ones, because surely everyone was always sur-veiled all the time. Quite possibly every human alive had a team of little drones following them. That was the only hope, in a way, in terms of lying low: that there were too many people to follow, too much data. And good reason by now to feel that he was not presently flagged as being of interest. He had his ID and his legend, and unless what had happened to him in his youth made the authorities look for him again for some reason, he should be okay. And he had not been that person for many years. Six or seven years. No, nine.

So he wore hats, and sunglasses, and grew a beard; and sometimes put a mouthgard in his mouth when out, and wore different kinds of cloth-ing. He tried to be four distinctly different people for the cameras, with the one that was really him the least often at large. See if the algorithms could sort that out; and given the eight billion people they were track-ing, it seemed possible he could slide under the radar. In any case he had to walk around, had to get out into the city. He couldn't hide in a room all day every day; he had tried that for a while, and it didn't work. It had almost killed him.

One day he saw a notice on a message board announcing a meeting of the 2,000 Watt Society. He looked it up and decided to go to the meeting. It was held in the back room of a little Italian restaurant west of the Haupt-bahnhof called Mamma Mia's. By the time the meeting started, about fifty people had jammed into the room. They looked like any other Swiss people, perhaps a little more bohemian in style, but not much. The Swiss were extremely regular in their appearance, but then on consideration Frank realized that this was basically true everywhere.

The meeting began on time, of course, and it had a schedule that was gotten through briskly. Frank's German was not up to the task, and this was Schwyzerdüütsch to boot, so he was completely lost and could only pretend to be comprehending, but no one seemed to notice. Their gut-tural looping sentences were calm, and they laughed pretty often. When they saw he was there, and that there were some other *Ausländer* there also, they summarized their proceedings in quick rough English. He liked the feel of the meeting, nothing dogmatic or virtuous about it, just people pursuing a project; something between a committee meeting and a party planning exercise. Like the local Swiss Alpine Club, no doubt, and in fact when he asked about that, he found that many there were in both clubs. Party planning—political parties—he wondered if the same word for both was the case in German also. *Partei*, yes. But birthday party? Maybe so. He wished he had brought along a translation earbud.

Back home he looked again at the society's online information. Started in Basel and Zurich about forty years before. The idea was that the total global energy generated by people, when divided by the number of people on Earth, came to about 2,000 watts per person. So the people in the society had decided to live on that much energy and see how it felt.

The 2,000 watts were to cover food, transport, home heating, and home utilities. When Frank saw the breakdown, he realized that his lifestyle was already well within the limits prescribed by the society. That made him laugh.

Swiss citizens in general used about 5,000 watts. This was compared to 6,000 in the rest of western Europe. Chinese citizens about 1,500. 1,000 in India. 12,000 in the United States. His country, the great whale in this as in everything, slurping down the world.

In Switzerland, their current usage per person cost about 1,500 watts for one's living space, including heat and hot water.

1,100 watts for food and "consumer discretionary."

600 watts for electricity, which included power for a refrigerator.

500 watts for automobile travel.

250 watts for air travel.

150 watts for public transport (trains, trams, subways).

900 watts for public infrastructure (Frank wasn't sure what that meant, but presumably in his case the cost of the library, the bahnhof, the sewer systems, and so on).

He considered the list for a while. The Swiss hoped to achieve the reduction from 5,000 watts per citizen to 2,000 mainly by swapping out their entire built infrastructure, to make it more energy efficient. They hoped that their economy would grow by 65 percent at the same time they were reducing their energy per person; and they wanted their people to live pretty much as they had before. No hair shirt, no saintly suffering. No Francis of Assisi mentality or behavior: this was Switzerland, not a monastery. Stolid burgher watch makers, cheese makers, made fun of by the rest of the world, or envied, or really both at once. In fact it was a bit mysterious how the Swiss had gotten so prosperous. Some still pointed to the deep past, including their mercenary soldiers and guards, their banks holding criminals' money, and so on; but it had to be more than that. Chemicals, pharmaceuticals, engineering systems, all the minutiae of daily life that the rest of the world couldn't be bothered to master—like Swiss watches in their day, but no one wore watches anymore, so they had moved on to something else. Manufacturing of almost anything was cheaper in China and India, so again, on to something else, or to the kinds of manufacturing that required exceptionally good quality control. On and on it went, with only 35 percent of their little country useful for agriculture, or even habitable by humans at all. It was strange.

And then there were the four language groups, the German speakers most numerous, then the French, then the Italians, then the Romantsch, who numbered only fifty thousand or so, and yet flourished in their little corner of the country. The Swiss were proud to assert that they had made Romantsch an official Swiss language in defiance of Hitler's raving about Aryan supremacy, and as far as Frank could discover, there was some truth to this, even though the defiance had been rather indirect and symbolic, as the Swiss had also been allowing the German and Italian militaries free passage across Switzerland at the time. Still it was a nice gesture toward language diversity, right when national governments in France and other countries were crushing their local dialects. The Swiss had always slanted against the grain, always pushed against the received wisdom that tended to wash over the rest of Europe in waves of intellectual fashion, everything from details of fashion to participation in world wars.

So, fine; the Swiss were mysterious. But this 2,000 watt project was a good idea. Frank already had it covered, he used an almost Bangladeshi amount of energy per year. He lived in an apartment or a garden shed, he didn't own a car and never rented one, he had stopped flying, he ate mostly vegetarian. There were websites on which one could calculate one's energy burn quite closely, using electricity bills and estimates of mileage on various transport systems and grocery lists. These calculators had existed from the start of the century or before, but as far as Frank could tell, no one used them. It was like avoiding the scale when you were overweight. Who wanted such bad news?

Scrolling around as he thought about it, he ran across one essay that said the people of the world could still be divided into roughly three groups of wealth and consumption, measured by their transport methods. A third of the world traveled by car and jet, a third by train and bicycle; the final third was still on foot.

He thought about that for a while. He walked a lot, it was easy in Zurich, a form of entertainment. That was true in many cities, as far as he knew. There were a few that were unwalkable, like Los Angeles, for which planners had struggled to invent new names, like conurbation or agglomeration or megacity; but in most cities walking still worked, at least in their central districts and their various nodes. Anyway it was no great deprivation to get back on foot, if you lived in a city like Zurich.

Of course it was good to get out of town sometimes, see something different. That meant trains and trams, but the watts used could be calculated there too; the 2,000 Watt Society had provided its members with lots of graphs to estimate individual use. Watts per kilometer even. Apparently he didn't travel much compared to most people. That felt right. If you were mentally ill your energy use inevitably dropped, because you couldn't put it together to live a normal life. He had gone to ground, he was living in a hole like a badger. Hibernating maybe. Waiting for some kind of spring to come.

In any case, strictly regarding standard of living relative to energy use, he was doing well. He was a comfortable badger. And it was interesting to think about life as a consumption of energy, it was now part of his project, his self-medication. One therapist had questioned it once, as possibly some kind self-punishment, but he didn't think so. He didn't feel his various forms of self-medication as more or less virtuous. Self-reliance was always a delusion, he relied on other people as much as anyone, he knew that. But it was interesting to try to do more with less. At the very least, it passed the time. And it kept him off some of the cameras.

He started going down the hill to a refugee aid center in north Zurich, to help with the free dinners offered there two nights a week. No more fraternizing with the refugees, that had clearly been a mistake, something beyond his capacities. But he could at least work, put his shoulder to the wheel, help turn the world. The organizers of these particular evening dinners were mostly Swiss women, and the workers who assembled to help them were from various charitable or aid organizations, or school groups, or churches, or people who were working off some kind of school or legal trouble by doing community service. The organizers only wanted first names from helpers, and they didn't ask questions. While he was there he helped set up tables and chairs and tablecloths, put out cutlery and cut donated cakes and pies into portions, and did a lot of kitchen and dining room clean-up. It was simple and calming, and there was time to sit or kneel by some of their guests and ask how they were doing, without getting involved in their lives beyond that. Some of the guests didn't want to talk, especially in English; some appreciated the chance; it was easy to tell which was which.

The space for these meals was some kind of civic hall, as far as he could tell, near the first bridge over the Limmat downstream from the Hauptbahnhof bridge. There were mini parks at both ends of this bridge, to give Zurchers the no doubt satisfying spectacle of their tamed river flowing under them, through the gateway created by one of their massively over-engineered bridges. It was indeed pretty mesmerizing. Frank watched for a long time as the water purled over a drop downstream from the bridge, like some kind of cake batter flowing in a giant mixer. One of the set-up crew for the dinner called this area "Needle Park," or so it seemed to Frank, as it had been said in Swiss German and he didn't know if he had caught the words right. Apparently a drug dealers' area back in the day, or even now, if Frank understood correctly. There was a needle dispensary nearby.

These days it was also a place where various refugees out of the camps for the day gathered before and after the free meals. Possibly something illicit was still being done in the two parks, he didn't know. Pairs of police would walk across the bridge occasionally, and sometimes they would stop and talk to people, but there was never any sign that they were confronting crime, or causing guilty parties to run and reveal themselves. In general the feel around Swiss police officers on patrol was completely different from what Frank remembered from his childhood. At home the presence of police meant trouble; something was wrong, guns might be involved, the big uniformed men were faintly menacing. Here in Zurich, and elsewhere around Switzerland, the police had the same vibe as the tram conductors, and they often carried similar looking scanning boxes. They seldom carried weapons, and there were about as many women officers as men. It seemed like they usually worked in pairs, a woman and a man, patrolling together doing something like outdoor marriage counseling. They did approach people, they did ask questions; but here too they resembled tram conductors, because Zurich's trams operated on the honor principle, and everyone bought tickets at kiosks or had annual passes, but conductors asking people to show their passes appeared on about one ride in every fifty. They wandered through the cars, and as people saw them they produced their tickets or their annual passes, and the conductors passed on with a nod. Very seldom were there freeriders, and many of these were tourists who had misunderstood the system.

Same in the park, or so it seemed. Legal? the passing police seemed to be asking just by their presence. Yes, legal. *Genau*, people would say in conversation with them, meaning *exactly*, a word the Swiss loved and used all the time, as in *okay* or *sure* or *yeah right*: exactly! and with a nod the police would pass on, the people too.

But as Frank spent more time in these two little Needle Parks, he began to see that there might be some people here functioning as nodes in some kind of refugee underground railroad. Groups of ten or a dozen people, looking like two or three families, would sit on a couple of benches, or on the grass if it was dry enough, and talk among themselves, looking around warily. They looked like they were from elsewhere; they weren't Swiss. They could have been from the Middle East, or South Asia, or Africa, or South America. Then someone who usually looked echt Swiss would approach them, and speak in their language—Frank usually couldn't hear well enough to tell more than the fact that it wasn't German or English—and the group would get up and follow that person away. Given there were surveillance cameras mounted on the streetlight posts, it had to be obvious to the authorities watching, or to their algorithms, that something was going on here. And yet it went on day after day.

He couldn't tell what it was. He couldn't speak the languages. English was the world's lingua franca, no doubt about it, and he heard a lot of that; but he couldn't join in the project of finding ways to help get these people to a place of greater safety, so he kept his distance. Any kind of aid offered on the spot was inherently suspicious. Best to stay uninvolved, to pitch in with the free meals and leave it at that.

Still, he began to go to the meals wearing a hat with the Zurich cantonal shield on it, the blue and the white. Then he would walk around the two bridge parks with children's down parkas stuffed in little bags, or little umbrellas collapsed on their stems into short squat cylinders, and if he saw foreign children shivering, he would approach the groups and say to the adults, *"Für die Kinder, sehr warm,"* and then quickly get away with a *"Danke mille fois"* said over his shoulder, this being the Zurchers' own sweet mash-up of German and French, particular to their city. He would leave them his little bundles and often on cold evenings the adult women in the group would nod gratefully. Off Frank would go. Jelmoli sold these jackets and umbrellas for ten francs each, so it was an easy way to help a little. Switzerland was cold and raw on winter afternoons when the wind swept down the lake and through the town; crossing a bridge often felt colder to him than Antarctica had felt. It was like Glasgow in that regard. A chill raw wet wind was far colder on the skin than a dry stillness of far lower temperature. And some of these people were unprepared for that.

One time one of the refugees in the park on the eastern side of the bridge was sick, and fell, after perhaps fainting. A quick little crowd surrounded him, and Frank joined them to see what he could do, his heart racing. He saw none of them seemed to have mobile phones, or maybe they were afraid to use them, and so he went to the little phone box on the bridge end and punched the button for the local 911, then asked for help in German, giving the situation and then the address, which he assumed they already had. He felt a brief sense of accomplishment in saying all that in German, then stood at the edge of the group watching, feeling so sick he worried that he too would keel over. When the ambulance arrived with its weird European siren, most of the people surrounding the stricken man disappeared. The emergency medical people approached and then a pair of police, both women; more refugees disappeared, until it was only the stricken one and Frank. Frank gestured the police over. One of them held a scanner in front of the man's face and clicked a photo and looked at the result. Possibly there would be an RFID embedded in the man's skin too, there to be read by the scanner.

Then the policewoman with the scanner glanced at Frank and stood up straight, holding the scanner toward him, gesturing at him with her other hand.

The other policewoman said "Nei," Swiss for *Nein*, and the one with the scanner returned to the stricken person. Frank nodded his appreciation to the one who had stopped her colleague, and when the emergency crew had the young man on a stretcher, Frank turned and walked away, shivering in the chill gray wind.

In the summers it was different. Some organization, maybe the Red Cross, maybe the Red Crescent, often set up a big open-sided tent in the western bridge park, and cooked food and gave out free meals. They were uninterested in finding out who they were serving, it was all anonymous on both sides of the table. It got hot and steamy on some days, which put Frank on edge, but he ignored that and did what he could to help. Set up, clean up. He didn't serve food, and he tried not to look at the people eating. It was too hot, too familiar. He didn't want to recognize what he was feeling, he looked around under the edge of the tent roof to see Zurich sights—the stone, the trees, the blue and white. Smell of bratwurst and beer. Red geraniums. Stolid medieval edge of the Rathaus, there upriver as far as he could see, visible through linden trees. Cold northern land, cool sober people, cool to the eye and the mind.

But not the people taking refuge here. The Fremdenkontrolle, the Stranger Control, a bureau of the police, estimated that there were now five million native Swiss and three million *Ausländer* within the borders. This ratio, one of the most extreme in the world, had caused membership in the various right-wing anti-immigrant parties in Switzerland to swell, and now they held a dozen or more seats in the Swiss legislature, led by the SVP, the Swiss People's Party. There were about thirty political parties in Switzerland, and all the ruling coalitions in the federal government were formed by majorities created by alliances between the central parties; center-right, center-left, with the more radical parties on each side just barely earning seats. SVP had even held a majority for a while, then after the heat wave it had lost popularity. Now they did better in the cantonal governments, but these had had their power shifted over the years to Bern's federal government—not entirely, but in national matters like this one, the federal government tended to get its way. The upshot seemed to be that there was a lot of pent-up anger in the anti-immigrant crowd, as they saw their country being "overrun," with nothing they could do about it, at the political level anyway.

Maybe it was this way everywhere. Now, whenever Frank saw small groups of obvious *Ausländer*, people from the global south or even the Balkans, walking in areas of the city where they might get accosted, he would strike up a conversation with them in English and walk with them. He saw it help once or twice. Racists were confused by mixed-race situations, so a white man with these dark-skinned people gave them pause. If it were a dark-skinned man with a Swiss woman it might make them angry, even though this was a common sight; but a white man with people of color was different, and anyway it took some evaluation before any racist could decide to get angry, so that created some time, and if one walked fast, staying on lit streets, which in Zurich meant any street not an alley in the Niederdorf, it would be enough to keep from getting assaulted, except by some guttural comments hurled one's way, comments designed not to be understood. So he walked with them when he could.

But then a day came when he saw a group of young Swiss men standing on the bridge looking at the big meal tent. Though it was hot, the air had gone dark; there were clouds building overhead, a thunderstorm was imminent. It would be a relief when it came. For now, it was getting hotter.

Big fat raindrops began to spatter the pavers outside the tent. All of a sudden they were in a downpour, and in that same moment the young Swiss men charged the tent throwing pavers and shouting. They were going for the servers as well as the guests, and people were screaming and going down, there was blood splashing from nowhere into reality, and seeing that, many among the refugees charged their assailants screaming with fury. Pavers gone, the Swiss thugs had nothing but their fists, but they too were enraged, and a donnybrook of punching and shoving and shouting erupted outside one edge of the tent. The Swiss who ran away were tackled or kicked in the back, the moves familiar and practiced from soccer pitches at their worst. The Swiss thugs then realized they had to retreat while still facing the refugee fighters, at least until they could get a clear route away. Some of them ran across a street right in front of a tram, which squealed down past them and gave them some cover for their retreat.

After that, wet people, screams, crying, blood on the ground and on the tabletops, which were now commandeered to lay injured people on something other than the cold wet ground. Police arrived, and clumps of quivering shouting people surrounded them to tell the tale of the outrage. The police teams went from person to person in the following hour, and Frank stayed, too upset by what he had seen to leave. He had to testify. But everyone the police interviewed they were also scanning, and when they were done talking to Frank, having done the same to him, they looked at each other. One of them was showing his scanner to his colleagues. These approached Frank and then surrounded him.

"Sorry," one of them said to him in English. "There is a warrant out for your arrest. Please come with us."

48

Babies crying at dawn. Already hot. People hungry. Sun over the hills like a bomb. Hot on the skin. Don't look that way or you'll be seeing white all morning. Shadows running off the west edge of the world. It'll be that way until the tent roof blocks the glare, about nine. By then it will be too hot to move. Better to sweat than not. Dust settles on the sweat and you can see little tracks of mud on your skin. Showers not till Saturday. It's my time of the month. Need that shower.

The dining tent opens at eight, still lit by the horizontal light. Long line of people at the entrance. People let the moms with little kids go first. Most people anyway. They're stacked up next to the entry, line collapsing, fretting. Unless you're starved, it's easier to hang back. Not to mention the right thing to do. After a while it's just a habit. You go where you've gone before. The group of women I usually join is there trying to stay normal, talk.

Inside the smell of eggs and onion and paprika. Big bowls of plain yogurt, my favorite. Load up on that and hope to last till nightfall. Skip the trials of the midday meal. It's too much sorrow to go through the dining hall three times in a day. People are exposed there, fretting and anxious, hot and hungry.

And bored. The same food, the same faces. Nothing else to do but eat.

The aid workers are from some northern country. They talk among themselves. Some are quiet and serious, others laughing and animated. Clean. They sweat but they're clean. I don't know where they come from. Sometimes I recognize faces, not just the handsome men, but a look, something in their look will catch my eye long enough to impress that face on me. After that I can't help seeing them. Not that they see me. When they work the serving line, dishing out food onto our plates, they make eye contact and ask if we want what they're serving, but very few of them really see us in any way that would register on them. It's one way to do their job without getting too sad. Even so they burn out pretty fast. Or maybe their deal is short. Either way, they come, they go. They aren't quite there and they aren't really real.

But it's important to refuse to get angry at them. You focus your feelings on what you can see, that's just the way people are. So there's a world out there of people who have put us in this camp. Not all of them specifically did it, but they're all part of it. They live in a world where this camp exists, and they go on. Anyone would.

But it adds up to us being imprisoned here, for nothing we've done except to live. Just the way it is. People know we're here, but there's nothing they can do about it. Or so they tell themselves. And in truth it would take a lot of doing to release all the people incarcerated in this world. So they don't. They focus their thoughts elsewhere and forget about us. I would do that myself. In fact I did do that myself. Only when things fall apart do you realize it can happen to you. You never think it can happen to you, until it does.

So then, some of these people volunteer to come to our camp and help feed us and do everything else that needs doing when you have eighteen thousand people stuck inside a fence unable to leave. Cleaning toilets, washing sheets, all that. And of course feeding us. Three meals a day. That's a lot of work. And yet here they are. Most of them are young, not all, but it takes a certain idealism, and that's mainly a young person's feeling. They are almost all younger than me now, but not so long ago, when I first came here, they were my age. And they learn things and see the world and meet other people like themselves and so on. And so they have to keep a distance from us, they have to or else they would become as unhappy as we are. At their best they are still indignant in our cause, and that's a stress on them. So they have to keep that distance. I know that.

But I still hate them for not seeing me. For looking me in the eye while they put food on my outstretched plate, and yet never seeing. I try not to but I hate them. Just as I hate everything else in this life.

No one likes to feel gratitude. Gratitude is recommended by the clerics but I say no one likes it, no one. Not even the clerics. They go into their trade in order never to be in a position to have to feel it. They receive our gratitude as they receive our pain, but they never have to give gratitude themselves. Or only in their professional capacity as our receptacles of feeling and of meaning, our representatives to God or whomever. No, I don't like clerics either.

When the sun is far enough to the west that I can do it without being scorched, I walk out to the northern perimeter of the camp to look at the hills. I should be back in the tent where we teach the children, and I'll get there eventually, but first I come out here. The hills still remind me of my home hills where I grew up, even though these are as green as limes. There's a ridgeline that looks just like the one I used to look at from our town when I was a girl. In late spring our hills would turn green also, not this wet green, but green enough: olive and forest green, in a dapple of furzes. I look through the links of the fence, which is like any fence anywhere, but topped with rolls of razor wire. Yes, we are prisoners here. They don't want us getting any idea that we're not, by way of looking at a fence that could be climbed. As for the wire mesh itself, it looks weak, like you could cut it easily, not with scissors maybe, but with tin snips for sure. No problem. But we have no tin snips in this camp.

So I lean against the wire mesh and feel it bell out under my weight. The bottom of the wire is dug into the earth pretty far, I can see that. Once upon a time it might have been possible to dig it up to crawl under it, with a spoon or one's fingers even. But now the dirt has flowed together and hardened. It would take some effort to dig under it, and worse, some time. They would see me. Still, I consider it every time I come out here. At sunset when I'm here I see the last of the sunlight pinking the ridgeline at the top of the hills, and I scuff at the dirt. No way. Maybe. No way. Maybe.

The sun goes down, the sky goes twilight blue. Then indigo. This is the 1,859th day I have spent in this camp.

49

In July of 1944, the United States government convened a group of seven hundred delegates, from all the Allied countries, to design the postwar financial order. They met at the Mount Washington Hotel in Bretton Woods, New Hampshire, and there, after three weeks of meetings, they published recommendations that when ratified by the member governments resulted in the International Bank for Reconstruction and Development, and the International Monetary Fund. The intended results of these new entities included the establishment of open markets and the stability of member nations' currencies.

An international trade organization was also proposed, but when the US Senate failed to ratify this part of the proposal, it was not founded. Later the GATT, the General Agreement on Tariffs and Trade, was established and took on the functions that the failed ITO would have fulfilled. Later the GATT was superseded by the World Trade Organization.

John Maynard Keynes, the chief British negotiator, also suggested at Bretton Woods that they found an International Clearing Union, which would make use of a new unit of currency to be called a *bancor*. The purpose of the bancor would be to allow nations with trade deficits to be able to climb out of their debts by calling on an overdraft account with the ICU that would allow them to spend money to employ more citizens and thus create more exports. Nations making use of their overdraft would be charged 10 percent interest on these bancor loans, which could not be traded for ordinary currencies, or by individuals. Nations with large trade surpluses would also be charged 10 percent interest on these surpluses, and if their credit exceeded an allowed maximum at the end of the year, the excess would be confiscated by the ICU. Keynes thus hoped to create an international balance of trade credit which would keep countries from becoming either too poor or too wealthy.

Harry Dexter White, the assistant secretary of the US Treasury and the chief American negotiator, said of this plan, "we have taken the position of absolutely no." As the world's biggest creditor and holder of gold by far, the US was in a position to enter the postwar period as the sole owner of the major global currency, the US dollar, which was to be backed by gold reserves. White proposed an International Stabilization Fund, which would place the burden of debt firmly on deficit nations; this later became part of the World Bank.

So at Bretton Woods, White's plan prevailed over Keynes's, and in the absence of the International Clearing Union and its bancor, postwar reconstruction and subsequent economic development was funded by the US dollar, which became the de facto global currency. The imperial coin, so to speak.

50

Mary returned to Europe, and from her base in Zurich began visiting the various central banks there, trying to improve on the unsuccessful meeting in San Francisco. In London she met with the chancellor of the exchequer, and some of the board members of the Bank of England.

In the days before that meeting, she read up on the history of the Bank of England, and saw that it was important in the financial history of the world. 1694: Charles II and William III had been borrowing money from private banks and not paying them back, or else levying taxes on all kinds of activities to be able to afford to make their debt payments, and thus making life more expensive for everyone, except for the royals involved, who were less and less liable for their profligate spending. So a Scottish merchant, William Patterson, proposed that 1,268 creditors lend the English king 1.2 million pounds for a guaranteed rate of interest of eight percent, and once William III signed off on that, a big piece of the system of the current world fell into place. The capitalizing of state power now had its roots in private wealth; thus the rich and the state became co-dependents, two aspects of the same power structure.

After that, the Bank of England became the mechanism by which the financing of the state apparatus was monopolized by a small group of wealthy tradespeople, and the shift from feudal land power to bourgeois money power was complete. The state from then on was always indebted to private wealth, and so relied on the good will of particular private individuals, who were unelected and unrepresentative of anyone but their own class, and yet were inserted right into the heart of state power. The Bank of England had also been founded in a state of emergency, during a war; but there was always an emergency that would serve when it came to finding reasons to perpetuate and extend state power. So whatever the law said, in practice the bank/state combination did what it pleased.

Naturally such a new thing was controversial. Tories of the time thought the Bank would lead to more parliamentary power, sapping the monarchy and leading to the mob; Whigs thought it was a mechanism by which the monarch would always escape paying his or her debts. Either way it was a rival to already-existing power bases, and was considered as such from the very start: an insulated enclave of power within government, made up of unelected rich bankers.

All this was ominous enough, but it kept occurring to Mary that these people's autonomous power might now actually put them in position to help enact a quick solution to the carbon problem, should they choose to join that effort. Even if they saved the world to save their privilege, maybe that didn't matter. Justice being not at this moment her first priority. So she went to London.

The Bank of England leaders were coolly unappreciative of her plan. Likely to cause inflation; could expose the central banks to currency-trading pirates; would create exposure to market pressure. Not sure how that could be avoided. When Mary reminded them that they had quantitatively eased trillions of pounds into existence when needed to save the banks, they nodded; their job was to save the banks. To quantitatively ease trillions of pounds into existence to save the world: not their job. That would take legislation.

The following week Mary went to Brussels. The European Central Bank, a much younger institution than the Bank of England, founded in the late twentieth century as a financial instrument of the European Union, offered to her representatives who were if anything even worse than the English. The group Mary met with was formed for the most part of German and French men. These were very sophisticated, intelligent, polite, and arrogant people. Their attitude toward Mary was dismissive in the extreme. For one, she headed an agency with no financial power and little legal leverage; the ministry was some kind of idealistic gesture to make people think extraordinary efforts were being made when really they weren't. And as an Irish woman she was doubly damned, more for being Irish than for being a woman; ever since Thatcher and Merkel and Lagarde, a few women had wedged their way into the top echelons of

European and financial power; Mary admired all these women, despite her hatred for Thatcher's politics, and of course none of them had managed to crack the top by way of being progressives. But Irish—no. A colony, a little country, one of the PIIGS, one of the many little piggie countries of Europe who had to pick up the crumbs of the big countries, and had no chance of achieving the gleaming polish of one of the big countries, which was really to say, Germany and France. These two old antagonists still fought for control of Europe, but it was a battle of two, the rest of the world was irrelevant, or at most instruments to be used. And somehow the little countries could never sort out their differences and band together to become a united front of their own. That kind of cooperation would be asking too much of nationalism and sovereignty. So the two big frenemies stood on the peak and regarded the rest with indulgent condescension at best, brusque command normally, and brutal arm-twisting at worst. Which of course was better than brutal military assaults, as in the past, but still not very satisfactory when sitting in a room with them. Thousand-euro suits: Mary did not fail to let her Irish disdain for such peacockery show. She could convey that disdain with a look while still being ostensibly polite, but of course it didn't help in terms of getting her what she wanted. She saw very clearly that the European Central Bank was entirely focused on price stability and increasing its power in the world to carry out that task. If they were asked to adjust the interest rate half a point to save the world, they wouldn't do it. Outside their purview.

The People's Bank of China on the other hand was a state-owned operation that held the most assets of any central bank on Earth, approaching four trillion US dollars; and although they were independent compared to most Chinese divisions of government, they were still ruled by the State Council. There was no point in talking to the Chinese bank heads; she would need to speak to the finance minister directly, and even better would be the premier and president, of course. In fact they were the strongest hope for her plan; they were not doctrinaire, not fixed on ideas from neoliberalism or any other political economy; practice is the sole criterion of truth, they said. Cross the river by feeling the stones, they said. If she could convince them of the worth of the idea, they wouldn't give a damn what the other banks thought.

But it would take a lot of central banks buying in to make this work.

Actually, thinking about that, she set Janus Athena to studying this. If the Chinese bank were to back it alone, could it work? No, J-A got back to her to say; no one bank could expose themselves to the market that way. Even China, even the US; these were just the biggest Lilliputians, in terms of any given entity trying to tie down the global economy. It would take a gang of them.

So, it wasn't going to happen. The bankers were useless. They would look at each other and see the mutual lack of enthusiasm in their peers, and hide behind that. If the world cooked and civilization fell apart, it wouldn't be their fault, even though they were funding the disaster every step of the way.

Something was going to have to make them do it.

The "structural adjustment programs" enforced by the World Bank on the developing countries caught in the debt crises at the end of the twentieth century set the conditions for what became the world order in the twenty-first century. These SAPs were instruments of the postwar American economic empire, which was unlike the older empires in that it did not insist on ownership of its economic colonies; it only owned their debts and their profits, no more than that. The best empire yet, in terms of efficiency, and the neoliberal order was all about efficiency, in its purest economic definition: the speed and frictionlessness with which money moved from the poor to the rich.

So there was a reason it was called the Washington Consensus. Its SAP requirements, made of any country that wanted a bail-out in the form of further loans, came only by adhering to the following conditions: a reduction in public spending; tax reforms, especially reducing taxes on corporations; privatization of state-owned enterprises; market-based interest and currency exchange rates, with no government controls on these; a set of strong investor rights, so investors could no longer be given hair-cuts (the long hair provisions, so-called); and the massive deregulation of everything: market activities, business practices, labor and environmental protections.

Even though these structural adjustment programs were widely criticized, and judged a failure by some analysts at the end of the twentieth century, they were the template for dealing with the EU crises in the small southern countries, and were inflicted on Greece in full to scare Portugal, Ireland, Spain, and Italy, not to mention the new EU countries from eastern Europe, at the prospect of what the EU (meaning in this case France and Germany) would do to them if they tried to create and hew to a line of their own. Join the EU, obey the European Central Bank; which meant, obey Germany and France. As Germany's economy was about twice the size of France's, what people around Europe took all this to mean was that Germany had finally conquered them all, no matter what it had looked like at the end of World War Two. Just as America had conquered the world by way of finance rather than arms, Germany had conquered Europe using the same methods—in some cases, using even the same capital. Because Germany had been very good at being a client state of America though the course of the Cold War. Now that the Cold War was over and Germany was in economic terms stronger than Russia, it could detach itself from the US a bit, cleverly pretending to be a client when it was convenient, but by and large pursuing its own course. This was obvious to everyone in Europe, but America's narcissistic myopia regarding the rest of the world didn't allow it to see that very well.

So a visit to Berlin was for Mary always a fraught venture. The bankers and finance ministers were a little less glossy and supercilious than in France and Brussels, but even more ominous in their burgher certainty that nothing could ever change. The Bundesbank, Germany's central bank, had been formed in the postwar period precisely to stabilize West Germany's status as a faithful and effective wingman for the American superstate, and given the incredible traumas of the two wars and the period between, it came as no surprise to Mary to learn that in the documents associated with the formation of the Bundesbank, the "preservation of the stability of the currency" was a "moral and legal necessity." The independence of the state bank was given a constitutional status that included the suggestion that currency stability be included among the catalogue of basic human rights. Life, liberty, and low inflation rates! Well, this was the country that had seen its currency blown to smithereens. Lose a war that you started and the conditions imposed on your surrender could include reparations you were in no position to refuse, that would damn your population unto the seventh generation. No wonder the Germans who had lived through that had said never again. "The economy creates public law," as the founding documents of the Bundesbank put it. That Mary saw this as exactly backwards was enough to give her pause; was it some kind of stealthy German acceptance of Marx over Hegel to say that first there is practice and then theory? She had no idea; she was not a philosopher, nor a historian. Just a working diplomat; but working diplomats tended to believe in cause and effect, in plan and execution. That's what governance was, that was bureaucracy, even economy. The laws defined the behaviors that were legal. Possibly it was chicken and egg, but never was it effect causing cause—that would scramble the definitions of the words beyond comprehension.

Anyway she was in Berlin, feeling oppressed by these Germans who had won by losing and then adapting to that loss, and trying again by different means. Maybe now they didn't want to rule the world; they only wanted to defend Germany by way of an active diplomacy, influencing Europe and the world as much as they could given their smallish population and economy. In which project they were doing a good job. So she, like them, tried to ignore their spectacularly awful history and focus on the moment at hand. Central banks, she urged them, quantitatively easing the world off carbon! Exertion of state sovereignty over the global market, by way of international cooperation of nation-states big enough to face down the market; even to alter the market. *To fucking buy the market.* She said this politely but urgently.

They stared at her. One of them wrinkled his forehead, a deep cleft appeared between his eyebrows. He spoke. China's central bank, the richest of them all, had four trillion US dollars in assets. All the central banks combined held about fifteen trillion in assets. The world's annual business, the GWP, was eighty trillion a year, and in the depths of the high-frequency dark pools, something like three trillion dollars got traded every day. Even admitting that these last were in some senses fictional dollars, it was still very clear: the market was bigger than all the nation-states put together.

Mary shook her head. Even if market and state were two parts of a single system, that single system was ruled by law; and the laws were made by the nation-states; they could therefore change the laws, that was sovereignty, that was where seigniorage and legitimacy and ultimately social trust and value resided. The market was constructed by, and parasitic on, that structure of laws.

The market can buy the laws, one of them suggested.

The market is impervious to law, another added. It is its own law, it is human nature, it is the way of the world.

Mary said, It's just a legal system. We change laws every day.

Central banks only exist to stabilize currencies and prices, to curb inflation and keep interest rates a viable tool for that.

Mary said, Together the central banks often advise their legislatures to change tax levels as needed to stabilize money. That means changing the laws.

Legislatures do what they want.

Mary said, Legislatures pass financial laws that the central banks tell them to pass. They're scared of finance, they let their quants write the laws in that realm. If you advised it, they would do it. Especially if you advised them to increase their own power over finance!

The Germans were practical people, their looks said. It was an idea worth considering, to see if it would help Germany.

This was not the worst response Mary had gotten, so she left the meeting feeling drained, and in serious need of a drink, following a session in the gym punching things; but not as depressed as she had been after some of the other meetings. Germans: they had seen the worst. They knew how bad it could get. Even though these were the grandchildren of the ones who had lived through it, or now mostly the great-grandchildren, there was still a cultural memory they could not escape, a memory that would last centuries. There was repression, of course, but wherever there is repression there is also the return of the repressed, often twisted and compressed by the repression into something even more dangerous to the self. Maybe that meant that Germans were focused on staying safe to a degree that had turned dangerous. They wouldn't be the first.

Then on to Russia. Russia's central bank was almost as much a state operation as China's. Half of its profits went to the Russian state, by constitutional design. It owned sixty percent of Sberbank, the country's largest commercial bank, and one hundred percent of the country's national reinsurance company. Their central bankers were very intent on protecting Russian interests above all. It made sense to Mary; and the men who met her were friendly. Any country that could produce Tatiana probably had some good in it. And Russians too had seen the worst that could happen. Their empire had imploded on them within living memory, and before that they had suffered the extreme trauma of the world war. They had reasons to hate the Germans, also to hate the Americans, to a lesser extent; really to hate everybody, you might say. Russia against the world: that had been part of their collective psyche for as long as they had been aware of the world. But their own problems absorbed most of their attention most of the time; they were a world apart, to some extent. So many places had been like that for most of human time, and were like that still—everyone living in the past of their own region's psyche to one extent or another, because they all lived in their languages, and if your native language was anything but English, you were estranged to one degree or other from the global village. Globalization was many things— including a reality, in that they all lived on one shared planet in which borders were historical fantasies—but it was also a form of Americanization, of soft power imperialism combined with economic dominance, in that the US still had seventy percent of the capital assets of the world secured in its banks and companies, even though it had only five percent of the world's population. So the globalization determined by physical reality could never be escaped, and would only become more prevalent as the biosphere problems got worse, while the globalization of American imperialism could not possibly last, as it was one of the main causes of the biosphere's problems. And yet the world's lingua franca was a permanent soft power.

So the two globalizations were at war, and both had to change; they had to be destranded and dealt with separately, but also understood for now as one, and dealt with together.

In the midst of these thoughts, feeling the useless spin of them, as in a bad dream, she got a call from Badim.

"What's up?"

"That man who kidnapped you?"

"Yes?"

"They've caught him."

"Ah! Where?"

"In Zurich."

"Really?"

"Yes. Down by the river, in Needle Park. He was helping at a refugee dinner, and they got attacked by some fascist group and he got in the fight."

"How did they know it was him?"

"DNA. And they've got him on some cameras, as usual. The DNA also links him to a death on a beach on Lake Maggiore, Swiss side. Someone got punched in the face and died. Looks like your man was the one who punched him."

"Damn," she said, feeling shocked. "Well, I'm on the train back to Zuri in the morning. I'll want a full report."

Mary spent that long trip home stewing. She didn't know what she thought or what she felt. A thrill of fear; a strong curiosity; a sense of triumph; a big touch of relief. Now at least she was safe. She wasn't going to wake up some night being strangled by that disturbed young man. That was good; but the thought of him imprisoned was also strangely upsetting to her. Should people suffering mental illness really be incarcerated? Well, sometimes they had to be. So it was quite a confusing mix of feelings to feel.

But there was no denying that whatever these feelings were, she was interested. Interested enough that she even somehow wanted to see this man again. And with him in jail, it would be safe. But why should she want it? She didn't know. Something about that night had snagged her. As of course it would.

As the train crossed Germany, she realized she was going to do it. She was going to go see him, whether she understood the impulse or not. That suggested to part of her that it might be a bad idea. Well, it wouldn't be the first time she had done something she knew was not smart. Always she had been prone to the rash act. She put it down to something Irish. It seemed to her that Irish women doing rash things was precisely how her people had managed to perpetuate themselves.

A phrase occurred to her: Stockholm Syndrome. Was that what this was? She looked it up. Conversion of hostages to a state of sympathy for their kidnappers. Generally regarded as a mistake on the hostages' part, or a psychological weakness, a result of fear, transference, and the hope to survive by turning belly up and exposing the throat (or other parts) rather than fighting their captor and getting killed.

But what if it wasn't a mistake? What if you had been forced, by being taken hostage, to focus for once on the reality of the other—on their desperation, which had to have been extreme to drive them to their own rash act? What if you saw that you might do the same sort of thing in the other's shoes? If that insight were to occur to you, in the immense protraction of time that occurred when taken hostage, you would then see the situation newly, and change somehow, even if much later. Possibly that change was, at least sometimes, the right reaction to what had happened.

No doubt it depended on circumstances, as always. In the original Stockholm situation, she read, a pair of bank robbers had held four hostages, three women and a man, in a bank vault that they too were trapped in, and over the course of a week many small kindnesses had passed both ways. By the time it was all over, and of course it had ended peacefully, with the kidnappers surrendering, there was a bit of sympathy established. The hostages had refused to testify at trial against the kidnappers. The syndrome named after them was said to affect about ten percent of the victims of kidnappings, and the kinder the behavior of the kidnappers, inside the fundamentally hostile act of holding a person against their will, the more likely the effect. As made sense.

There was also a Lima Syndrome, she read, in which the kidnappers developed for their hostages a sympathy so strong that they let them go.

What if both syndromes occurred at once? Surely that was how it would sometimes happen, in a kind of symmetry: two people, both suffering from different kinds and degrees of trauma, recognizing in a moment of high stress a fellow sufferer. Wasn't that right?

So hard to say about these things. Trying to put a name on a confusion of feelings. A wrong feeling that nevertheless got felt. Many psychologists doubted the reality of Stockholm Syndrome. It had never made it into the DSM handbooks. It was pop psychology, a journalistic term, a fiction.

Well, but. Every possible thing happened, eventually. In moments of high stress strange things happened. Probably it was stupid to try to put any kind of label on these events, any kind of explanation. Syndromes, psychology in general: bollocks. It was just that one time, every time. In this case, just Mary and that young man, in a kitchen for a very intense couple of hours. Not that different from a bad date you felt obliged to ride out. Well, no. That pistol, that moment of fear—a jolting spike of fear for her life—she had not forgotten that, or forgiven it. She never would. Nothing was quite like that. But nothing was quite like anything.

In Zurich she left the Hauptbahnhof and crossed the river to the tram stop and got on the next 6 tram and rode it up the hill to Kirche Fluntern. Wearily she walked along Hochstrasse to her building and let herself in. Bodyguards still stood outside, pleased to see her. Up the stairs, feeling wasted. Poured herself a glass of white wine on ice and slurped it down fast. She didn't like living alone, but she didn't like living with bodyguards either. Probably she should have invited them inside; it was cold out. But at times like this she didn't want to talk to anyone. It would have been impossible, she was too confused. She would have been snappish, even if she had tried to be polite. As it was she took a quick shower and fell into bed, still stewing. Luckily sleep swept her off.

In the morning she went first to her office and got through the necessary things. Then she looked at the report that Badim had worked up on the young man. Frank May. He had indeed survived the great India heat wave; he had been right in the area worst hit, while there as an aid worker. At the time he had been twenty-two years old. And his DNA had been found at the murder site on Lake Maggiore, on a chunk of wood used as the weapon. Probably that would be charged as manslaughter. But they had him dead to rights, it looked like. She sighed. The other stuff was trivial in comparison, but the problem was that it made a pattern, added up to a repeat offender. Meaning more time in prison.

She looked up where he was—Gefängnis Zurich, the Zurich Jail. After a call to make sure of visiting hours and his availability, she walked to the tram stop. While waiting for the next tram she inspected the stuff in the kiosk shop. What did you bring a prisoner? Then she remembered she was visiting her kidnapper. She didn't buy anything.

The jail was located on Rotwandstrasse, Red Wall Street. The nearest tram stop was Paradeplatz. She got off and walked to the street; no red walls in sight. It could have been repainted, or knocked down eight hundred years before. The jail was obvious; a three-story concrete building, extending for most of a block. Institutional; tall windows in deep embrasures, obviously not meant to be opened.

She went in and identified herself. They called in; the prisoner was willing to meet with her. There was a big meeting room. First she had to leave her phone and other stuff in a locker, then go through an X-ray machine, as in airport security. After that she was escorted by a guard down a hall and through two doors that unlocked and opened automatically. Like an airlock in space stations, she thought. Inside this building, a different atmosphere.

And it was true. It looked different, it smelled different. The Swiss almost always displayed a little stylishness in even their most institutional institutions, and that was true here too—a blue wainscotting line, a big room with lots of widely separated tables with chairs, potted plants in the corners and some Giacometti imitations gesturing toward the ceiling in their usual elongations. But it smelled of ozone and power. Panopticon. A pair of guards sat behind a desk on a small dais by the door for visitors. Another guard came in the door to the prisoners' side, escorting a slight man who moved as if hurt.

It was him. He glanced up at her, smiled briefly, uncertainly, confused to see her; a fear grin; and looked back at the floor. He gestured at one of the tables, walked to it. She followed him, sat down in the chair across the table from him. The table was clearly called for. When they were seated, the guard who had escorted her in left them and walked over to chat with the other guards.

She looked at him in silence for a while. After a single startled glance, as if to reaffirm that it was her, he looked at the table. He looked withdrawn. He had lost weight since that night in her apartment, and he had been skinny then.

"Why are you here?" he asked at last.

"I don't know," she said. "I guess I wanted to see you in jail."

"Ah."

Long silence. Now I've got some of my own back, she didn't say—barely even thought. Now I'm safer than I was before. I won't get shot in the street. Now the tables are turned, you're the one being held against your will, I'm free to go. And so on. It wasn't feeling like good reasons had brought her here.

"How are you doing?" she asked.

He shrugged. "I'm here."

"What happened?"

"I got arrested."

"You were at a refugee dinner, I heard?"

He nodded. "What else did you hear?"

"I heard that they got attacked by some hooligans, and you went to their defense, and you were still there when the police came, and they were looking for you for something that happened at Lake Maggiore."

"So they said."

"What happened at Lake Maggiore?"

"I got mad at a guy and hit him."

"You hit him and he died?"

He nodded. "So they say."

"Are you some kind of a . . . ?"

He shrugged. "Must have got lucky."

"Lucky?" she repeated sharply.

He squirmed. "It was an accident."

"Okay, but don't joke about it. From now on, what you say may have an impact on your legal status."

"I was just talking to you."

"Practice it with everybody."

"No jokes? Really?"

"Will that be a hardship for you? I don't recall you making many jokes when you visited my flat."

"I was trying to be serious then."

"Do that now too. It could make a difference."

"In what?"

"In how long a sentence they give you."

The corners of his mouth tightened, he swallowed hard. No joke there.

"This person who died, did you hit him with something?"

"Yes. I was holding a piece of driftwood I found on the lakeshore."

"And that was enough to kill him?"

"I don't know. Maybe his head hit something when he went down."

"Why did you hit him?"

"I didn't like him."

"Why didn't you like him?"

"He was being an asshole."

"To you or to others?"

"Both."

"Were these others Swiss or *Ausländer?*"

"Both, really."

She regarded him for a long time. Apparently theirs were the kind of conversations that included many long silences.

Finally she said, "Well, it's too bad. It means they've got you for multiple things now. So . . . Well, I'll put in a word for you, if you want."

"That I was a good kidnapper?"

"Yes. That's already been put on the record. I mean, I reported it as a kidnapping, so now I can't exactly say I was having you over for a nightcap. They don't have any three strikes type laws in Switzerland, as I understand it, but you do have this latest fight you were in, to add to the death and what you did to me. That will have an effect on the judgment. If I weren't already on the record saying otherwise, I would consider telling them we were just having a night."

He was startled. "Why?"

"To help shorten your sentence, maybe."

He continued to look surprised. She too was surprised. Who was this man, why should she care? Well, because of that night. He was obviously damaged. Something wrong in him.

He shrugged again. "Okay." Then suddenly his look turned dark. "Do you know how long it might be? My sentence?"

"I don't." She paused to think about it. "In Ireland I think what you did would get you a sentence of a few to several years, depending on circumstances. Then there's time off for good behavior and so on. But Switzerland is different. I can look into it."

He stared through the table, down this unknown abyss of years. "I don't know how long I can do it," he said quietly. "I already can't stand it."

Mary pondered what she could say. There wasn't much. "They'll give you work," she ventured. "You'll be let go out to work. They'll put you in therapy. It might end up being not that much different from how you were living before."

This earned her a quick fierce black look. Then he was staring at the table again, as if unhappy she was there.

She sighed. In truth there was little encouragement to be had in a situation like his. Well, he had made his choices and here he was. If they had been choices. Again the question of sanity came to her. All the horrid violent crimes in the world, worse by far than anything this man had done—weren't they all prima facie evidence of insanity? Such that any subsequent punishment became in effect punishing someone for being ill?

Or in order to keep the community safe.

She didn't want to be thinking about these things. She had bigger fish to fry, she had a busy day. But there he was. Stuck, jailed, miserable. Possibly insane. Not just post-traumatic, but damaged by the trauma itself in some way even more crippling than PTSD. Some kind of brain damage, from overheating or dehydration or both, damage which had never healed. It seemed quite possible; everyone else on the scene had died.

Well, who knew. He wasn't going anywhere. She had things to do, and it would be easy enough to come back.

"I'm going to go," she told him. "I'll come back. I'll see what I can learn, and talk to your lawyer. Have you got one?"

He shook his head. "They assigned one to me."

He looked completely hopeless.

She sighed and stood. One of the guards came over.

"Leaving?" the guard asked.

"Yes."

She was the one in power now. Briefly she touched him on the shoulder, almost as he had first touched her on Hochstrasse. She was partly there for revenge, she felt that now. He was warm through his shirt, hot; feverish, it felt like. He shrugged her off like a horse shivering away a fly.

51

The thirties were zombie years. Civilization had been killed but it kept walking the Earth, staggering toward some fate even worse than death.

Everyone felt it. The culture of the time was rife with fear and anger, denial and guilt, shame and regret, repression and the return of the repressed. They went through the motions, always in a state of suspended dread, always aware of their wounded status, wondering what massive stroke would fall next, and how they would manage to ignore that one too, when it was already such a huge effort to ignore the ones that had happened so far, a string of them going all the way back to 2020. Certainly the Indian heat wave stayed a big part of it. They could neither face it nor forget it, they couldn't think about it but they couldn't not think about it, not without a huge subconscious effort. The images. The sheer numbers. These recalled the Holocaust, which had left a huge hole in civilization's sense of itself; that was six million people, but an old story, and of Jewish people killed by Germans, it was a German thing. The Palestinians' Nakba, the partition of India; on and on these bad stories went, the numbers always unimaginable; but always before it had been certain groups of people who were responsible, people from an earlier more barbaric time, or so they told themselves. And always these thoughts obtruded in an attempt to stave off the heat wave, which was now said to have killed twenty million. As many people, in other words, as soldiers had died in World War One, a death toll which had taken four years of intensely purposeful killing; and the heat wave had taken only two weeks. It somewhat resembled the Spanish flu of 1918 to 1920, they said; but not. Not a pathogen, not genocide, not a war; simply human action and inaction, their own action and inaction, killing the most vulnerable.

And more would surely follow, because they all were vulnerable in the end.

And yet still they burned carbon. They drove cars, ate meat, flew in jets, did all the things that had caused the heat wave and would cause the next one. Profits still were added up in a way that led to shareholder dividends. And so on.

Everyone alive knew that not enough was being done, and everyone kept doing too little. Repression of course followed, it was all too Freudian, but Freud's model for the mind was the steam engine, meaning containment, pressure, and release. Repression thus built up internal pressure, then the return of the repressed was a release of that pressure. It could be vented or it could simply blow up the engine. How then people in the thirties? A hiss or a bang? The whistle of vented pressure doing useful work, as in some functioning engine? Or boom? No one could say, and so they staggered on day to day, and the pressure kept building.

So it was not really a surprise when a day came that sixty passenger jets crashed in a matter of hours. All over the world, flights of all kinds, although when the analyses were done it became clear that a disproportionate number of these flights had been private or business jets, and the commercial flights that had gone down had been mostly occupied by business travelers. But people, innocent people, flying for all kinds of reasons: all dead. About seven thousand people died that day, ordinary civilians going about their lives.

Later it was shown that clouds of small drones had been directed into the flight paths of the planes involved, fouling their engines. The drones had mostly been destroyed, and their manufacturers and fliers have never been conclusively tracked. Quite a few terrorist groups took credit for the action in the immediate aftermath, and demanded various things, but it has never been clear that any of them really had anything to do with it. That multiple groups would claim responsibility for such a crime just added to the horror felt at the time. What kind of world were they in?

One message was fairly obvious: stop flying. And indeed many people stopped. Before that day, there had been half a million people in the air at any given moment. Afterward that number plummeted. Especially after a second round of crashes occurred a month later, this time bringing down twenty planes. After that commercial flights often flew empty, then were cancelled. Private jets had stopped flying. Military planes and helicopters had also been attacked, so they too curtailed their activities, and flew only if needed, as if in a war. As indeed they were.

It was noticed that none of the experimental battery-powered planes had been brought down, and no biofuel-powered planes, also no blimps, dirigibles, or hot-air balloons; no airships of any kind. But in that period there were so few of these, it was hard to tell at the time if they had been given a pass or not. It looked like they had, it seemed to make sense, and the surge in airship manufacture, which had already gotten a small start, has never ceased to this day.

The War for the Earth is often said to have begun on Crash Day. And it was later that same year when container ships began to sink, almost always close to land. Torpedoes from nowhere: a different kind of drone. It was noticed early on in this campaign that ships often went down where they could form the foundations for new coral reefs. In any case, they were going down. They ran on diesel fuel, of course. Loss of life in these sinkings was minimal, but world trade was severely impacted. Stock markets fell even lower than they had after Crash Day. A worldwide recession, a feeling of loss of control, price spikes in consumer goods, the clear prospect of the full-blown depression that indeed followed a few years later... it was a time of dread.

Two months after Crash Day a group called Kali, or the Children of Kali, issued a manifesto over the internet. No more fossil-fuel-burning transport. That was about twenty to twenty-five percent of civilization's total carbon burn at this point, and all of it vulnerable to disruption, as the manifesto put it.

Next up, the Children of Kali (or someone using that name) told the world: cows. Later that same year the group announced that mad cow disease, bovine spongiform encephalopathy, had been cultured and introduced by drone dart into millions of cattle all over the world. Everywhere but in India, but especially in the United States and Brazil and England and Canada. Nothing could stop these cattle from sickening and dying in a few years, and if eaten their disease could migrate into human brains, where it manifested as Variant Creutzfeldt-Jakob disease and was invariably fatal. So to stay safe, people needed to stop eating beef now.

And indeed in the forties and ever after, less beef got eaten. Less milk was drunk. And fewer jet flights were made. Of course many people were quick to point out that these Children of Kali were hypocrites and monsters, that Indians didn't eat cows and so didn't feel that loss, that coal-fired power plants in India had burned a significant portion of the last decade's carbon burn, and so on. Then again those same Indian power plants were being attacked on a regular basis, so it wasn't clear who was doing what to whom. Scores of power plants were being destroyed all over the world, often by drone attacks. Power outages in those years were most common in India, but they happened everywhere else too. The War for the Earth was real, but the aggressors were nowhere to be found. It was often asserted they were not in India at all, and were not even an Indian organization, but rather an international movement. Kali was nowhere; Kali was everywhere.

52

Our Sikkim became a state with fully organic agriculture between 2003 and 2016, aided by the scholar-philosopher-feminist-permaculturist Vandana Shiva, an important figure for many of us in India. Of course Sikkim is the least populated Indian state, with less than a million people, and the third smallest in economic terms. But it grows more cardamom than any place but Guatemala, and one of its names is Beyul Demazong, "the hidden valley of rice," beyuls being sacred hidden valleys in Buddhist lore, including Shambala and Khembalung. Sikkim is precisely this kind of magical place, and its version of organic agriculture, one aspect of permaculture generally, became of interest across India in mid-century, as part of the Renewal and the New India, and the making of a better world. Here again Vandana Shiva was an important leading public intellectual, combining defense of local land rights, indigenous knowledge, feminism, post-caste Hinduism, and other progressive programs characteristic of New India and the Renewal.

Important also has been the example of Kerala, at the opposite end of India, for its crucial innovations in local government. A state blessed by location on the southwest coast of India, with a long history of interactions with Africa and Europe, home to fabled Trivandrum, Kerala has long been governed by a power-sharing agreement between the Left Democratic Front, which was very influenced by the Indian Communist Party, and the old Congress party, which though out of favor at the national level, always kept a fairly substantial level of support in Kerala, as representing the party of independence and Gandhi's *satyagraha*, meaning *peace force*. The LDF has been the dominant party in Kerala since Indian independence a century ago, and one of the main projects for the party has always been devolution to local governance, closing in on the goal of what one might call direct democracy. In Kerala now there are pan-chayats for every village, then district governments that coordinate these panchayats and oversee disputes and care for any necessary business at the district levels; and above them is the state government in Thiruvanantha-puram, looking after such business as involves the whole state together. The focus on local government has been so intense and diligent that there now total 1,200 governmental bodies in Kerala, all dealing with issues in their particular area whatever it might be.

Strangely, perhaps—or perhaps it is not so strange—it is often said of both Sikkim and Kerala that they are exceptionally beautiful places, and both attract more tourists than most Indian states. But in truth almost every Indian state is beautiful, if you take the trouble to visit it and look. It is a very beautiful country. So the answer to the success of these two states cannot be put down entirely to the beauty of their landscapes.

Although treated very poorly for a long time by various invading inter-ests, including the Mughals, the Rajas, the British, globalization, and corrupt national governments from every party, but especially from the Congress party in its waning days of power, and then the Hindu triumph-alist BJP party, both now disgraced for their connections to the heat wave, India continues to rise above all and do its best. So in the New India, in the time since the heat wave, there was a complete revamping of priori-ties, and if at all possible the introduction of India-based solutions to the problems afflicting us. Here Sikkim and Kerala are in their different ways great exemplars of what can be done by good governance and India-first values and practices. Of course the solutions to our ills can't come entirely from within India's borders, big though it is compared to many coun-tries, both geographically and in terms of population. No matter how big, all countries now need the whole world to be behaving well. Still, India is big, and big enough to provide human, agricultural, and mineral resources for a great deal of rapid upgradation.

What has been of interest is to see if the advances pioneered in the states could be rapidly scaled to the whole country. Not just Kerala and Sikkim, of course, but also states like Bangalore, the so-called Silicon Val-ley of India, which has led the way in that regard by founding a great number of engineering colleges. Now Bangalore, the Garden City, third largest in India, is thriving as a global hub for IT, and innovations there in the creation of the so-called Internet of Land and Animals might well help the process of full modernization of the Indian countryside. And then of course there is Bollywood in Mumbai, and the enormous mineral resources all up and down the spine of the country. Even taking coal off the table, as we have, the sheer physical wealth of India is unsurpassed; and in a world powered by solar power, India is indeed blessed. More sunlight energy falls on India than on any other nation on Earth.

So the important thing now is to join up past and future together, and join up also all the diversity of people and landscapes that is India, into a single integrated project. The biggest democracy on Earth has huge problems still, and also huge potential for solutions. We need to learn our agriculture from Sikkim, governance from Kerala, IT from Bangalore, and so on and so forth; each state has its contribution; and we must then put them all together to work for everyone in our country. When we have done that, then we will provide an example for the world that it needs; and also, having solved the contemporary problems of life in a democratic fashion for one-seventh of all humanity, there are that many fewer people for the rest of the world to worry about.

53

I zing and I ping and I bring and I bling. Freed to self in the heart of the sun, I banged around in there for a million years before popping out the surface and zipping off. Then eight minutes to Earth. In the vacuum I move at the speed of light, indeed I define the speed of light by my dance.

Into Earth's atmosphere, exhibiting aspects both wave and particle but not either, I am a four-space conceptualized as an hourglass shape in three-space, where time and space cross in the human mind. Hitting things again, slowing down, breaking off brothers and sisters trapped in things, all the same, all without mass, all spin one, and so bosons, not fermions; one three-hundred-sixty-degree turn and I'm back where I began in that regard, while for fermions it takes seven hundred and twenty degrees of turning to get back to its original position; fermions are strange!

I am not strange, I am simple. Banging into atoms and moving them as I too move, simple as Newton, bang bang bang, the atoms of the atmosphere move more than they would have without my kick in their pants. That's heat. Until I kick into something that captures me and I stop moving on my own. That or I might bounce off something I hit and head back out into space, become the light in the eye of some lunatic observer, looking up at the big blue ball and seeing me bang something in their retina. A blue pixel of the utmost granular fineness, so much so that the wave I make is easier to detect, maybe even easier to imagine. The wave/particle duality is a real thing, and both at once is hard to think, hard to see. It doesn't really work, four-space in your three-space mind, no. I am a mysterious thing, massless but powerful. There are more of us than there are of anything else. Well, perhaps that's not true. We don't know about dark matter, which should be called invisible matter, we don't know them or what's going on there. Presumably its individual constituent elements are something like me, but maybe not; no one knows. All that stuff flies around as if in a parallel universe slightly overlapping ours, maybe just waves, in any case gravity works on it for sure, because its very existence has been revealed to us by its gravitation effects. If we are similar to those dark-matterinos, it is in the way that light and dark are similar. Two parts of a whole, perhaps. I am visible, I embody light itself; dark matter is not actually dark, it is invisible, and we don't know how or what it is. Our absent self, our shadow, our twin. Although there are a lot more of them than us, maybe. A mystery among all the other mysteries, we fly through each other like ghosts.

But me, me, me, me, banging Earth, bouncing back up into the eye of an observer on the moon, then off again, immortal, immutable, I bang bluely on and on and on, and in the process, passing by just this once in a cloud of my brothers and sisters, we hit Earth and light it up, and the gas wrapping the planet's hydrosphere and lithosphere gets warmer from our touch. My brothers and sisters follow me and they continue that warming.

What am I? You must have guessed already. I am a photon.

54

Mary hunkered down in the wan light of winter Zurich, meeting daily with her team to plot their next moves. The whole baker's dozen gathered in the mornings to share the news and plan the day, the week, the decade. It was like a war room now, yes, and their work felt like war work. Not war, however; there was no opponent, or if there was, it consisted of fellow citizens with a lot of money and/or passion. If it was war, they were outgunned and on the defensive; but really it was mostly just discursive struggle, a war of words and ideas and laws, which only had brutal death-dealing consequences as a derivative effect that could be denied by aggressors on both sides. It was a civil war, perhaps, a body politic punching itself over and over. In any case, war or not, it had that same besieged awful feeling of existential danger, of stark emergency that never went away. The large number of people living rough in Zurich, busking for change, looking for work, or worse, not looking for work; that never got ordinary. It was not a Zuri thing.

Nevertheless, Mary's life settled into a routine that was, she had to admit, almost pleasant. Or at least absorbing. Busy; possibly productive. There were worse things than having a project in hand that felt crucial. She stayed in Zurich. When its winter gloom got to be too much for her, she took a Sunday train up to one of the nearby Alpine towns, popping up through the low cloud layer that banked on the north side of the Alps and made winter life in Zurich so sunless and gloomy. There had been eight hours of sunlight in the first fifty days of this year, they told her, scattered over several days; it was like them to quantify a thing like that. But on a Sunday she would take a break and train up into brilliant snowy light. The ski resorts were too crowded, and Mary was not a skier; but there were many Alpine towns that had no skiing to speak of, being surrounded by sheer cliffs: Engelberg, Kandersteg, Adelboden, some others. In these towns she could walk on cleared snow trails, or snowshoe them, or just sit on a terrace, soaking in the chill brilliant light, the small crystalline sun in its big white sky. Then back down into the gloom.

In Zurich she met with her team and with outside helpers and antagonists. The fossil fuels industry's lawyers were getting more and more interested to learn how much they could extort from the system if they chose to sequester their asset. This was by no means a trivial topic, indeed it was a crucial matter, and Mary entered into it with great interest. To a certain extent it felt to her like she was negotiating to buy off terrorists who had explosive vests strapped around their waists, and were saying to her and to the world at large, *pay us or we blow up the world*. But this was not exactly right. First, if no one would buy their product, their explosives wouldn't go off, and they couldn't blow up the world. So their threat was hollowing by the day, which explained why they were talking to her at all; their leverage as terrorists, with the biosphere as their hostage, was lessening. And as the Ministry for the Future was one place where they might be able to negotiate a settlement, they were coming to Zurich to bargain.

Then also, they were not terrorists. It was an analogy, maybe a bad one, or at least a partial one. Civilization needed electricity, and it was citizens who had powered themselves on these fossil fuels for the last couple of centuries. The owners of these fuels were sometimes private individuals who had gotten fantastically rich, but many times they were nation-states that had claimed ownership of the fuels found within their boundaries as assets of the state and its citizenry. These petro-states controlled about three-quarters of the fossil fuels that had to stay in the ground, and they too were seeking compensation for the losses they would incur by not selling and burning their fuels.

So it was a mixed bag in that respect. ExxonMobil was big, yes, and held more in assets than many nations in the world; but China, Russia, Australia, the Arab states, Venezuela, Canada, Mexico, the US—these were bigger in carbon assets than ExxonMobil and the other private companies. And they all wanted compensation, even though all of them had agreed in the Paris Agreement to decarbonize. Pay us for not ruining the world! It was extortion, a protection racket squeezing its victims; but the victims were the national populations, so in effect they were putting the squeeze on themselves. Or their elected politicians were putting the squeeze on them. So the situation was bizarre, both hard to define and constantly changing.

So meetings kept happening. Mary kept on pushing the idea of creating the money to pay for decarbonization and all the necessary mitigation work. As the weeks passed and the global economy went from recession to depression, Dick almost convinced her that taxation might be a good enough tool to do the job all by itself. What counted, he said, was differentials in cost and benefit. So in that sense, meaning in effect the bottom line in accounting books, taxes were both stick and carrot. On the bottom line, whether a person's or a company's or a nation's, there was no difference between stick and carrot. They were both incentives. And if nations levied taxes they got the money themselves, rather than having their central banks print new money; they took in rather than shelled out. So if you did something that was stiffly taxed, and progressively taxed the more you did it, that tax was a stick to beat you away from that activity; then, if you didn't do that activity, you thereby dodged the tax and didn't have to pay it, and your bottom line showed that lack of payment as a plus in the ledger. So avoiding the action that got taxed created a carrot.

Mary saw Dick's point, but at the end of every discussion, didn't actually agree with him. Maybe it was just psychological rather than economic, but people liked to be paid for doing things more than they liked avoiding having to pay for something. There was a mental difference between carrots and sticks, no matter if the numbers were the same in a ledger. With the one you got fed, with the other you got hit. They simply were not the same. She would make this point to Dick over and over, and he would respond with his crazy economist's smile, his acknowledgment of economics' fundamentally alien nature, the way it was a view from Mars, or a helpful but clueless AI. Which last was more or less the case.

Also, in terms of making a carrot: the oil industry had equipment and an expertise that could be adapted from pumping oil to pumping water. And pumping water was much easier. This was good, because if they were going to try to pump some of the ocean's water up onto Antarctica, the pumping effort was going to be prodigious. Even if they confined their efforts to draining the water from under the big glaciers, a lot of pumps would be needed.

So the oil industry needed to stop pumping oil, for the most part, but they could be hired to pump water. Or to pump captured carbon dioxide down into their emptied oil wells. Direct air capture of atmospheric CO_2 was looking more and more like an important part of the overall solution, but if it scaled up to the size of the problem, it was going to produce an enormous amount of dry ice, which had to go somewhere; pumping it underground was the obvious storage location. In some ways, these reverse pumping actions were easier and cheaper than pumping up oil; in other ways they were a bit harder and more expensive. But in any case they made use of an already-existing technology which was very extensive and powerful. If the industry could be paid to do something with its tech to help in the current situation, all the better.

The fossil fuel lawyers and executives looked interested when this was proposed to them. The privately owned companies saw a chance of escaping with a viable post-oil business. The state-owned companies looked interested at the idea of compensation for their stranded assets, which they had already borrowed against, in the usual way of the rampant reckless financialization which was the hallmark of their time. Paid to pump water from the ocean up to some catchment basin? Paid to pump CO_2 into the ground? Paid how much? And who would front the start-up expenses?

You will, Mary told them.

And why? When we don't have it?

Because we'll sue you if you don't. And you do have it. You can pay the upfront costs of the transition, and if you invest in that, we'll pay you in a guaranteed currency that is backed by all the central banks of the world to increase in value over time. As an aspect written into the currency itself. A sure bet no matter what happens.

Unless civilization crashes.

Yes. You can short civilization if you want. Not a bad bet really. But no one to pay you if you win. Whereas if you go long on civilization, and civilization (therefore) survives, you win big. So the smart move is to go long.

Go long. She found herself saying that a lot. Bob Wharton called it the Hail Mary pass. It was a little strange to be saying "go long" to middle-aged men, so sleek and smooth in their wealth; sexually satisfied, said their aura, such that saying to them "go long" might be enough to trigger what she had once heard described as a hard-on in the heart. Well, if she was enticing them to expand confidently into the world, to make an intervention into the body of great mother Gaia, so huge and vivid and dangerous, then fine. So what. Mary had no illusions that she herself, a harried middle-aged female bureaucrat in a limp toothless international organization, represented for these men any kind of stand-in for the Earth mother; but she was a woman, perhaps even the fine remains of a woman, as the pirates of Penzance had said about Ruth so hard of hearing. She could push that a little, and she did. Little sadistic lashings of what was really just Irish contempt for any form of pretense, of which there was a lot in these meetings. These men were often quite disgusting, in other words, but there was a higher purpose to be pursued.

Then one day Janus Athena came into her office. This person was the opposite of all that bankerly masculine erotic charge; the project for J-A was to efface gender, to exist in that very narrow in-between, the zone of actual gender unknowability. Which itself was a new gender, presumably. Mary wouldn't even have believed that zone existed if it weren't for Janus Athena standing before her every day, completely unknowable in that aspect of self, or to put it better, neither masculine nor feminine as any kind of dominant. This could only be the result of an extensive and canny effort.

Mary regarded her mysterious AI expert with her usual curiosity. She wanted to say, J-A, what gender were you assigned at birth, if any?

But this would be to break the social rules which either J-A or Mary, or society at large, had imposed on her. It would be to intrude or to interfere with J-A's project. Probably J-A got asked this pretty often, one way or another, but Mary wasn't going to. Acceptance of the other and their project; this was crucial.

"What have you got for me, Janus Athena?"

"The AI group is making open source instruments that mimic the functions of all the big social media sites."

"So people can shift over to this new set?"

"Yes. And it will protect their data for them using quantum encryption."

"Then China probably won't let their people use them."

"Maybe not. China is under huge pressure to change, so, unclear how that will play out. For everyone else, using these sites means they'll control their data, rather than it being used and mined. That privacy can then be a resource to them. They can sell their personal data if they want. That plus the security of encryption, and the public ownership of these sites as a commons, should be enough to entice every user on the planet to shift. Publicize it, make it easy, set a date, be ready to handle the influx, boom."

"How many do you think will shift?"

"Maybe half. After a few years, everybody."

"So, the decapitation of Facebook."

"And all the rest like it."

"Replaced by a system owned by its users, in effect."

"Yes. Open source. A distributed ledger. The Global Internet Cooperative Union. GICU."

"Is that a good name?"

"Is Facebook a good name?"

"Better than GICU."

"Okay, think of a better one. Then if it works, it will serve as the operating platform for ICU."

"Which means," Mary prompted, playing along.

"International Credit Union. A people's bank. The team has set that up too. Lots of bank mirroring, and credit unions are already a thing. This won't be quite like a credit union, because it would be an open network of people who make a distributed issuance of credit, issuing carbon coin fractions to each other on proof of good action on carbon. People deposit their savings and create new value in a customer- and employee-owned distributed ledger. Their bank, as one function of their YourLock account. It invests mindfully as a group mind, a kind of planetary mind, that has to always be funding biosphere-friendly activities. Also, a place to go if everyone removes their deposits from current private banks at the same time. Those banks are so over-leveraged that they will immediately crash. Then individuals have to have a safe harbor. For-profit banks will go running to central banks to ask for bail-out, and legislatures will panic and agree to let central banks create however many trillions the central banks recommend. That's been the template so far. So, for any planned attack on private banks, best to have a safe harbor ready. Then you can tell the legislatures to approve central banks' bail-out QE, but only on condition of buying equity in them."

"Nationalizing the banks, you mean."

"Yes, but that puts it too simply. Central banks take possession of private banks, having saved them from bankruptcy or death. Good as far as it goes, but legislatures also need to take possession of their own central banks, by increasing political control of them. It's a double action. You need both."

"Will it work?"

"Is it working now?"

Mary sighed. "Point taken. Can you get the future into the names somehow? Since we're the Ministry for the Future?"

"The internet interface name could be, I don't know: Children, Own Your Data. COYD?"

"That's terrible."

"Some people are not good at names."

"I've noticed." As for instance, the names Janus and Athena had nothing to do with each other. But that was neither here nor there. "Let's make the name a game. Get the whole gang down to a dinner at Tres Kilos and see what they come up with."

"As long as you're picking up the tab. Pitchers there are ridiculously expensive."

"Right."

"And so, what about these projects?"

"Oh yes," Mary said. "Let's try it. Something has to be done. We're still losing."

That night at Tres Kilos, over dinner and a couple pitchers, they came up with some ideas for names for the Facebook replacement, which Mary scrawled down on her napkin: DataFort, EPluribusUnum, WeDontChat, OnlyConnect, A Secure and Lucrative One-Stop Replacement for Your Many Stupid Social Media Pages, TotalEncryption, FortressFamily, FamilyFortress, HouseholdersUnion, Skynet, SpaceHook, WeAretheWorld-WeArethePeople, PourquoiPas, Get Paid To Waste Time!

"Maybe we're still looking," Mary concluded as she read the list on her napkin. "Although I do kind of like WeDontChat."

55

La Vie Vite! It was a time.

The *gilets jaunes* shifted the model for how to proceed, away from May 68 or any fainter impressions of the Commune or 1848, not to mention 1793, which it has to be admitted is now like a vision from ancient history, despite the evident satisfactions of the guillotine for dealing with all the climate criminals sneaking off to their island fortress mansions. No, modern times: we had to get out into the streets day after day, week after week, and talk to ordinary people in their cars stuck in traffic, or walking past us on the sidewalks and metro platforms. We had to do that work like any other kind of work. It wasn't a party, it wasn't even a revolution. At least when we started.

But soon we saw that people wanted to talk to us. They all knew they were being used, that they were just tools now. I myself was a kid, the main thing that got me out there was how much I hated school, where I had always been made to feel stupid. I was slotted into the bottom classes early on and my life was sealed at that point, on a track to servitude, even though I knew I had real thoughts, real feelings. So the main thing for me in that initial break was to get my ass out of school. Although parenthetically I have to admit that I later on became a teacher.

Something then caused us to all converge on Paris. In France, that's where you go. No one had to direct us. It was Trotsky who said the party is always trying to keep up with the masses. Strategy comes from below and tactics from above, not the reverse, and I think that's what happened here, some trigger or combination of triggers, the extinction of some river dolphin, or another refugee boat going down offshore, who knows, maybe just lost jobs, but suddenly we were all headed to Paris together, often on foot when the highways jammed. Of course once we got there we couldn't take on the police or the army, that would be suicide, we had to overwhelm them with numbers. And eventually there were so many of us we couldn't be contained, everything ground to a halt. At that point problems immediately jumped out of the pavement and hit us in the eye. Some were simply logistical, a matter of food and toilets. Others were ideological. The younger we were the more we wanted. Older people were hoping just to make things a bit better. So the traditional infighting began, but I have to say this was mostly a discursive battle, we weren't like Spain in Franco's time, killing each other or watching the Russians kill us. It was France on its own now, and really we are the country of revolution. Now we had to show what could happen in our time. So we took over the city, Paris was ours by way of sheer bodies jamming the streets. And of course some of us had read about the Commune and realized if we didn't win decisively we would be hunted down and killed, or at best jailed for life. So at that point it was win or die, and we buckled down to making it work as an alternative system of life, a kind of commons that was post-capitalist, even post-money, just people doing what it took to keep everyone fed. And I must say, so many Parisians came out and helped us, cooked food, provided rooms, manned the barricades in every way, that again we had to realize that it wasn't just those of us in the streets, it was all France, maybe even the world, we couldn't tell. But for sure what happened then was the most intense and important feeling I could ever live in this existence. Here's what it was: solidarity. We could see so many others with us, all on each other's side. Paris was a commons, France was a commune. So it felt. Later that proved to be impressionistic to that time, but while it lasted it was amazing.

But also exhausting. To live without habits, making it up day by day, trying to get a shower, a meal now and then, find the right way to pitch in, it's much more work than if you're just a wage slave. Much more. But people felt it was important, all over they were dropping what they had been doing and joining the fray and giving it all they had, and it felt right. Somehow we kept finding the ways to give it our all. This we felt was a French thing above all, a kind of political improvisation that our whole history and even our language made us good at, if we could figure it out and pull it off.

Help came from weird places. When the internet was killed the union of proof-readers, historically anarchist which is of course very funny, came out of their tiny niches in the publishing industry and plastered the city with posters—posters on walls, as if the world was still real! And we realized that social media actually meant many things we had forgotten, and could be taken back under our own control, at least sometimes. Simply talking was the strongest social media of all of course, it was obvious once we rediscovered it, but those posters made the city itself our text, as it had been more than once before.

But under all that, the right was regathering their forces. And in fact we didn't have the logistics set up to keep it going forever. We didn't have a good plan to change government itself, and we argued with each other about how to proceed. A movement without leaders is a good idea in theory, but at some point you have to have a plan. How to make one wasn't so clear. State power is a mix, first of the government proper in all its parts, but then also the military, finance, and the population's support, and you need all of these working together to make any lasting progress. In our case, supporters in the populace began to complain they couldn't get to their usual bakery, it wasn't open when they got there, and so on. And if there's no plan, if there's nothing that follows the moment of occupation, or the moment of the Commune, then we're stuck in helplessness and a drift back toward the center. And as someone said, in France the center has neither a left nor a left. So we were in trouble.

And so the police waited until after midnight one night, and charged us. Pepper spray and men with giant shields, like Roman soldiers in a nightmare. I had a paving stone to throw at them, but at the last moment I couldn't, because a vision of a wounded man came to me, of what it would feel like if that heavy stone in my hand hit me. So I threw it at the ground itself, I was weeping with fury that I couldn't fight properly, and then I joined all the rest who lay down and forced them to drag us off to their vans. They beat us with batons and pepper sprayed us right in the face, the spray was amazingly painful, my whole face convulsed and tears poured out of my eyes and nose and mouth, even out of the top of my head it felt like. But all through that I kept thinking fuck it I don't care, I refuse to care, if they kill me here and now at least I'll have died for something I believe in. And in the end they just tied our arms behind our backs and dragged us off. There were so many vans.

After that it was all over but the shouting, which of course has never subsided. No one agrees on what happened or what it meant. But I know that a lot of what we did mattered, and it was supported at the time by ordinary Parisians, especially the women of the city, they were the real organizers when it came down to it, not the people at the microphones. And now of course a lot of us are back in yellow vests talking to people driving by in the roundabouts, and there is a lot of support for what we say. One driver when the traffic had stopped leaned out his window and said to me, Look it's all about how we treat the land, the revolution will happen there. Another man said I don't own my kids' teacher, I don't own my doctor, I don't need to own my house. I just want to pay the collective for it, not some landlord.

So maybe someday the solidarity will overcome the splitting. I hope so. During the occupation I didn't want reform, I wanted something entirely new. Now I'm thinking if we can just get the fundamentals working, it would be good. A start to something better. I don't like to think of this as giving up, it's just being realistic. We have to live, we have to give this place to the kids with the animals still alive and a chance to make a living. That's not so much to ask.

Of course there is always resistance, always a drag on movement toward better things. The dead hand of the past clutches us by way of living people who are too frightened to accept change. So we don't change, and one hard thing now is to go through a time like that, like ours during Paris, two hundred days of a different life, a different world, and then live on past that time in the still bourgeoisified state of things, without feeling defeated. For a time everything seemed possible, you felt free. You feel things so intensely when you're young, and really it's the first time I spoke to the world, the first time I wasn't just the stupid kid in school, but a real person with a real life. Those seven months made me, and I'll never forget it, never be the same. I only hope to live long enough to see it happen again. Then I'll be happy.

56

The International Criminal Court was separate from the UN, founded by its own international treaty to prosecute crimes that were somehow outside of nation-state court systems' jurisdiction. It was targeted toward individual crimes, and could be effective only when a combination of factors came into an unlikely conjunction. The US and several other big countries had withdrawn from the court's jurisdiction after negative rulings against their citizens.

The World Court, more properly named the International Court of Justice, was a UN-sponsored body, so that all member states of the UN were theoretically bound to follow its rulings; but it was designed to adjudicate state-to-state disagreements, and all UN agencies were forbidden to bring cases before it. The Ministry for the Future was an agency created in the implementation meetings of the Paris Agreement, but the Paris Agreement had been brokered under the umbrella of the United Nations; so the ministry was forbidden to bring cases to the World Court.

What it came down to was that to prosecute a state, you needed to be a state, and not the UN or an individual. More than that, you had to be a state that had agreed to be within the World Court's jurisdiction. To prosecute an individual outside a national court system, you needed to convince one of the international courts to take your case, and historically this had been hard to do. In short, these two international courts, both located at The Hague, were not well-designed as places to litigate climate justice.

Tatiana was getting more and more frustrated by this situation. If the ministry had been founded to bring legal standing and therefore lawsuits on behalf of the people of the future and the biosphere itself, fine, she said, fine fine fine: but in what court?

Tatiana concluded one of her reports to Mary with this:

—*If legal action has to occur in the various national courts, even though the crime is global, and committed by those same states whose courts would be asked to judge against them, the actions will not prevail.*

—*If we convince a signatory nation to go to the World Trade Organization and file a complaint to the effect that destroying the biosphere contravenes WTO rules against predatory dumping and the like, this is so tenuous and slow as to be useless.*

—*Meanwhile the fossil fuel companies keep pouring vast sums into buying elections, politicians, media, and public opinion. Even when they come to the table to negotiate with us, they never stop these other aggressive actions. Because the best defense is a good offense.*

—*This last is true for us as well as for them.*

Mary read this and ground her teeth. Tatiana was one of her fiercest warriors. They were now at atmospheric CO_2 levels of 463 parts per million.

The crash in insect numbers put every ecological system on terrestrial Earth in danger of collapse. Collapse—meaning most of the species currently on Earth dead and gone. The surviving species subsequent to this event would be free to spread in all the empty ecological niches, spread and evolve and speciate, so that in twenty million years, maybe less, maybe only two million years, a differently constituted array of species would fully re-occupy the biosphere.

Mary replied to Tatiana's memo with an instruction: Pick the ten best national cases from our point of view. Do what you can to help them.

The inadequacy of this, the futility of it, made her shudder. Hoping to counter this desperate feeling, she convened her nat cat group. Natural catastrophes: it was possibly a contradiction in terms, maybe the group should be called anthropogenic catastrophes, but anyway, them. Also the infrastructure and ecology groups. She needed to get away from legal abstractions.

They spoke enthusiastically of carbon-negative agriculture, clean energy, fleets of sailing ships, fleets of airships, carbon-based materials created from CO_2 sucked out of the air and replacing concrete; thus direct air capture of CO_2, a necessary component of the drawdown effort, would provide most construction materials going forward. Cheap clean desalination, clean water, 3-D printed houses, 3-D printed toilets and sewage, universal education, vastly expanded medical schools and medical facilities. Landscape restoration, habitat corridors, ag/habitat combinations—

"Okay!" Mary said, chopping short their flood of suggestions. She could see that the people in these divisions were feeling a little neglected. Finance had filled her head for too long. And as Bob Wharton had just said, you could literally fill a medium-sized encyclopedia with the good new projects already invented and waiting to scale. "Admitted; there's no end to the good projects we could fund, if we had the funds. But what should we be telling national governments to do now?"

Bob said, "Set increasingly stringent standards for carbon emissions across the six biggest emitting sectors, and pretty soon you're in carbon-negative territory and working your way back to 350."

"The six biggest emitters being?"

"Industry, transport, land use, buildings, transportation, and cross-sector."

"Cross-sector?"

"Everything not in the other five. The great miscellaneous."

"So those six would be enough."

"Yes. Reduce those six in the ten biggest economies, and you're hitting eighty-five percent of all emissions. Get the G20 to do it, and it's essentially everything."

"And how do you get reductions in those six sectors?"

Eleven policies would get it done, they all told her. Carbon pricing, industry efficiency standards, land use policies, industrial process emissions regulations, complementary power sector policies, renewable portfolio standards, building codes and appliance standards, fuel economy standards, better urban transport, vehicle electrification, and feebates, which was to say carbon taxes passed back through to consumers. In essence: laws. Regulatory laws, already written and ready to go.

"This sounds like a litany," Mary observed.

Yes, they told her, it was standard analysis. US Department of Energy in origin, quite old now, but still holding up well as an analytic rubric. The EU energy task force had done something similar. Really there were no mysteries here, in either the nature of the problem or the solutions.

"And yet it's not happening," Mary observed.

They regarded her. There is resistance to it happening, they reminded her.

"Indeed," she said. They were caught in a maze. They were caught in an avalanche, carrying them down past a point from which there would be no clawing back. They were losing. Losing to other people, people who apparently didn't see the stakes involved.

She walked down to the park fronting the lake. She sat on her bench and stared at the statue of Ganymede, holding his hand out to the big bird. She watched the white swans beside the tiny marina, circling about, hoping for bread crumbs. Beautiful creatures. Bodies so white against the black water that they looked like intrusions from another reality. That would explain the way water rolled off them, the way the light burst away from them, or perhaps right out of them. Not creatures of this world at all.

It wasn't going to happen from the top. The lawmakers were corrupt. So, if not top-down, then bottom-up. Like a whirlwind, as some put it. Whirlwinds rose from the ground—although conditions aloft enabled that to happen. People, the multitude. Young people? Not just congregating to demonstrate, but changing all their behaviors? Living together in tiny houses, working at green jobs in co-operative ventures, with never the chance of a big financial windfall somehow dropping on them like a lottery win? No unicorns carrying them off to a high fantasy paradise? Occupying the offices of every politician who got elected by taking carbon money and then always voted for the one percent? Riot strike riot?

She didn't know if her failure to imagine that bottom-up plan working was her failure or the situation's.

Then Badim appeared before her.

"Mind if I join you?"

"No. Sit." She patted the bench by her. She supposed he had been able to consult her bodyguards to find out where she was. This was a little disturbing, but she was glad to see him.

He sat beside her and regarded the view. "Who was Ganymede again?" he asked, regarding the statue.

"I think he was one of Zeus's lovers."

"A gay lover?"

"The ancient Greeks didn't seem to think about things in those terms."

"I guess not. Didn't Zeus rape most of his lovers?"

"Some of them. Not all. If I recall right. I don't really know. In Ireland it wasn't a school topic. What about in India?"

"I grew up in Nepal, but no. Greek mythology was not studied."

"What about Hindu mythology? Don't they have gods behaving badly?"

"Oh yes. I don't know, I never paid it much attention, but the gods and goddesses seemed like a family of, I don't know. Distant ancestors. Very heroic and noble, very proud and stupid. It made me wonder what the people were like who told each other these stories in the first place, as if they were interesting stories. Like Bollywood musicals. So melodramatic. I was never interested."

"What interested you?"

"Machines. I wanted my town to be like the towns in the West, you know. Clean. Easy. Full of shiny buildings and trams. And cable car lifts, for sure. I had to walk four hundred meters up and down to get to school and then back home, every day. So. I wanted it to be kind of like Zurich, actually. I wanted to move into the present. I felt like I was caught in a time warp, stuck in the middle ages. You could see on the screen shows what the world was like now, but for us it wasn't like that. No toilets, no antibiotics, people died of diarrhoea all the time. In general people were sick, they were worn out, they died young. I wanted to change that."

"You were lucky to want something."

"I don't know. Wanting something can make you unhappy. I was not happy. I lashed out."

"Happiness is overrated. Is anyone ever really happy?"

"Oh, I think so. It looks like it, anyway." He gestured around them.

Zurich, so solid and handsome. Were the Zurchers happy? Mary wasn't sure. Swiss happiness was expressed by a little lift at the corners of the mouth, by thumping down a stein after a long swallow. Ah! *Genau!* Or by that little frown of displeasure that things were not better than they were. Mary liked the Swiss, their practicality anyway. Undemonstrative, stable, focused on reality. These were stereotypes, sure, and in the part of their lives hidden except to themselves, the Swiss were no doubt as melodramatic as opera stars. Italian soap opera stars, another stereotype of course, those were all they had once they started thinking about groups, they simplified to an image, and then that image could always be turned to the bad.

"I'm not so sure," she said. "People who have it all, who don't want anything, they're lost. If you want something and your work gets you closer to it, that's the only happiness."

"The pursuit," he said.

"Yes. The pursuit of happiness is the happiness."

"Then we should be happy!"

"Yes," she said unhappily. "But only if we get somewhere. If you're pursuing something and you're stuck, really stuck, then that's not a pursuit anymore. That's just being stuck."

Badim nodded, regarding the great statue curiously.

Mary thought he had come down here to tell her something, surely; but he had not done that. She watched him for a while. Nepal had fallen to a Maoist insurrection that had killed thirteen thousand people over a period of about ten years. Some would call that a lot; others would say it wasn't so many.

"I notice things happening out there," she said to him. "Davos gets seized and the happy rich folk put through a reeducation camp. Glamping with Che. Then all those planes going down in one day."

"We didn't do that!" he said quickly. "That wasn't us."

"No? It killed the airline industry, more or less. That was ten percent of the carbon burn, gone in a single day."

He shook his head, looking surprised she would even think such a thing. "I wouldn't do that, Mary. If we had any such violent action in mind, I would confer with you. But really we're not in that kind of business."

"So, these so-called accidents happening to oil executives?"

"It's a big world," he said. Ah ha, Mary thought. He was trying not to look uneasy.

"So," she said, "what did you come down here to discuss? Why did you find me and come down here?"

He looked at her. "I do have an idea," he said. "I wanted to tell you."

"Tell me."

He looked off at the city for a while. Gray Zurich. "I think we need a new religion."

She stared at him, surprised. "Really?"

He turned his gaze on her. "Well, maybe it's not a new religion. An old religion. Maybe the oldest religion. But back among us, big time. Because I think we need it. People need something bigger than themselves. All these economic plans, always talking about things in terms of money and self-interest—people aren't really like that. They're always acting for other reasons than that. For other people, basically. For religious reasons. Spiritual reasons."

Mary shook her head, unsure. She'd got enough of that kind of thing in her childhood. Ireland had not seemed to have benefited from its religion.

Badim saw this and wagged a finger at her. "It's a huge part of the brain, you know. The temporal lobe pulses like a strobe light when you feel these emotions. Sense of awe—epilepsy—hypergraphia..."

"It's not sounding that good," Mary pointed out.

"I know, it can go wrong, but it's crucial. It's central to who you are, to how you decide things."

"So you're going to invent a new religion."

"An old one. The oldest one. We're going to bring it back. We need it."

"And how are you going to do that?"

"Well, let me share some ideas with you."

57

Next season I was back down there again, helping with the seawater pumping experiment, even though it was obvious to all of us that it was a crazy idea. Ten million wind power turbines? Thousands of pipelines? Not going to happen. It was a fantasy cure.

But someone had to try it. And the project had one more season of funding. So we got the pump intake back through the sea ice into the water. Then we followed the pipeline up the big white hill. It was laid right on the ground, because snow or ice was a better insulator than air, and warmer too. Still, a big part of the total energy budget was for heating the pipes to keep the water liquid on its way to its destination. The rest of the energy was simply to move the water uphill. And water is heavy, and Antarctica is high. So, whatever. An experiment or an exercise in futility, depending on your view.

There were people proposing to generate energy from ocean currents. The Antarctic Current runs around the continent like a belt, clockwise as seen from above, and of course it gets channelized through the Drake Passage; if electricity could be generated from that faster section of the current, great. But none of us thought it would work. The sea eats everything you put in it, and the size and number of turbines that could spin up enough electricity to do the job was off the charts.

Then there were those who still held the dream of space-based electricity. Russians for the most part. They had used their Molniya orbit for communications satellites for a long time—this being an almost polar orbit, in an elliptical shape that brings it close to Earth twice a day. So the Russians were putting up satellites with solar panels and microwave transmitters to send power down to Earth. Microwave collection stations were to be located by the Antarctic pumps and heating elements, and electricity thus beamed down from space to help power the warmed seawater uphill and inland, even during the long night of the Antarctic winter.

Maybe, we said. Although the truth is that solar power from space is not likely to work very well. Capture, transmission, reception, all problematic.

Even if a sufficient power supply was found, people would still be required at the upper ends of the pipelines to oversee the water getting poured out up there. So this season we tried that part too, and it was a weird sight to see. Typical polar plateau scene, Ice Planet Zero, a sastrugied white plane to every horizon, domed by a dark blue sky very low overhead, stupendously awesome, you feel like the Little Prince and have to pinch yourself from time to time, also do some Pete Townshends to keep your hands warm, it's fucking cold that goes without saying—and then there's this pipeline like some bad dream from Alaska, their oil pipeline I mean, a nightmare. But there it is.

When the water pours out of the end of this pipe, it steams madly in the dry polar air and then sploshes down onto the ice and runs away, just as we had planned it, having aimed the nozzle down a slight hill. But the tilt of this so-called hill was about two meters in every kilometer, as good as we could get in the region. So we got surprised by how fast the water froze. Maybe we shouldn't have been, most of us had tried the old Antarctic trick of taking a pot of boiling water outside and tossing it into the air to watch it steam and crackle and freeze to ice bits before it hits the ground—it's an experiment that never ceases to amaze. Always good for a laugh. But pouring it out in quantity, as from a fire hose or a sewer outflow, we thought it would take longer.

Not so. In fact, newly frozen ice stacked just a few meters from the end of our pipe, creating a low dam that checked the water's progress, making downhill no longer down, so that the unfrozen water began to flow back toward the pipe outlet, and then past it in the other direction. Oh no!

We hustled down to the little ice dam and started trying to break it, which worked about as well as you might expect. In the midst of our fuss, as we yelled directions at each other, not desperate but maybe a little panicked, Jordi shouted, Hey I'm stuck! Help!

He was standing in what had been ankle deep water near the outlet, which was now ice that had him stuck in place, with more water flowing over his boots all the while. Help!

We laughed, we cursed, we tried to cut him out, nothing worked. He was not in imminent danger, but on the other hand we couldn't free him. And in the race between rising ice and the newly emerged water sliding slickly over it, the ice was winning. It actually takes quite a bit of thermal energy for water to turn into ice, the process actually warms things a little, weird though that seems, but at 30 degrees below zero that warming is a hard thing to detect, or rather it has little to no effect on anything. The frost steaming up from the mess fell on us like Christmas tree flocking, and it began to look like we couldn't get Jordi out without chopping his feet off, which gave me an idea. Pull him out of his fucking boots, I said. Leave his boots.

This was easier said than done, but luckily he was wearing NSF's white rubber bunny boots, rather than the more tightly fitted mountaineering boots that most of us had on, so we were able to stand by him and give him something to hold on to while we all pulled him up and out of the boots, him cursing as he got his feet up and out in the ridiculously cold air, after which we had to half-carry him to the heated dining hut. Someone had shut off the water quite a bit before, I don't know who, I have to admit it hadn't occurred to me, we were in the midst of an experiment, I don't like to cut those short. Anyway, Jordi's boots are still stuck out there, NSF will not be pleased, they will ding us for it.

So Jordi was saved, but the problem remained—water was going to freeze fast enough to create a problem in getting it to flow across the surface of the ice. The pipe outlet would have to be whipping back and forth like a hose on the driveway with too much pressure in it—which maybe could be arranged, but yikes, how to control that, it's kind of non-linear. A sharper gradient would help too, although on the polar plateau those are not easy to find.

So we closed down for a week, and rebuilt one outlet to emerge under pressure and snake back and forth like a windshield wiper, see how that went. And when we tried it again, water came out of the pipe and flowed downhill and froze along the way, and finally pooled pretty well away from the outlet, where it mounded and the new water shifted and flowed around it and slid down yet again. And we got better and better at mapping where it might ultimately pool, and staying out of those pools while it happened.

The estimate we came away with was that we could deposit about a meter of water per year on any part of the polar plateau and still have it freeze successfully. More than that and we would be exceeding the capacity of the air and ice in combination to chill the water. So we would need a wide spreading zone; at a meter thick, that would be about a third of Antarctica.

Not going to happen, no way in the world. We had firmly established a trifecta of impossibilities: not enough energy, not enough pipe, not enough land.

So the straight-up seawater pumping solution wasn't going to work. It was a fantasy solution. The beaches of the world were fucked.

A lot of us didn't want the beaches of the world to be fucked. As we sat in our little habitats, like mobile homes half filled with insulation, we would gather around the table looking at maps and talking it over. World maps, I mean.

The endorheic basins of the world, meaning basins where water does not drain to the sea, were many in number. And many of them in the northern hemisphere were dry playas, where water had existed at the end of the last ice age but dried out since, partially or all the way. The Caspian Sea had been helped to dry down to its current level by people, the Aral Sea even more so. The Tarim Basin was completely dry all on its own, Utah's Great Salt Lake was the remnant of a much bigger lake from the past—on and on it went, mostly in Asia and North America, and the Sahara. Of course there were people living in some of these places, but not many of them, given the problems of desertification, or disasterated shorelines in the case of the Caspian and Aral. If you added up their volume of empty available space, it was considerable. A lot of seawater could be relocated there, in theory. We ran the numbers; well, it would do for a meter or two of sea level rise. But then all those basins would be full, and you'd be back to the unworkabilities of Antarctica.

No. We needed to go back to the plan to pump water out from under the big glaciers, to drop them back on rock beds to slow them down. Slawek had been right all along. It was the only thing that was going to work. We had been following the money, taking it where we could get it and doing what they asked us to do with it. The billionaires and oil companies and Russia even the NSF had said, Pump seawater back onto Antarctica, cool idea! Do it! But we were the glaciologists, so we had to guide the process if it was going to have any chance of success. Give expert advice, guide the money where it needed to go.

So, the next Antarctic season we were back. We went every year anyway, of course, but for once this was a really good reason to go. Mostly we just like it down there and are looking for excuses. Science! Is the ice on Antarctica five million years old or fifty million years old, huge argument! That kind of thing. Pure science. This was much more applied. It felt good.

This time we went to the Pine Island Glacier above the WAIS, the West Antarctic Ice Sheet. Pine was running narrow and fast into the sea, right next to Pine Island. It had been a study site for many years, so NSF had the logistics in place for getting a camp installed there.

Although in fact they weren't experienced in getting as much stuff there as we needed. We needed almost as much as McMurdo itself, not really, but NSF had decided that if we were going to give this idea another test we had better make it a good test, or we wouldn't know what our results meant. Our first try long ago was now worthless as data, because no one had ever followed up on it to find out why that hole had cut off and gone dry.

So we borrowed the ships that resupplied McMurdo at the end of every summer, and a couple of Russian icebreakers, us having been so stupid as to neglect icebreakers for decades and then build only a couple of wimpy ones, really suitable only for the Arctic, where there's practically no ice left anyway. But the Russians love icebreakers, and they sent a couple of their monstrous beasts south to help us bash an open-water lane to Pine Island, where we could land stuff and drag it over and up onto the glacier, using the same snow tractors that had been dragging fuel and gear from Mactown to Pole for many years. Without too much in the way of fuck-ups we were established on the Pine Island Glacier outlet about a month after Winfly had opened Mactown for the season. A very impressive logistical effort.

This was my twenty-fifth visit to the Ice, and while on Pine Island I was going to break the six-year mark for living on glaciers, which was not the craziest of icehead records, but respectable. My wife has never been happy about this, but I like glaciers. And here I was again, out on Pine Island, which is really just a bump in the ice, entirely submerged by it, white sastrugi to the horizon in all directions. It has to be admitted that all Antarctic glaciers look much the same. The Dry Valleys are fantastic, but 98 percent of the continent is not like them. It's ice on ice, for as far as the eye can see.

So, we deployed the equipment to our first borehole site and got to work. It was like dragging a village over the ice, some kind of Baba Yaga thing, the monster tractors pulling trains of four or five huts in a row behind them from spot to spot. We started between the Hudson Mountains and Pine Island itself; this was a perfect pinch point in the glacier's fall, a place where if we could thump it back down on its bedrock, it would slow for sure. So we circled the wagons and got to work.

It's cold in Antarctica, yeah. You forget that back in your university office in Louisiana or Pennsylvania or California or Ohio, or wherever you winter over. Even when your home in the world feels colder than Antarctica, like in Boston, you still forget that the Antarctic cold is really a true, deep cold, so cold it burns. Then after a few days in it you forget about it again. Of course a wind will remind you, and you don't want to go out poorly dressed, or get frostnipped or even frostburned, or get to the point where if you flick your cold ear it breaks off and falls to the ground, but for the most part, it's just the way it is. Cold. Keep a good selection of gloves and deploy them as appropriate. Humans evolved in ice ages, and properly dressed are good in the cold. Just deal.

So, the ice borers were still as simple as showerheads. Slow but effective. In the old days we burned a lot of fuel to get the showerhead's water hot. Now solar panels helped to power the heaters. The meltwater under the showerheads gets suctioned and pumped out of the hole and reheated and used again, with the excess piped a distance away to freeze somewhere else. To both sides of the Pine Island Glacier are regions of ice moving slowly, so the water could be dumped there. In truth it was such a small volume that where we dumped it didn't matter. We could have let it go into the ocean, it wouldn't even have registered there. We could almost have drunk it as our drinking water.

We also began to make use of the microwave energy beamed down to us from the Russian satellites in their almost-polar orbits, to power the pumps and the showerheads when the sun went down. Test whether we could make that work, so we could keep the system going year round.

That first experiment had at least taught us this, that you probably wanted to choose a homogenous block of ice within the flow of the glacier, so that the movement of the glacier wouldn't deform the hole as we drilled it. We had chosen our site partly for that reason—it was a single block that was forty kilometers long and extended all the way across the glacier. Still 130 kilometers upstream from the ice shelf, and a couple hundred meters above sea level. Perfect.

We had done the calculations as to how many drill holes we needed, and how far apart they should be and so on. It was pretty audacious; suck enough water from the underside of the glacier for the whole block of ice that we were working on to lose its water cushion and crash back down onto bedrock, hopefully with a mighty squeal and maybe a crashing sound, as of tires braking on asphalt followed by car hitting wall. Even if that only happened in our heads, it did seem like it was going to be palpable when it happened.

Slawek's numbers were holding good. Amount of water lubricating the bottoms of Antarctica's glaciers roughly sixty cubic kilometers. Not insignificant, a clear cube of ice about four kilometers on a side and the same high, so, half as tall as Everest—yes, a lot of water for humans to pump. But not outside the zone of what we already pump every year.

Still, a lot of water to pump, but that's for all of Antarctica. Around the circle of the continent, about fifty glaciers dump the majority of the ice now rushing into the sea, with a few being the major contributors. Starting from McMurdo and running clockwise, you need to stick the Skelton, the Mulock, the Beardmore, the Carlyon, the Byrd, the Nimrod, the Lennox, the Ramsey, the Shackleton, the Liv, the Axel-Heiberg, the Amundsen, the Scott, the Leverett, the Reedy, the Horlick Ice Stream, the Van der Veen Ice Stream, the Whillans Ice Stream, the Kamb Ice Stream, the Bindschadler Ice Stream, the MacAyeal Ice Stream, the Echelmeyer Ice Stream, the Hammond Glacier, the Boyd, the Land, the Hull, the deVicq, the Murphy, the Haynes, the Thwaites, and our Pine Island; then around the Peninsula to the Drewry, Evans, Rutford, Institute, Möller, and Foundation ice streams; then the Support Force Glacier, the Blackwall Ice Stream, the Recovery Glacier, the Slessor Glacier, the Bailey Ice Stream, the Stancomb-Wills Glacier, the Vestraumen Glacier, the Jutulstraumen, the Entuziasty, the Borchgrivinkesen, the Shirase, the Rayner, the Beaver, the Wilma, the Robert, the Philippi, the Helen, the Roscoe, the Denman, the other Scott, the Underwood, the Adams, the Vanderford, the Totten (biggest of them all), the Dibble, the Francois, the Mertz, the Ninnis, the Rennick, the Tucker, the Mariner, the Priestly, the Reeves, the David, the Mawson, and the Mackay.

That's actually seventy-four. So, sixty cubic kilometers sucked out from under seventy-four glaciers. Okay, not so bad!

Especially compared to 3,600 cubic kilometers, right?

So we were melting and casing twenty boreholes into Pine Island Glacier. After that we'd pump up all the water that we could. It didn't seem so bad! In the same realm as wells up in the world, draining fossil water for farms all over the Ogdalilla and other places set atop irreplaceable groundwater resources. Can be done! Solves all problems!

All right, it doesn't solve all problems. But let's not get picky. If sea level rises even a meter, all the beaches in the world are gone, and seaports and coastal infrastructures and salt marshes and you name it. And as Hansen and his team pointed out in their 2016 paper, if the rate of rise doubles every ten years, quickly you are fucked, all the coastal cities of the world devastated, damage in the quadrillions, if you think you can put a price on it. What's the monetary value of human civilization? Trying to answer that question proves you are a moral and practical idiot. Well, economists make such calculations all the time, but that's their job, and they think it makes sense. In this case, better just to throw up your hands and say civilization is effectively a fiscal infinity, a human infinity.

Tacking down the Antarctic glaciers won't completely stop sea level rise, of course. But if we could get the big ones back to one-tenth of their current speed, like the good old days, that would help a lot. Best to do it in Greenland too—the ice there is going even faster, and even though it has only a tenth the ice of Antarctica, that's still a seven-meter sea level rise, if all of it were to liquify. So fine, do them too; Greenland is easier, it's basically a stone bathtub with narrow cracks in it. Tack down those suckers falling through the cracks, and you've got a somewhat stabilized sea level, rising at about a millimeter a year—meaning a thousand years before it rises a meter. Enough time to draw down enough CO_2 to get back to 350 parts per million—hell, enough time to start another ice age if you want to!

Basically, the sea level rise problem gets solved. Beaches still in existence.

So, someone asked tonight in the mess tent, is what we're doing down here geoengineering? Who the hell knows! What's in a word? Call it Glacier Elevation Operations, Based on Estimates of Godawfulness Gobsmacking Interested Nations' Goodness: GEO-BEGGING. Call it whatever you want, but don't immediately clutch your pearls and declare we can't predict the unintended consequences, we are sure to create backlash effects so bad they overthrow the good we intended, etc. There are some things man was not meant to know—my ass! We are meant to know everything we can find out. So get over that whole wimpy line of objection. And I'll tell you what the unintended side effects of slowing down the glaciers of Antarctica will be: nothing. Nada. No side effects whatsoever, and the beaches and coastal cities of the world will stay out of the drink.

So if this works, and it looks like it will, I think we'll be doing it. Our team is relatively small. The project is expensive but not that expensive. Like drilling a few wells anywhere else, pumping water anywhere else, plus keeping everything warm. And getting the stuff and people here in the first place. That will be expensive, yes, and dangerous, but a quick calculation of the cost of this operation of ours multiplied by something like a hundred, or even a thousand—that gets us to ten billion dollars. Well, probably it will be more expensive than that, everything is when you actually get into it. But whatever—call it fifty billion dollars. This is such a bargain!

So, I've been waking up every morning excited to get out there. Actually this is always true for me in Antarctica. Jamesways, mobile homes, tents, Mactown; get up and try to stay warm through breakfast and prep, gear up, out the door and bang—freezing sunny cold air, like a wallop of vodka and a slap to the face. Eyes streaming from the double whammy. Give a hoot and head to the line of snowmobiles. Get one of them started, off on the track leading to whichever hole I'm working on. The snowmobile roads are flagged, and the flags tell me what the wind is up to that day. If it's windy then it's going to be cold, that's the law in Antarctica.

That's why they're always saying, it's not the cold, it's the wind! Although when it's windless, it's still cold. But with a wind you are going to freeze your butt. The wind just ransacks you, no clothing is enough. You'd have to be in a spacesuit to be shut of it, and even then you'd be cold.

Today I made the rounds and found that every hole was working. Pumps pumping, heating elements keeping the water liquid. The lines from each pump feed into a bigger pipeline, like a piece of the Alaska pipeline lying right on the ice, with repeater heaters every kilometer, and joints where you can get into the line and run a pig if you need to melt an ice clog. The pipeline is running slightly uphill, which I think is an interesting decision, as you could actually just let it run downhill into the ocean, the addition to sea level rise would be so trivial, well below detection level. But people like to be neat.

It's a funny place to describe, Antarctica. You can't really get it across to people, what it looks like and what it feels like. I think the clean dry air makes for some optical illusions that add to the feeling that things don't look quite real. Often it's just snow and sky. That is very common. A sky clean of clouds. Around the coast there are clouds sometimes, more often than on the polar plateau, but still: blue sky, white snow under it. Sometimes smooth snow, sometimes sastrugi. Sastrugi and even completely flat snow look different when you're looking toward the sun and when you're looking away from it. When looking toward it, there's a glare off icy snow, a blaze off snowy snow. Then when looking away from it, the white goes somehow dark—which is a strange thing, maybe due to the polarization of your sunglasses, but very noticeable when you turn around and look sunward again. The contrasts are more than the eye can deal with. If there are clouds, like the fleets of low marine puffballs that float above the ice near the coast, then their shadows on the snow appear black; a flat snow expanse can suddenly look hilly or peppered with black mesas. More contrast than the mind can handle, now why is this beautiful? I can't say. Maybe a lot of people wouldn't say it is. But I like it.

Looks like tomorrow we'll be able to run the final tests and declare the job done. Then in about a year we should be able to tell how it's working in terms of slowing this beast down. Maybe five years to be safe. Although people will be in a hurry to declare one way or other. But if you're going to be putting billions into it, and training a bunch of crews, you'd do better to be sure. Actually, one of the choke points in the supply chain, in terms of getting this done, if they decide to do it, is simply people. It takes a certain expertise. On the other hand if we're shutting down all the oil operations, as we really ought to do, then that's a lot of people out of work. And the work in question isn't that different. Some of them might even regard this job as easier. Simple stuff, although colder. But if you're working in Saudi Arabia, maybe you like the idea of cold. And if you're working in Alaska, maybe it makes no difference. Yeah, that part will probably work out. We've only got fifty people here now, and you could do it with thirty—almost half of us are here to study the other half, or do science while we have the camp here, in the usual way. Again, scale that up and the number of people is still trivial.

We'll have a party tomorrow to celebrate.

I'm sorry to report that this is the last entry in this file of Dr. Griffen's laptop. He took a last inspection run on February 6, and on the way back to camp he took a shortcut that left the snowmobile road, which is clearly flagged. Visibility was good, so no one is sure why he did that. Usually he stuck to the flagged road, just like the rest of us. It's normal practice on glaciers, where crevasses can be covered with old snow and completely invisible. And taking the shortcut wouldn't have saved him much time anyway. So we are all mystified.

We were in the dining hut waiting for him, and when about an hour had passed, our mountaineers got worried and went out to check. Another thing we do is usually travel in twos. But Dr. G wasn't always meticulous about that, and since the entire work space of the project was within sight of camp, none of us were. I mean you'd go to check on a pump or whatever by yourself, and to your tent by yourself, to the bathroom and so on. Just normal.

So when the mountaineers couldn't find him in a quick tour of the wells, we sent one crew up the pipeline to see if he had gone to check the spillway and had a snowmobile problem. The rest of us made the circuit looking for snowmobile tracks leaving our road system. It was all we could think to do, since we could literally see the whole surface of the glacier, and no sign of him. That in itself was deeply worrying.

It was Jeff, one of our mountaineers, who followed a snowmobile track to an unobtrusive hole. He then came back and got Lance, the team's other mountaineer, looking grim. They went out to near it and then stuck in some ice belays and roped up, and walked over to look in the hole. We watched from the dining hut, standing around silently. Lance belayed Jeff, who descended into the hole. He disappeared and was gone about, I don't know, twenty minutes. It seemed longer. Finally he reappeared and climbed out and stood there, then walked back to Lance. They conferred; Lance put out an arm and held him by the shoulder. They hugged. They looked our way, saw us. Jeff shook his head. We understood it as clearly as if he had said it: Dr. G was dead.

That was a bad night. We sat around the dining hall, stunned. Jeff sat by the stove looking grim and distant. Of course when he got back to us we asked him what had happened, but he could only say what was obvious. Dr. G had driven off course, his snowmobile had dropped into a hidden crevasse. He went on to say the snowmobile was stuck about twenty feet down, and Dr. G's body under it. Jeff didn't say what condition the body was in. He didn't want to inflict that on us, I guess, or didn't want to compromise Dr. G's privacy, I don't know. Sophie and Karen were crying, and we all cried from time to time. Dr. G had given most of us our jobs, and several of us our educations, our careers. He was one of the old iceheads, the ones who keep coming back, the true Antarcticans. Jeff didn't cry, nor did Lance. Lance was concerned about Jeff, he didn't care about Dr. G. Beakers were always doing stupid things and getting killed, that's why every team in Antarctica is required to have mountaineers as part of the group, to keep them safe. But you can't lead them around on leashes, and if they break protocol you can't stop them, even if you insist on protocol at the start of every deployment, go through the drill with the utmost seriousness, which truthfully many mountaineers are not good at mustering. So Jeff and Lance were grim, but not because they felt responsible. They were going to have to go down in that crevasse the next day and get Dr. G's body into a bag, and haul him up to the surface, and get the body on a sled and back to camp. A plane from Mactown would be in tomorrow, if the weather allowed. They already knew about it there; their voices sounded shocked, sympathetic, the usual.

We sat around drinking. Damn, we said. Why? we said. Some of us talked about the bathtub graph. People doing dangerous things make mistakes when they're first learning it, and then when they've known it forever. These were the two periods with higher rates of accidents, while the in-between was a stretch of low accidents. So the graph looked like a bathtub. People who fly planes fit that graph really well. Working on glaciers is somewhat the same.

That's how scientists talk at times like that. Maybe everybody. Faced with a death, with a friend suddenly disappeared from the world, the mind shies away from the shock of it, the incredulity. Why? Can't we have a do-over? Pop back just a few hours, do it differently?

No.

So we sat there and drank.

Well, Sophie said, at least he died saving the world.

No! Jeff said. No! It was a mistake!

Even then he didn't cry, though his face went red and he looked furiously distraught. We huddled in a mass around him and sobbed or didn't. These feelings come on you or they don't, the timing is weird. Lots of people dissociate in moments like that, and it only hits them later. Sometimes so much later you can't believe it, I know this myself—for me, once it was literally twenty-one years between the death involved and me feeling it. Twenty-one years, I swear. But on this night most of us cried, all but Jeff. We were distraught.

After that we pulled ourselves together and cleaned up the place and discussed plans more quietly. Nothing to be done. Finally we gave up and went to bed, reluctantly, as it seemed too normal, it seemed like giving up on any chance of things changing or the world going back in time. Just had to give up and go to bed; we were going to have to deal with a lot of shit the next day. And there was no point in drinking any more. It wasn't going to do any good. Our leader had made a simple but deadly mistake. The world would go on, but for us it would never be the same.

58

The usual view of liberation theology locates it in South America in the latter part of the twentieth century. The phrase was invented to describe this Latin American phenomenon, so it's fair enough to think that's what it refers to.

But in Spain we think there was an earlier example of a young idealistic Catholic priest, helping his people in defiance of the church hierarchy. No doubt it has happened many times without anyone noticing it outside the community affected. Of course the situation with young priests has gone wrong so many times. But maybe more times, the young idealistic man, trying to do good in the world, intense, devout, isolated, put out there in a community of poor people, people suffering in so many ways, just trying to make ends meet, to hold it all together, and their church supposed to be part of that effort—when some of these young men get confronted with that situation, in all their belief and their desire to help, their trust in the church, quite a many of them must have fallen in love with their people and worked furiously their whole lives to do everything they could to serve them.

In this particular case in Spain, the young priest was named José María Arizmendiarrieta. Born and raised in the Basque part of Spain, he took arms in the Spanish civil war on the Republican side, then got captured by Franco's soldiers. It's said that he then had a sort of Dostoyevsky moment, in that he was condemned to execution and scheduled to be shot, but in his case was spared by a bureaucratic oversight, as they failed to show up and get him on the day in question, no one knows why. Let's say God had a plan for him.

After that he took holy orders, perhaps feeling his life was meant for something, and he was sent to Mondragón in 1941, when he was

twenty-six years old, as part of an attempt by the Franco regime to pac-
ify the Basque people, who were still rebellious in the aftermath of the
Republic's defeat.

At first his congregation was not impressed by him. He had only
one eye as a result of the war, he read verses in a monotone, he seemed
distant and tentative. One can wonder if he was shell-shocked, or a bit
on the spectrum as we would say now. It took him a few years of quiet
listening to his people to come to a determination of how he might help
them best. Before the war the area had supported some light industry
which had not returned. Father José María wondered if they could start
something up again, and as part of that, he helped them to organize a
polytechnic school, now known as Mondragón University. Soon after
opening, it provided enough engineering support to bootstrap the expertise to begin a few manufacturing businesses again, starting with paraffin
burners. And on his suggestion, and with his help, these were organized
from the start as employee-owned cooperatives. This mode of organization was in the Basque tradition of regional solidarity, a manifestation
of that precapitalist, even pre-feudal gift economy of the ancient Basque,
which goes back as far as can be determined, into the time before written
history.

Whatever the explanation, these cooperatives thrived in Mondragón,
and a complex of them has been growing there ever since. Eventually
they included the town's banks and credit unions, also its university and
insurance company. These worker-owned enterprises became a kind
of co-op of co-ops, which now forms the tenth largest corporation in
Spain, with assets in the billions of euros and yearly profits in the millions.
The profits don't get shifted out as shares to shareholders, but are rather
divided three ways, with a third distributed among the employee-owners,
a third devoted to capital improvements, and a third given to charities
chosen by the employees. The wage ratio between management's top salary and the minimum level of pay is set at three to one, or sometimes
five to one, or at most nine to one. All the businesses and enterprises
adhere to the cooperative principles formalized later by the larger worldwide cooperative movement, of which Mondragón is somewhat the jewel
in the crown: open admission, democratic organization, the sovereignty
of labor, the instrumental and subordinate nature of capital, participatory

management, payment solidarity, inter-cooperation, social transformation, universality, and education.

This list is worth studying in some detail, but not here. Taken together,
if these principles were to be applied seriously everywhere, they would
form a political economy entirely different from capitalism as generally
practiced. They make a coherent set of axioms that would lead to a new
set of laws, practices, goals, and results.

How this has worked out in Mondragón is open to interpretation. The
system has been enmeshed in the world economy all along, and it had to
make adjustments when the European Union formed, as well as continuous adaptations to the markets and countries it existed in. There are those
who say it could not succeed outside of its Basque context, that Basque
culture makes it possible; this seems unlikely, but there are many who
don't want to consider that an alternative to capitalism, more humane,
what you might even call a Catholic political economy, not only is possible, but has existed and thrived for a century, and is still going strong.

There have also been moments of crisis, as when recessions struck just
at the moment that certain critical cooperatives had expanded, or when a
manager absconded with an immense amount of money, causing severe
cash flow problems. Still, the place makes a good living for its people, and
creates a culture that is mostly loved by those who perform it. There is
solidarity and esprit de corps, and even in a world of intense competition,
it makes a profit most years, enough for over a hundred thousand people
to make a living from it and to give back to the general culture.

There are other such enclaves around the world, and systems that while
not as distinctive and whole, are yet somewhat like it. They survive, sometimes they thrive. The question is, to put it in the dominant vocabulary of
our time, could they scale? Are they a way out, a way forward, a step along
that way?

We think so. For us, the project is to spread the system throughout
Spain. For everyone else, maybe the world. But this is our contribution.
We give you Mondragón.

59

I was in my apartment in Sierra Madre, which is a little town lined by tall palm trees, wedged between Pasadena and Azusa, set right at the foot of the San Gabriel Mountains, which tower over that part of LA like a brown corrugated wall, pretty ugly if you ask me. It was kind of a blessing when the smog got so thick you couldn't see it, which could happen even from only three miles away. None of the ring of mountains backing LA are good-looking. But they do form quite the wall, as we found out that day.

Luckily I was an active kayaker before I broke my arm, and I still had my kayak. My apartment was a granny flat over a garage, its separate entrance something I valued a lot, as I didn't have to bother my landlord going in and out, and he usually never saw me and thus never got a chance to scam on me. It was kind of a mercy rental on his part, or so I thought before his intentions became clear, as I couldn't afford Hollywood any-more, no doubt obvious when I gave him my clichéd young-aspiring-actress-currently-waitressing shtick. And he let me store my big stuff in his garage below my studio, which was really just a storage shed with a bathroom in it, tacked onto the flat roof of his garage. So when the atmo-spheric river hit, I was one of the few people in the city with watercraft on hand.

It rained hard all that first night, the timing was part of the problem—by the time everyone woke up things were already bad, and it just went downhill from there. Literally so for the water, and therefore everything else. But it wasn't your ordinary flood, or maybe it was, I don't really know, but in our case, the water came roaring down off the side of the San Gabriels onto us. It was terrifying to see what the dawn revealed that day. The mountains are ten thousand feet high over Sierra Madre, and they were catching all that rain, which was falling as hard as in a hurricane or something, and then it was all rushing down those vertical ravines onto the streets that poked up into the ravines, now all of them whitewater rapids brown with mud, and filled with boulders and shrubs and pieces of all the houses that were coming apart farther up the street.

I looked out my front door and saw cars floating down the street on a brown wave about three feet high that covered everything. I could see the water was already coming into my landlord's house, and rushing hard toward the 210. Everywhere I looked was a big sheet of brown water! I shouted to my landlord but he had already left without informing me, very typical. In fact I couldn't see anyone anywhere except for a family on top of their SUV, getting taken for a sideways ride and looking desperate.

I got down my outside stairs and sloshed through brown water to the garage side door. Inside I found the power had gone out, of course, so it was a struggle to get the big garage door open. I managed to pull it up from the inside, and a sloosh of water flowed into the garage, a little wave about a foot high. But there was my kayak and I grabbed it and the paddle off the wall, wriggled into the skirt and got into the kayak and took off into the street.

That was a crazy moment, realizing the streets were all flooded and my kayak was the only way for me to get around. So much water! And brown as hot chocolate. And it was still raining cats and dogs too, so it was hard to see very far, and hard to believe what you could see. Later I heard the whole LA basin was flooded, all the way from the Hollywood Hills down to San Clemente, past Irvine where I grew up. Orange County was just as bad as LA, which makes sense given they are the same coastal plain with the same backing mountains. Of course there are some high points here and there on the plain, as everyone found out that day. Palos Verdes sticking up near Long Beach of course, and a few inland neighborhoods on lines of low hills like Puente Hills and Rose Hill, and back where the freeways meet around San Dimas. But for the most part LA is just one big coastal plain, and on that day, a big brown lake. In lots of places the elevated freeways were the only flat surfaces that stuck up out of this new lake, so, with no other place to go, the freeways were where people went. There were still some cars up there, but none of them were moving, and as the need for space to accommodate people got greater, a lot of cars were shoved over the side into the drink.

I kayaked under the 210 through a very scary underpass, and paddled around getting people off roofs and over to the freeway, where onramps served as boat docks. A lot of people were zipping around on motorboats they had kept in their driveways, also some kayaks like mine. We were doing all we could to help, some people were really desperate, especially if they had kids, and it was hard to keep them from tipping my kayak over in their panic. My arm started to hurt where it had broken, and I kept feeling a sense of unreality that this could all be happening at all, it was too much like a cheesy disaster film, but whatever, I must have gotten cast at last, and besides the fear on people's faces and in their voices kept reminding me that no, this was real no matter how weird it was. And my arm hurt, kayaking is just very bilateral, you can't do it one-armed, but I just kept saying Fuck it and kept paddling.

One thing about the LA basin being so huge was that even though the whole thing was flooded, it was never flooded very deep. Lots of taller rooftops stayed above the flood level, although many other buildings had collapsed into the water. Most palm trees were knocked down, it was a shock to see and a danger to navigation. One of many! Sometimes I had people hanging on to the back of my kayak and the current would take us toward a floating tree or a car and I would have to paddle insanely to get away from them, my arm hurting and the people hanging on and not always kicking in the most helpful ways. And all the old washes crossing the plain were now revealed again as fast places in the flow, it was spooky to see those currents, dangerous too, and it was never obvious which way water would be moving on any given street, because it depended on where the nearest wash was, since the water got sucked toward those, and the streets were almost flat. A whole river network was being revealed or half-revealed by currents in the streets, north south east west, they could be running in any direction. Orange Grove Avenue was on enough of a tilt to be like a water slide running south, then the old sunken part of the Pasadena Freeway just west of it ran in the other direction, it was crazy. Sepulveda was scary fast, I was told, the other kayakers all said Stay off Sepulveda, it's like class 8! And the rain kept pounding down on us. Hard rain, in LA, for hour after hour? It was like Noah's flood! And it looked like it could go on for forty days and forty nights too, why not?

So, ten million people stranded on all the high points left sticking out, and no food to speak of. Rain pounding down for hour after hour. Lots of little boats but nothing big, and nothing organized. All the freeways packed with very wet people. It was never colder than about seventy degrees, although that feels cold when you're wet and it's windy, but cold wasn't the issue. Nor was the flood like some do-it-right-now-or-die emergency, where you get an hour of total danger followed by relief. That became clear as it went on. So, not like a movie, not at all. Which was impressing me more and more. Here I was helping people, all of us wet and scared, and my right bicep just screaming, and I kept thinking This is real, this feels good, why again are you trying to be a fucking actress? Oh sure, some people had gotten caught somehow and drowned, it was inevitable given the number of people and the power of water, water is a force of nature, you can't resist it if it gets you, but for the most part people were on rooftops or on the freeways, and it was more a matter of evacuating everyone before they starved than anything else. If you didn't get drowned right away then it was just a matter of holding on and waiting for relief.

So as the day wore on I joined a crowd on the roof of a restaurant and they fed us spaghetti. They had broken through the roof and were using the restaurant's food supplies and big pots from its kitchen, and cooking over an open fire a group of men were tending, set on a big sheet of corrugated metal they had pried off of something, with another sheet as a roof over them to keep the rain off. Chances seemed good they would burn that restaurant down, but on the other hand they could always take their roof sheet away and the rain would drench any fire that wasn't nuclear, it was still pouring down so.

After a while there my arm felt better and I went back on the water, fueled up and ready to help more people. There were fewer of them in the later afternoon, people were either dead or had gotten to a high point of some kind. So I joined the other boaters and we made some street sweeps. It was really fun to ride down Orange Grove Avenue, I have to say, running the brown flow almost as fast as a car, but you had to stay sharp, because sometimes an easy flow would head under a freeway bridge or the like and it could quickly become desperate getting out of that current before you got sucked down and killed. People shared knowledge of these danger zones, that's how I heard about Sepulveda, no one's phone was working but some people had GPS devices with maps saved on them, and they were happy to share their orienteering news, and a lot of people out had local knowledge as well, so we paddled around and the motorboats zipped around, often wasting gas without thinking that they weren't going to be able to refill anytime soon, but after a while they remembered, if they didn't run out, and so most of the action as the sun went down was kayakers and a few rowboats and the like, even some sailboats with their sails down and people in them paddling along awkwardly. Little flotillas like human water bugs on the great lake of LA.

And what was occurring to me over and over again as all this was happening was, Hey: I hate LA. I was born here and I know it well, and have even read or been told some of its history in school, and I really do hate it. The truth is, after World War Two this place went from a sleepy little spread of villages to the ten million people here now, and during that time the developers were getting rich making ticky-tack suburban neighborhoods, that and putting in the freeways, which cut the plain into a hundred giant squares, and all of it crap. No plan, nothing good, no parks, no organization, no plan of any kind. Just buy some orange grove and subdivide it and tear out the trees and build a bunch of plywood houses, and then do it again, over and over. It happened in a snap of the fingers, and it was never anything but stupid. And that's what we've been living in ever since! And more than a few of us trying to live out a remake of the movie *La La Land*. It was double stupid.

So as we were paddling around in our kayaks, people were saying to each other, This whole fucking place is gone! Everything is going to have to be torn out! *The entire city of Los Angeles is going to have to be replaced.*

Which was great. Maybe we could do it right this time. And I myself am going to find a different job.

60

Spring came and Mary began to swim again from the Utoquai schwimmbad, first once or twice a week, then every day. Then tram back up to the office. She gave the final nod to Janus Athena's YourLock, and J-A posted the website address to the internet and they watched it go through its unobtrusive birth, a slow week as it turned out, as it was just one spike in the endless interference patterns of discourse. Then people began to share the news that you could transfer everything going on in the rest of your internet life into a single account on YourLock, which was organized as a co-op owned by its users, after which you had secured your data in a quantum-encrypted cage and could use it as a negotiable asset in the global data economy, agreeing to sell your data or not to data-mining operations out there who quickly saw the new lay of the land and began to offer people micro-payments for their data, mainly health information, consumption patterns, and finance. The royalties for being oneself in the world machine were not insignificant, a kind of lifetime annuity, small but useful. And so people began to make the shift, and one day that tipping point arrived where a non-linear shear occurred, like an earthquake, and suddenly everyone had a YourLock account and would henceforth be conducting their internet life by way of it. A whole new internet ecology, the much-hyped but previously vaporwaresque Internet 3.0.

This was news, of course, remarked on everywhere. But on the other hand, when Mary went down to the lake in the mornings to swim, everything looked the same; and this was true everywhere. Global revolutions these days were strange, Mary thought, being as virtual as everything else. And of course in the virtual world it had indeed caused an uproar. What did it mean? Who owned this new system? It was open source, some said, no one owned it. People working in the gift economy had made it, which meant maybe just people playing around. So who profited from it? Other people said its users were its owners and thus made whatever money it made, mostly, as always, by way of advertising fees. It was somewhat like a credit union, perhaps, inserting itself into the social media discourse space. As with a move from bank to credit union, instead of the company using the consumer, the consumer used the company, and owned it too. What did the company per se get out of it? Nothing, because a company was nothing. It was just an organization devised to help its employee-owners, nothing more. Like any other company, in the end. If you thought that was what they were.

All this was going over very poorly in China, where the stance had always been that the Chinese Communist Party was precisely a company in that mold, owned by its consumers and only in existence to advance their welfare. So they kept the new YourLock site outside their Great Firewall, as they had so many previous Western internet companies. But the Great Firewall was riddled with holes, and although some argued that most Chinese netizens were completely absorbed and happy in their Chinese space, it did seem to be true that many of them kept links to multiple accounts they had going all over the world. The internal migrants in all the big cities of China, sometimes called the billion, were still an exploited labor force to the point of wanting some outside leverage over the *hukou* system that had made them illegal for moving to the cities; and the prosperous middle class was always interested to slip some assets offshore. So in some sectors of the Chinese populace, the site was also being taken up in significant numbers.

Days passed, and as far as Mary could tell in Zurich, the impact of all this internet turmoil was minimal on daily life. Possibly it was the global revolution that internet advocates had been calling for since the beginning of the internet, but as no previous manifestation of this poorly defined revolution had ever come to pass, unless one counted the great privatization of the late 1990s, no one could say for sure what it would look like when it happened. Indeed the internet's earlier rapid colonization and capitalization of the mental life of so many people had occurred in a similarly invisible fashion, so Mary wasn't sure people even knew what they were wishing for when they postulated an internet revolution.

But her team knew—or they were imagining it. Now everyone who signed up for YourLock and started using it was also helping to sustain it, by hosting their part of a blockchained record of its history from its beginning. A distributed ledger: it was only by way of work given for free (meaning not just the labor but the electricity), by many millions of people, that this new organization could function at the level of the computing required. Even if that worked, Mary wasn't sure it was going to represent a net gain in terms of a sustainable civilization. Probably it would depend on what this new network was used for, or on what people did in the physical world. As always, the decisive moves were still to come. Possibly it was true that they would happen first in the realm of discourse, then afterward in the realm of material existence.

She tried to focus on that latter part of life. Morning swim in the Zurichsee, its temperature creeping up as spring turned to summer. Tram and trudge to work, trudge home. Take a weekly tram ride down into the city center and the Gefängnis, to visit Frank May. This was some kind of duty.

He seemed to be doing all right. The Swiss prison system was typically Swiss—practical, benign, a kind of community college dorm that you couldn't opt out of. Frank spent his days out around the city doing public work of various kinds, from street cleaning to nursing assistant, depending on the need and that month's schedule. He was either calmer and happier than on the night they had met, or else subdued and depressed—Mary didn't know him well enough to be able to tell. Possibly a bit of both, if that was possible. Other people: if they didn't want to share with you, you had no way to tell. When she visited, he regarded her curiously, not surprised anymore that she was showing up, just perhaps a bit discomfitted, or mystified. But not enough to ask her why she came. If he had, she wouldn't have been able to answer him very well. In her head she staged conversations with him that were completely unlike what really happened when they were together. Tramming back up the Zuriberg she would watch the other blue cars of the tram, bending ahead and behind in the switchbacking S turns that the tracks made to get up the slope, saying to the Frank in her mind, If you would just ask me, I would say to you, I visit you because I want to rest easier, because I am helping you to rest easier. My conception of the world going well is a world in which even you look at it and feel it possible to rest easier. A world in which you gave yourself a break, and forgave all the rest of us our sins, and forgave yourself too. And in these mental conversations he would often nod and say, Yes Mary, I feel better about things. Your stupid ministry has put its shoulder to the wheel and helped to shove the cart out of the ditch. Although it's not out yet, not by a long shot. Because the ditch was eating the road.

Nothing remotely like that ever passed between them in their actual meetings.

She kept track of Badim's informal work in private meetings away from the office, in the pattern they had established. They didn't meet often, nor was there any way to communicate in the office that wasn't subject to surveillance, so pretty often it was a matter of handwritten notes left on her desk, never direct messages but rather lines attributed to Rumi or Kabir or Krishnamurti or Tagore; she didn't know these poets' work, and wasn't sure if the quotes were real or made up. *The gods are in disarray. It is the theory which decides what one can observe. A great comet will appear in the sky tomorrow. Look to windward.* These phrases, as gnomic as Nostradamus, were only meant to tell her that things were happening, it was time to meet again. Or so she assumed. If there were specific messages encoded in them, she wasn't getting them.

So she kept reading the news. Two days after a note had appeared on her desk that said only *riot strike riot*, she read that Berlin, London, New York, Tokyo, Beijing, and Moscow had experienced simultaneously, in the very same hour no matter the local time of day, teacher and transport worker strikes. This caused chaos in the streets and in the markets. Already the past year's chaos had been sufficient to cause a massive drop in most of the stock markets, and they had never really recovered from Crash Day, so that was low indeed. The bear of bears. Of course the slack was soon enough taken up by risk-seekers looking to buy low and sell high later, but the sense of panic didn't go away, the sense that bubbles were about to burst all over the place. The striking workers in the big cities returned to work, but before the situation had settled, the seemingly endless drought afflicting the Middle East, Iran, and Pakistan suddenly intensified into another killer heat wave, this though it was still only May. But the high pressure that sat on the area had jacked temperatures briefly up into the wet-bulb 35 zone, mainly this time a matter of sheer high temperature rather than humidity, and at the same time some cities there were running out of water. Refugees from the area were pouring across Turkey into the Balkans, also north into Armenia and Georgia and Ukraine and Russia, also east into India. India, a refuge from heat waves! But the Punjab was also caught in a drought, so India had sealed its border with Pakistan, already militarized and easy to close. Disaster all around. Pakistan threatened war, Iran threatened war. Something like ten million people were on the move and in imminent danger of dying. The humanitarian aid programs were overwhelmed, as were the national militaries.

Esmeri Zayed, her refugee division head, told her that if the current refugee population were a country, it would have about the same population as France or Germany. A hundred million people were out there wandering the Earth or confined in camps, displaced from their homes.

In the midst of this situation, an atmospheric river struck southern California, and though its winds were not as forceful as the winds of cyclones or hurricanes, its rainfall was at least as intense, and longer lasting. It looked like it might be something like a repeat of the catastrophic winter that had struck California in 1861–62, arriving several hundred years earlier than would have been expected by the US Army Corps of Engineers, as they had labelled the earlier storm a thousand-year storm, but of course all those probabilities were useless now. The tall mountains hemming in the LA basin had caught the truly torrential rain and poured it down onto the mostly paved surface of the basin, and the devastation was universal. Initial estimates pegged the death count to a remarkable low of seven thousand or so, but the infrastructure damage dwarfed anything the Angelenos' much-feared earthquake would have done to them. Actually there were scientists warning that the weight of that much water might trigger that very earthquake. The Big One, right in the middle of a mega-storm! Only in LA, people said, feeling shivers of schadenfreude, tinged with regret: the world's dream factory was being destroyed before their eyes. No more Hollywood faces to haunt the global unconscious; that age was over. Restoration costs for the damage they were seeing on the raindrop-spattered images would cost more than thirty trillion dollars, Jurgen estimated.

So now one could imagine that the American people might support action on the climate change front. Better late than never!

But no. Already it was becoming clear that LA was not popular in Texas, or on the east coast, or even in San Francisco for that matter. In fact, no place that was not LA cared about it at all. The dream factory for the world, universally unpopular! People had not liked those dreams, perhaps. Or had not liked having their dream life colonized. Or maybe they just didn't like being stuck in traffic.

In any case, California's government, one of the most progressive in the world, and the US federal government, one of the most reactionary in the world—both were making efforts to help. Love it or hate it, LA was important to them. And really, Mary thought, keeping the death count down to seven thousand was an amazing accomplishment of civil engineering and citizen action, also rapid deployment by the US Navy and the rest of the military, and the quick actions of the citizens themselves. The initial rush of the flood had been the most fatal part of it, and after that it was just an accumulation of small accidents. So it was an admirable emergency response. Really the US was in many respects the gold standard for infrastructure, a brick house in a world of straw; those stupid raised freeways, built strong enough to withstand the Big One, had served as refugia for the entire population of the city, and the subsequent evacuation had proceeded successfully. A very impressive improvisation.

Despite LA's uneven popularity across the world, it was for sure immensely famous. The dream factory had accomplished that at least. Many people all over the world felt they knew the place, and were transfixed by the images of it suddenly inundated. If it could happen to LA, rich as it was, dreamy as it was, it could happen anywhere. Was that right? Maybe not, but it felt that way. Some deep flip in the global unconscious was making people queasy.

Despite this sense that the world was falling apart, or maybe because of it, demonstrations in the capitals of the world intensified. Actually these seemed to be occupations rather than demonstrations, because they didn't end but rather persisted as disruptions of the ordinary business of the capitals. Within the occupied spaces, people were setting up and performing alternative lifeways with gift supplies of food and impromptu shelter and toilet facilities, all provided or enacted by the participants as if in some kind of game or theater piece, designed mainly to allow unceasing discourse demanding the official governments respond to the needs of their people rather than to the needs of global capital; and the governments involved had to face either siccing their police and militaries on their own people, or waiting out the occupations for what could be months, or actually changing in the ways demanded. Time to dismiss the people and elect another one! as Brecht had so trenchantly phrased it.

Meanwhile flying was still much reduced, except for an increase in battery-powered short flights, and an immense surge in airship construction. Ocean trade was disrupted; millions were out of work; millions were in the streets. Online, people were joining YourLock and abandoning the other social media sites, now called the predatory social media. So many people were withdrawing their savings from private banks and depositing them in credit unions and alternative cooperative financial institutions that another financial crash not only was happening but was the deepest in over a century. The banks had all been so over-leveraged for so long that what that actually meant had been lost; so now, in a crisis as big as this one, most of them had been brought to their knees, and were stumping to their governments' central banks to squeal for salvation. This time the governments' treasuries, although still in the hands of financial industry veterans, found they could do nothing like what they had done in the 2008 bail-out; that crash was looking minor compared to this one, and because of the 2020 recession, awareness of what was happening, and why, was much higher. It was a different time, a new structure of feeling, a new material situation. Already people were saying this was bigger than 2020, bigger than the Great Depression, maybe the biggest economic crash ever—because it wasn't just economic. The whole damn merry-go-round had spun off its flywheel and was disintegrating as it fell.

So Mary called up the heads of the various central banks around the world and got them to agree to gather to talk things over yet again. Many of them wanted her to come to them, and she almost agreed to hold the meeting in Beijing: the Chinese would be key to any solution. But the Chinese were punctilious about joining international meetings, and would go anywhere the meeting was held, she judged; they didn't care so much about national prestige that they would refuse to join a settlement being worked out elsewhere. The puffy nation in that regard would be the United States, but Mary was pretty sure Jane Yablonski would also come to the meeting wherever it was. And the Bank for International

Settlements' annual meeting in Basel was coming up soon anyway. So she told them to please come to Zurich too, right after the BIS meeting. Air travel being now so fraught, specially arranged stealth military flights would bring them all to Switzerland; or now, quite a few would fly in on airships.

Hosting a meeting of a dozen or twenty of the most powerful people on Earth, which meant also accommodating large staffs for each of them, was a big job, but one that the Swiss were used to performing. For this meeting there were too many people for the ministry offices to hold, so they held the meeting in the Kongresshall, down by the lake.

The morning they convened, the broad picture windows spanning the south wall of the big room provided them with the pathetic fallacy in full measure: a spring storm lashed the Zurichsee, with low shifting gray clouds dropping black brooms of rain onto the silvery lake surface, the windows running wild with deltas of rainwater kaleidoscoping this view. Nothing unusual in that, no LA-like climate apocalypse here, just Zurich spring weather as usual; but still very fitting, given the mood in the room, which was one of grim virtue. They would weather this storm, they said to each other while looking out at it, and all the stark emotion in the room was heightened by the dark metallic sublimity of the rain-lashed whitecaps on the lake, the sound of the wind ripping through the flailing trees.

Mary brought them to order. She reminded them of the meetings she had had with them over the past few years, in which she had urged them to create a new currency of their own collaboration, based on carbon sequestration, and exchangeable on currency exchanges; money like other money, but backed by the central banks working together, and securitized by the creation of really long-term bonds, bonds with a century pay-out at a guaranteed rate of return large enough to tempt anyone interested in fiscal stability. In essence, as she had been saying, creating a way to invest in survival, to go long on civilization, as opposed to the many ingenious ways that finance had found to short civilization, thus in the process shifting most of the surplus value created in the last four decades to the richest two percent of the population, making those few so rich that they could imagine surviving the crash of civilization, they and their descendants living on into some poorly imagined gated-community post-apocalypse in which servants and food and fuel and games would still be available to them. No way, she said to the bankers; not a chance that would happen. Shorting civilization and imagining living on in some fortress island of the mind was another fantasy of escape, one of many that rich people entertained, as ridiculous as retreating to Mars. Money was worthless if there was no civilization to back it, no civilization to make things to buy—things like food. So even if the central bankers were regarding their task in the very narrowest terms, as stabilizing prices and helping the employment rate, and more than anything else, preserving the perceived value of money itself—to do that now, they had to leave their usual monetarist silos, and regard themselves as what they were, the not-so-secret government of the world. In that capacity, something more was now called for than merely adjusting their fucking interest rates.

Yes, they were shocked by her bluntness, by her disgust for their timidity. These Irish! they were thinking; she could see that. But they were also paying very close attention to her; they were transfixed, the storm outside forgotten. Now the storm was in the room, in the form of one angry intense middle-aged woman.

Well, she had to remember; when a meeting got hot it was usually going badly. This was a calculated risk, getting their attention by lashing them a bit; now she had to calm it down. So she did that. The last time she had asked them to do this, she reminded them, they had refused her. Now, she told them, the situation was different. It was so much worse it could scarcely be believed. And as the current representative in their midst of all the future generations to come, she was going to have to insist that they act. She was (remembering what Dick had told her about letting them invent the instruments) open to their suggestions as to how best to act. Possibly the Bank for International Settlements could be brought out of its twentieth-century time capsule and used as the instrument at hand for this. But act they must. Because civilization was trembling on the brink. They were going down.

Here the pathetic fallacy of an ordinary Zurcher spring storm helped bring her point home. The wind was really howling now, the air was black though it was just before noon, the whole lake was slamming into the windows and blurring the view, then the wind clarifying it with a blast, time after time—it was almost as rainy as Galway.

The new Chinese minister of finance, who served as the head of their central bank and was also at the same time a member of the standing committee, and thus one of the seven most powerful people in China, stood up to speak. A woman who had learned her English at Oxford, it sounded like, and she had a cheerful relaxed manner, as if they were discussing history, which Mary supposed they were. She pointed out that Mary had not visited China in her tour of the central banks, nor had she herself been finance minister at that time, so she had not been part of the earlier unenthusiastic response Mary had just described. In fact, in China the national banks were always trying to throw their weight around as vigorously as they could to help China's economy, and they would be happy to join any international effort that they felt could help in a way that was good for China and the world. Indeed, it sounded to her as if what Mary was asking for was precisely the kind of thing that the Chinese government did all the time.

True enough to be a discussion point, Mary replied. But no matter which national or surpranational (with a nod to the European Union's central bank head, and the BIS head) model they referenced or preferred, now she urged them to consider again something new and fully international: a carbon coin, a digital currency backed by a consortium of all the big central banks, with open access for more central banks to join; these coins to be backed by long-term bonds created by the consortium, and shored up against financial attacks by speculators who were sure to attack it. Defended by all the central banks working together, they would be able to repulse successfully any entities that tried to hamstring their new system. Indeed, if the central banks blockchained not just the new carbon coins but all the fiat money that existed, they could probably squeeze parasitic speculators right out of existence. The best defense being a good offense.

The crucial banks, Mary thought privately, were the US, the ECB, and China. Germany and the UK were also important, also Switzerland itself. The more the merrier, of course, as always; but the big three were crucial. Even if it were just those three, they could probably go it alone; although if they were in, Mary was sure others would join.

So, right now, although the new Chinese finance minister thought she was being positive by cheerfully comparing the proposal to ordinary

Chinese practice, this wasn't actually helping much with the others; they were looking skeptical that becoming more like China was really the answer for this moment. China was debt-laden, opaque, oligarchic, authoritarian. Even granted the modifying Chinese characteristics always referenced, they were avowedly socialist, even Marxist. What that really meant no one knew, not even the Chinese, but their financial practices were constantly offending ordinary Western norms and sensibilities, so it hadn't been a very diplomatic move on the part of the Chinese finance minister to suggest to them that by necessity they were now having to become more Chinese. But looking at her, Mary didn't think that this new finance minister was really very regretful about that. Her look was amused, but in the way a hawk might be amused, something hard to imagine. She had a fierce edge.

On the other hand, all central banks were undemocratic technocracies, not that dissimilar to China's top-down system. They were run by financial elites who did what they felt was best without consulting even their own legislatures, much less the citizens of their countries. As institutions they were in fact specifically designed to function outside any legislative or democratic whims, the better to keep the financial ship of the world steadily sailing on into the great west of universal prosperity—for the elites first, and everyone else if they could be accommodated without endangering the elites on the first-class deck. So an invitation to become more undemocratic, if couched diplomatically enough, would not be entirely unwelcome to this crowd. It would be a matter of how one phrased it.

Phrasing was also important when showing the stick. First the carrot, which she felt was the best way to lead: do it, she told them, and you are the saviors of the world, staving off chaos and allowing the huge resources of humanity and the Earth to be brought to bear on the greatest crisis in history. People would be writing about them, analyzing them and copying them, even celebrating them, for centuries to come; and a model would be built by them here and now which could be adapted to deal with any future crises of similar dimensions. Thus the carrot.

The stick: if they didn't do it, Mary and her team could arrange the whole thing to happen through YourLock accounts as a distributed ledger coin, created and given by people to each other. This would cut hard into any power central banks might be said to have. Then also, the Ministry for the Future had allies within every relevant legislature, and Mary's legal team had prepared detailed advice for governments to introduce new legislation that would expand legislative control over the central banks, giving them mandates and responsibilities to mitigate climate change proactively, as opposed to just responding to the financial risks reactively. The new mandates would require central banks to create a digital currency and manage the exchange rate of it, using all the mechanisms at their disposal. In short, Mary was prepared to start a movement worldwide in which governments put their central banks on leashes and directed them to act in ways governments wanted. The great example of how effective this nationalization or internationalization of the national banks could be was the takeover of the Bank of England by the British Treasury during the Second World War. Britain had commandeered the Bank of England to properly guide capital where it was needed to win the war. The same could be done again with climate change, if the relevant legislatures felt it was necessary. Country-appropriate laws were ready to be introduced by sympathetic powerful politicians in every country.

That's what we'll do if we have to, she concluded. She was being blunt again, as at the start; she was slipping into that certain Irish rhetorical mode that was so often useful, the one that said No more fucking around, reality has struck—said with blunt disdain for any naiveté or cowardice that refused to admit the obvious facts. That mode was a mode she liked.

But of course, she went on silkily, I don't think a total takeover of central banks by governments, or replacement by a new people's currency, will be necessary. Sharing for a moment a single basilisk glare with the Chinese minister, who clearly was enjoying her presentation. Renminbi was Chinese for *people's money*, after all. The situation we're facing is unprecedented, she went on, and its causes are clear, and now we have to act, and so we will.

Are the causes so clear? asked Jane Yablonski sharply. I'm not so sure!

Mary let Badim and the rest of her team make the case. She had asked them to prepare a kind of group presentation that ran around the table in cause/effect mode, describing each aspect of the problem in turn. Of course the causal chains ran in all kinds of directions, it was a cat's cradle, but she could make that point at the end; for now, three minutes each to describe the problem: climate change caused by carbon dioxide and methane released to the atmosphere; knock-on effects very close to releasing vastly larger quantities of CO_2 and methane, now cached in the Arctic permafrost and the ocean's continental shelves; oceans unable to uptake more CO_2 and heat; rate of extinctions already as high as at any time in Earth's history, in terms of actual speed of extinctions per century, thus set now to match the Permian in terms of total percentage of species gone from the land, which was ninety percent; subsequent to that coming extinction, inevitable famine, dislocation, and war—possibly nuclear war—leading to the destruction of civilization; impossibility of insuring against such an eventuality, or clawing back from it. Irreversible and unfixable catastrophe.

Thus, ultimately, as a result of all these converging factors, Mary concluded at the end of her team's presentations, they were facing the impossibility of stabilizing inflation rates and employment rates as the climate heated up. The specific principal tasks that central banks were charged with could no longer be fulfilled if the climate emergency got out of hand. In other words, central banks would fail in their principal tasks if they did not save the civilization that had charged them with those tasks. And although it was true that full employment would always remain a key objective for them, she finished, it wasn't such a victory if the remnant of humanity that survived the crash ended up working as scavengers and peasant farmers. That wasn't the kind of full employment that the world had in mind when central banks were created.

She saw that Yablonski and the Europeans were offended at this final sarcasm, and she pondered for a moment simply shouting suddenly in their faces, or taking her shoe off and pounding the table Khrushchev style. Or throwing a chair through the picture window and letting the storm pour in over them. Sudden fury at their mulishness: Fuck your inflation rates! she wanted to shout. Do the job that only you can do!

And judging by their faces, it was possible that her own face held all these sentiments and imagined actions and curses, perfectly visible in the way she was looking at them. The power of the eye. Not Medusa, turning them to stone or killing them with a strike from a snake on her head— instead, she fervently hoped, some kind of electrical jump-start, applied by jumper cables that had her two eyes as the contact grips, leaping the gap from mind to mind. Yes, she was very close to losing it.

Then she saw that the Chinese finance minister was grinning broadly, not even trying to hide it. She checked her cheat sheet; who was this woman again? Madame Chan. Daughter of a finance minister from the generation before. A child of the Party hierarchy, like Xi had been, and so many others. Mary liked her look.

In the following days the representatives of the central banks kept meeting in the same room, and while the Zurichsee provided emotional guidance by way of a succession of brilliantly sunny days, marked by tall clouds like galleons, sailing over the lake as if carrying vast treasure, the central bankers finally invented a proposal they could all agree on. It was as bold as anyone could want, and Mary felt that none of the central bankers there would have touched the plan with a ten-foot pole if all of them weren't in it together to take the heat sure to follow. They would issue together a single new currency, coordinated through the BIS: one coin per ton of carbon-dioxide-equivalent sequestered from the atmosphere, either by not burning what would have been burned in the ordinary course of things, or by pulling it back out of the air. They promised to establish a floor in the value of this carbon coin, which exposed them to great danger from speculators trying to scare money out of the plan; and they foretold a rise in the value of the currency over the coming decades. By doing these things they made this investment a sure thing, assuming civilization itself survived. That by itself would guarantee a certain large amount of capital from many different sources looking for just such a sure thing. Pension funds, small national reserves, big corporate assets, really anyone responsible would want the security involved, especially now that there was no security anywhere else. In essence it was like throwing a life ring to drowning people. It could overwhelm the system, actually, if everyone grabbed at once; but carbon had to be sequestered to create the coins in the first place, so if there was a mad rush to do that, it wouldn't be a bad thing. And the central banks could always adjust upward the amount of carbon saved that would be required to earn a coin, creating derivative complications of all kinds and giving them more control knobs. Getting the certification teams for the sequestrations up and running was going to be a crazy effort. In fact, at the end of the agreement they all lent some fiat money of the ordinary kind, pooled into a fund administered through the BIS, which would be enough to pay for this new bureaucracy of verification that would have to be created to certify that carbon was really being sequestered. This was a bureaucracy so vast no single bank could afford it, nor of course the ministry, not even close. It was almost a full employment plan all by itself.

So it was a total program. Mary's team wrote it up in detail, in consultation with the bankers on hand and their staffs, taking all their suggestions and folding them in, and then in the end, after each bank had consulted with its government back home, they announced it, and offered the first tranche of carbon coins for purchase. Began to disburse them too; and the trade price for them held, even rose a little.

Then nothing happened.

This, Mary thought as the days and weeks after the meeting passed, was beginning to look like a pattern. They were only really doing things to try to ameliorate the situation they were falling into *after* it was too late for those things to succeed. They kept closing the barn door after the horses were out, or after the barn had burned down. At that point their actions, which a few years or decades earlier might have been quite effective, weren't enough. Maybe even close to useless. Over and again it was a case of too little too late, with nothing stronger anyone could think of to apply to the worsening situation.

If this were really true for something physical, like the Arctic's permafrost melt, or the ocean's acidification past the point of life at the bottom of the food chain surviving it, or the Antarctic's ice sheet collapsing fast— then they were fucked and no denying it.

And yet there were still people fighting tooth and claw. And it could be that it was only in the realm of the social that they were so far behind the curve of the moment. Anyway people were fighting.

Although not just for the good, but also against the good—fighting tooth and claw to forestall their efforts, to hamstring them. Thus in effect there were people trying to kill every living thing on Earth, in some awful genocidal murder-suicide. Here they were, walking a tightrope over the abyss, and these fuckers were jumping all over the balancing pole they were holding, doing their best to cast them all down to disaster and death.

"There will always be idiots," Badim intoned as Mary cursed another manifestation of these people.

But it was worst than that. "There will always be assholes," she said viciously.

Badim said, "Focus on all the people still fighting for the good. There are many more of them than the other kind."

Then one month after the carbon coin announcement, a bomb went off in their offices on Hochstrasse.

It happened at night, with no one in the building; perhaps that had been the bombers' intent, but there was no way to tell. The Swiss police who accompanied Mary to see the wreckage were taut with apprehension, and they were not so much apologetic as they guarded her, as stolidly disapproving. Blaming the victim being one of the errors that law enforcement people were so often prone to.

They advised her to accept more police protection. In fact they insisted on it. And it was true that the sight of their offices, solid stone Swiss buildings, built to last a thousand years, blown open, shattered, their interiors visible from the street, a black shambles that would have killed them all had they been inside, was indeed shocking to her. So she agreed.

It turned out that what they had in mind was not like being arrested and let out for the day with protection. No, her guardians were suggesting, or requiring, that she go into hiding. They had facilities to make that work for her, safe houses not far away, fastnesses in the Alps. Or anywhere in the world she liked; but she needed to hide for a while. Threats on her life were very much in play. They insisted. She got to choose the where, a little, but not the how.

If she left Switzerland, she would be leaving her team. And Zurich.

So she chose the Alps.

61

Negative reactions to news of biosphere collapse are not uncommon. Grief, sorrow, anger, panic, shame, guilt, dissociation, and depression are frequently seen responses to news of global climate catastrophe. These negative reactions can sometimes become extreme enough to be labelled pathological.

One pathological reaction, a form of avoidance, has been called The Masque of the Red Death Syndrome, after the story by Edgar Allan Poe. In the story, a group of privileged aristocrats, isolated in a castle on a peak above a countryside devastated by a plague, stage a masquerade to distract themselves, or to display indifference or defiance to their eventual fate. They arrange the rooms of the castle such that each room is illuminated by light stained a different color, and then, having dressed themselves in costumes including masks and dominoes, they parade through the castle dancing to music, eating extravagant meals, and so forth. A silent masked stranger then appears and stalks through the party, and few readers are surprised when this stranger turns out to be death itself.

The syndrome is thus an assertion that the end being imminent and inevitable, there is nothing left to do except party while you can. The late middle ages' dance of death, *danse macabre* in French, *Totentanz* in German, is an earlier example of this response, in this case associated with the Black Death; it is likely to have been one of Poe's inspirations.

Even more extreme pathological responses to biosphere collapse are possible, and have been observed. Some who feel the end is near work to hasten it, or worsen it. Their position seems to be that if they're going to die then the world must die with them. This is clearly a manifestation of narcissism, and has been named the Götterdämmerung Syndrome. Hitler in the last days of World War Two has become the canonical example of this response. Hatred of the other is also quite obviously manifested in such a reaction.

The name for this response comes from Wagner's opera *Götterdämmerung*, which ends with the old gods of the pre-Christian Norse mythology destroying the world as they die, in a final murderous and suicidal auto-da-fé. A folk translation of this word into English has it as "the God-damning of the world," although this makes use of a false cognate and the German actually means "the twilight of the gods," and is Wagner's German neologism for the Norse word *Ragnarok*.

The Götterdämmerung Syndrome, as with most violent pathologies, is more often seen in men than women. It is often interpreted as an example of narcissistic rage. Those who feel it are usually privileged and entitled, and they become extremely angry when their privileges and sense of entitlement are being taken away. If then their choice gets reduced to admitting they are in error or destroying the world, a reduction they often feel to be the case, the obvious choice for them is to destroy the world; for they cannot admit they have ever erred.

Narcissism is generally regarded as the result of a stunted imagination, and a form of fear. For the narcissist, the other is too fearful to register, and thus the individual death of the narcissist represents the end of everything real; as a result, death for the narcissist becomes even more fearful and disastrous than it is for people who accept the reality of the other and the continuance of the world beyond their individual end.

Even the night sky frightens the narcissist, as presenting impossible-to-deny evidence of a world exterior to the self. Narcissists therefore tend to stay indoors, live in ideas, and demand compliance and assent from everyone they come in contact with, who are all regarded as servants, or ghosts. And as death approaches, they do their best to destroy as much of the world as they can.

The phrase *Götterdämmerung capitalism* has been seen. This marks a shift, possibly inappropriate, from psychology to sociology, and is therefore outside the purview of this article; and is in any case self-explanatory.

62

Sibilla Schmidt, officer in charge. We took subject Mary Murphy under protection 7 AM June 27. My team for this mission: Thomas, Jurg, Priska, me. Priska did her best to make M feel comfortable. M was clearly not happy.

Her ministry was thought by some in the intelligence division to have been involved with the hijacking of Davos, so Jurg suggested to Thomas that she shouldn't be so picky now. Of course he was professionally polite and did not say this around her. I told him to keep such thoughts to himself from now on. We're a good team, rated best in the Swiss federal secret service's internal evaluations. Of course our branch came in for heavy criticism after the Davos incident, and many people felt we must have been involved in that somehow, that we were complicit in it, or allowed it to go on longer than it should have. Always an obvious pleasure in criticizing us, especially when coming from the political figures we protect. The *Spasspolizei*, they call us, the killjoys, but they don't usually reject our services when required. Switzerland is a very free and safe place compared to most.

Old European joke: in heaven the cooks are French, the police British, the engineers German, the lovers Italian, while in hell the cooks are British, the police German, the engineers Italian. The Swiss are in the joke somehow too, but I forget as what. Maybe the same job in both heaven and hell? Schedulers? Bankers? Security force? I can't remember. Maybe it's a Swiss joke.

We moved M in one of our vans, bulletproofed, road bomb resistant, darkened windows, secure comms. Priska and Thomas made perfunctory compliments on how light M traveled. We put her in back with Priska, the rest of us sat in the middle seat and up front. Jurg drove. Highway to

Bern, no incidents. Bern to Thun, then up the western shore of the Thun-ersee. Up twists and turns into Heidi Land, big wooden houses lined with red geraniums, green alps rising to the dark cliffs of the Berner Oberland. I prefer Graubünden.

In Kandersteg we left the highway and took the track up to the Oeschi-nensee, using their private service road beyond the upper cable car termi-nal. On the way Priska explained to M why Kandersteg was a backwater, pointing around us; basically because of no skiing, simply because all the town's surrounding slopes are cliffs. Only way out of the box canyon is through the old train tunnel to the Rhone, one of the oldest tunnels of all. So it's quiet, like all the alpine canyons too steep to ski. There are far fewer paragliders than skiers.

Drove by sheep on an alp and M said it looks like Ireland, if you don't look up. Ridge to south very tall.

Arrived at the Oeschinensee, 10:40 AM. Substantial round lake under tall cliff, a curving wall of gray granite at least a thousand meters high, all in a single leap up from the lake, very dramatic. Lake an opaque blue, sign of glaciers somewhere overhead.

Priska told M about the Oeschinensee, explained to her that it was rare to have a lake this big this high, because all the alpine valleys had been so smoothed by the giant glaciers of the ice age that no ribs of rock remained to hold water. So no ponds or lakes until you got down to the giants of the Mitteland. But here a landslide had fallen off the cliff overhanging the valley, creating a blockage that eventually filled with a natural reservoir of melted snow. The lake has no outlet stream, Priska said, because its water seeps through the landslide and comes out in a big spring partway down to Kandersteg. I didn't know about this either, and found it interesting, but M just nodded, too distracted to be interested.

We drove by the mountain hotel on the lakeshore to its second build-ing, reserved for us. Family owners aware of our situation, helping as they have before.

One of the SAC huts above the lake was being cleared for our use, without a fuss, so that it would take a couple of days. So now we stayed in the lakeside hotel's second building. Two days were spent walking around the lake with M. She refused to stay in building, and I felt the walks were secure enough, confirmed that with Bern. Paths around the lake partly forested, rising into alps above treeline. M always paused to inspect the wooden statues in forest, carved from tree trunks not cut down. Primeval figures, beast faces, local folklore, the Böögen, etc., all looming in shadows among trees. Higher up the trail runs through a krummholz, trees small and gnarled. Typical Berner Oberland, Priska said. As we walked she told M about the Alps, mostly things we all know, but Priska knows more.

The cliff backing the lake really is very dramatic. Every day scraps of cloud hung partway up the cliff, showing how tall it was. M said the cliff was all by itself taller than the tallest mountain in Ireland. Its height, and the strange pastel cobalt color of the lake, gives the area the look of bad computer imagery or a painted paperback cover, too improbable and fantastic to be real. I definitely prefer Graubünden.

M spoke with the hotel's owners one evening. They are now middle-aged; I met them once when they were younger, when my parents brought me here. Now their children are grown enough to run the place. The son will be fifth generation to own and run place, they told M. She commented on how unusual that was, and they nodded. They felt lucky, they said. They like it here.

M asked if one could hike all the way around the lake, gesturing at the cliffs. Priska shook her head at this. There's a crux, the son replied, pointing across the lake. It can be done, but there's one ledge, pointing at a green line crossing the cliffs about halfway up. It can be done, he said, but I only did it once, and I was young. I wouldn't do it again. There's one spot where it's too narrow for comfort.

As so often, M said.

The son nodded, said There's always a crux.

Next day we went up to the SAC Fründenhütte, now emptied for us. 5:20 AM departure. A thousand-meter ascent, hard in places. M was tired from the start, and seemed frustrated we hadn't stayed at lake. Orders from Bern, I told her. Standard procedure. The real safe house here is the SAC hut.

Six hours steep uphill walk, always on trail. Cable handholds lining one steep wall section. M maybe suffering from altitude. She was slow and quiet.

Fründenhütte was imposing, a big stone box faced with red and white chevroned window shutters. Strange to see in such high remote location, as always with SAC huts. Each more unlikely than the next, it's a game they play to amuse their fellow climbers and hutkeepers. This one located on a rise in the bed of a glacier now gone, the remnant ice still hanging on at the top of the basin. An old terminal moraine extends in low curves to each side of the hut. Photos on the hut's dining room wall show the Fründengletscher in 1902, a big ice tongue almost reaching the hut, looming over it. Then four more photos through the years, two aerial, showing ice recession. Now just a scrap of grayish white, pasted up there under the cliff at the low point of the ridge.

M rested that afternoon. I suspected a touch of altitude, and gave her Diamox. Later she made a few phone calls on an encrypted connection we made for her. After that she napped, woke in time for sunset. Strong alpenglow against clear sky, some high clouds to east also pink. M said she could work by phone and hide here for a while very happily. A good sign.

Another good sign was her appetite that night. Cooks did raclette and rosti, salad and bread. Hutkeepers a middle-aged couple, with a pair of young assistants. They led M to a dorm room, all they had, which meant she had an entire matratzenlager to herself. She laughed to see that. The single mattress extended down the whole length of the room, with numbers on the long headboard marked for twenty sleepers, a duvet and pillow for each. She took two pillows and said good night. 9:10 PM.

Next day, hut empty except for us and the hutkeepers. M breakfasted in dining room, did her e-mail and made calls on encrypted lines provided to her. Then drinking coffee on patio overlooking lake, 1,200 meters below. The Alps are big, she remarked to Priska.

Later that day she asked to go for a walk, and we led her up to the foot of the Fründengletscher, some six kilometers up basin. Less steep than the ascent to hut had been the day before. Priska explained why this was when Mary remarked on it: instead of going up the side of a big glacial U valley, as we had yesterday, we were now walking along the bottom of a smaller higher U valley. Rock-strewn floor of a hanging valley, much less steep than where it falls into the bigger valley below. Same as always. Less moss and lichen and alpine flowers the higher we ascended, until bare rock, probably under ice until just a few years ago. By midafternoon we reached the foot of the glacier, which was mostly covered with black rubble fallen off the ridge, but also cut by white vertical melt incisions, making the glacial ice visible, and in the deepest parts of cracks, quite blue.

It must be depressing, M remarked. You can really see the glaciers are melting.

It's bad, Priska said. Maybe not as bad as the Himalayas, where the melt is their water supply. Still, it changes things here too. We lose some water, some hydro power. And it feels wrong. Like a disease. Some kind of fever, killing our glaciers.

Even so, the remaining wall of this glacier's foot stood about fifteen meters overhead. Getting onto the glacier proper would involve climbing a lateral moraine, then crossing the gap between moraine and ice. Possibly a job for crampons going up the ice itself, unless a good level bridge of rock or ice were found. Not on this day's program.

Hiked with M back down basin, seeing better just how steep our ascent had been. Pleasant evening at hut.

We all woke at 2:46 AM to a very loud roar and clatter. We rushed to M, prepared for trouble, Jurg with pistol ready. To windows to look out, but a moonless night, nothing to see. The sound had ended, nothing more to hear or see. Avalanche, Priska suggested. No, rock fall, said one of the hutkeepers; not snow but rock. Rock for sure, he said, the noise had been so loud. It had lasted perhaps thirty seconds. The hut was set on its little rise of rock, well away from the cliffs flanking it, so the hutkeeper said we shouldn't be in any trouble from rock fall or the run-outs that sometimes happened.

Back to bed for most of us, but Jurg and Priska and I stayed up for a while outside M's room, sitting on floor not sleepy. Thomas and one of the hutkeepers went out to have a look around, came back reporting a new mass of rock now lay just west of the hut. We alerted Bern, wondered what was going on, if we had been attacked. Waited to hear back from Bern about incident, get their take on possibility that hostiles had located M and sent something her way.

At dawn we went out and saw it; a new rockslide, yes. It had come off the steep ridge to the west of the hut. The run-out across the basin floor had reached almost to the hut. Immense boulders of schist and gneiss and granite now stood tall on the basin floor. Contact between different kinds of rock was always a weak point, Priska said. The biggest chunks had rolled the farthest, in the usual way. One boulder, almost as big as the hut, lay only about twenty meters from it. Looked like a rough statue of the hut itself. It would have crushed the hut if it had run into it with any momentum at all. One more roll of this big dice, in other words, and boom, we would have been crushed.

I conferred with my team, in Schwyzerdüütsch so M couldn't understand us. This is too much of a coincidence, I told them. I have a bad feeling about this. We are now code red. We moved into that protocol.

Bern agreed. Code red for sure. Get ready to leave, they said. We'll get back to you with an evacuation plan as soon as we have it. Cover must be blown.

We considered that. If cover was blown, it would be dangerous to extract by helicopter. Drone attacks were all too possible. Of course the hut itself vulnerable as well. Bern said the plan for us would be ready within the hour.

Long before that Priska proposed her own plan. I called Bern and ran it by them. They took it in, put us on hold, got back to us fast. Do it, they said.

We told M: We must leave.

Again? she cried.

Again. Bern thinks your location may be somehow known to hostiles.

Do you really think someone could trigger a rockslide as big as that?

Possibly yes. The hutkeepers say the cliff there had an overhang. Could have fallen naturally, but if the overhang got hit by a missile, maybe not even an explosive missile, just an inert mass hitting at speed, the cliff could have come down. That would bury all signs of a missile, and look like an accident. The rockslide could very well have crushed the hut. It just missed. Couldn't have been sure about it until trying it.

It couldn't just have been a coincidence?

Cliff falling now, after all the centuries of standing there, right when you are here? On that day, it falls?

It could still be a coincidence, she said. That's what coincidences are.

Thomas shook his head. They've seen something in Bern, he told her. They don't think it's a coincidence.

All right, M said, looking more and more disturbed. Where to now?

We told her our plan.

63

They came for her just after midnight, knocking as if to wake her, but she hadn't slept a wink. The whole hut dark and chill, her guardians hushed and nervous. This is when the climbers always leave, Priska told her reassuringly, to get up high before sunlight starts the rock fall.

Priska and Sibilla took her into one of the washrooms and ran a wand over her body as she stood shivering in her underwear. Then the wand all over the clothes she was to wear, everything she was going to take along, which was hardly anything. They had asked her to leave her phone at the hut; it would be conveyed to her later on. Same with her clothes. They thought she was clear of tracking devices, they said, but it was best to be sure, and leave behind everything not needed for this day.

The hutkeepers outfitted her with warm clothing, with climbing boots and crampons, and an outerwear suit like a pair of overalls but lined with down. A kind of spacesuit, it seemed to her. Also a climber's helmet, and a harness to wear around her waist and thighs.

I don't think I like this plan, Mary said.

It will be all right, Priska said. The Fründenjoch is not that hard.

I don't like the sound of that, Mary said. She knew that not that hard, when talking about the Alps, was Swiss-speak for fucking hard. And she knew that *joch* meant pass. Meaning probably the low point at the head of their basin, up there above the glacier they had visited the day before. There had been a notch up there in the cliff over the ice, the cliff which was a walled-off section of the crest of the great Berner Oberland. Once you knew to look for it, the notch was visible even from the hut. But the day before, she had seen that the black rock below the notch had looked completely vertical. Not that hard—right!

At 2 AM they went out into the frigid night. There was no moon, but illuminated by stars the basin walls glowed as if with a black interior light of their own. Their headlamps speared the night and illuminated variable circles and ellipses of rough stone ground ahead of them. Mary was roped up between Thomas and Priska, with Sibilla and Jurg on another line beside them. They all had headlamps on their climbing helmets, so no one looked at each other as they spoke.

After a couple hours of walking up the stony slope, Mary huffing and puffing and warming up throughout her body, all except for her nose, ears, toes, and fingertips, they came to the foot of the glacier remnant. After that they scrambled up the left lateral moraine, which was composed of loose boulders held poorly in ice-crusted sand. Then Mary had to focus on getting up onto the ice side of the glacier itself. The slope of white ice they were proposing to ascend was tilted at about forty-five degrees, maybe more; it was therefore crampon work, a hard little climb. She had never done anything like it. They sat her down and helped her strap the crampons onto her boots, and handed her an ice ax, and after that, when she kicked the ice of the glacier, the front points of her crampons bit into it very nicely. With a good kick it became like standing on the step of a ladder, a step which was really just her stiff-bottomed boot, stuck in place. Rather amazing. Up the side of the glacier she went, kick, kick, kick. With the rope extending up from her to Thomas, who was already up on the flatter top of the glacier, it was almost simple.

Then she was walking on the glacier's top with the rest of them, and feeling her crampon's downward-pointing points stick into the ice surface with every step. Sometimes she sank a bit and then stuck, piercing a layer of hard snow. That was firn, Priska told her. Good to walk on. In fact it was quite strange. Mary found she preferred the bare ice, where she stuck instantly with each step, remaining almost a full crampon tooth's height above it. She had to free her feet with little jerks at each step, then step a little high when moving her feet forward, or she would catch a spike and trip. After she stepped and stuck, she couldn't have slid her foot even if she wanted to. That was reassuring. The boots they had equipped her with were a little too large for her feet, she thought; sliding around inside the boots was the only give in the whole process.

None of this was comfortable for her, it wasn't her kind of thing. She wondered if the whole adventure was even necessary, but didn't want to ask about that, as it would sound like a criticism. And if she was in danger, they were in danger, and yet of course they were sticking with her, it was their job. So she did what they told her to without any commentary of that sort. The fact that they thought this was necessary was quite frightening, actually, if she allowed herself to think about it. Which she didn't. She focused on the ice underfoot, on her breathing.

They crunched up the high glacier at a steady pace. The creak and scritch of their crampon straps and points were the only sounds. Once they heard the clatter of a rock falling. Other than that, windless silence. Black sky, filled with stars. Milky Way almost setting in the west, like a noctilucent cloud. Thomas was following a series of flags that flew from wooden sticks, set into cans filled to the top with cement. Mary shuddered to think of carrying the cans up here, they had to weigh twenty kilos at least, but now they stayed put, and guided the way up to the pass. Priska said they were threading some crevasse fields. They would move the flags when the glacier's ice moved, she said, although that wasn't such a problem on this one, which had melted up to a kind of minimum remnant, an almost stationary ice field. Priska and Sibilla tried to point out some crevasses to her as they passed them, but Mary couldn't see them. Slight depressions in the snow blanketing the ice; maybe firn rather than ice; she would have walked right over them. A bad idea. Walking by one of the flags, she saw by her headlamp glare that both cans and flags were painted orange. By starlight they were gray.

After two hours of this ascent, they came to the top of the glacier and faced the black rock of the pass. There was a short and steep-sided gap between the ice of the glacier and the black rock. This gap was the bergschrund, they told her. Famous for presenting a problem in getting from glaciers onto headwalls, sometimes a terrible problem. Happily this bergschrund had a staircase of sorts hacked into the ice side under their feet, its shallow irregular steps much punctured by crampon tips, leading down to the black rock and ice blocks junking up the bottom of the dark little ravine. Tricky work, once they got down there; Priska and Thomas actually took her by the hands and watched her footwork with her, and she took their support gratefully. I'm fifty-eight years old, she wanted to tell them. This isn't my kind of thing. I'm a city gal. From the bottom of the slash it was eerie to look up and see just a narrow band of stars overhead.

On the rock side of the bergschrund they climbed by spiking their crampon tips into cracks in the rock. This seemed like a bad idea, but in fact her boots stuck on the rock even more firmly than they had into ice. And the rock wall proved to have setbacks in it, almost regular enough as to have been cut by and for people, though Priska said they weren't.

Then they topped the wall, and were hiking scratchily, as if on tiny stilts, over almost flat slabs of black rock, leading them slightly up between vertical black walls to each side of them. It looked like a roofless hallway, carved by Titans. Priska, in full tour guide mode despite or because of the surreal weirdness of it all, told her that joint faults in the rock had allowed the glacier, when it had been so high as to cover this entire section of the ridge, to pluck and shift loose blocks out of this passage, pushing them probably to the south, as they would see shortly. Now the missing fault block made a break in the ridge, the notch they had seen from below, as squared off as if drawn by a plumbline and carpenter's level. Very surreal. Not to mention the thought of an ice sea so high that it had covered this part of the range, and presumably all the rest of the Alps, all except for even higher ridges and peaks. Just another Swiss alpine pass, Thomas and the others seemed to suggest by their attitudes, but Priska was obviously proud of it. Each pass in the Alps had its own character, she said. Most were well known since the middle ages, or perhaps long before, back to the time when people had first come to these mountains thousands of years ago. Like the Ice Man, found emerging from a glacier in a pass to the east of here. He had crossed his pass five thousand years before. Or failed to cross, Mary thought but did not say.

So they walked through the notch of the Fründenjoch. It was like passing through a hallway from one world to another. It only took about five minutes. They were just thirteen meters short of a three-thousand-meter pass, Priska said. People often jumped up to pretend to touch that three-thousand-meter height above the oceans. Swiss people, Mary thought. She couldn't have jumped an inch off the ground.

When they came to the far end of the notch, dawn was flooding the Alps to the south. The raw yellow of morning. A new world indeed. Alpenglow stained east-facing peaks pink; slopes facing the other points of the compass were mauve or purple or black. The ice below them was a rich creamy blue, the sky overhead a clear pale gray, tinged by the yellow light in the air. Peaks extended to the horizon in all directions, and to the south another great range paralleled the one they were crossing. Below them a long sinuous glacier was flanked by black lateral moraines. The Kanderfirn, Priska told her. Not bare ice but firn, which gave it its velvety look. A dark turquoise velvet, very strange to see.

Directly below them lay a drop of empty air, then a slightly tilted mass of bare ice, which served as a kind of terrace, making a long run down to the firn much farther below. The last drop at the end of the ice terrace to the firn was invisible to them from their angle, indicating something obviously steep: a cliff. Mary gritted her teeth as she looked down at it. It was a long way down, and she was already tired, her calves and Achilles tendons aching, an exhausted feeling in her muscles everywhere.

She said nothing. She followed them, step by step. Down, then down, then down again. They helped her in places. Hiking on rock in crampons was awkward as hell, but it was solid too, in that her boots kept sticking in place, not the worst sensation to have, given what a slip might do. Priska confirmed what she felt in her feet; you could spike right into a crack on purpose, and hold firm in a way that bare boots wouldn't have. That meant the pressure on her ankles was tremendous, and often she found herself right at the edge of her strength, right on the brink of a fall, feet tilted this way and that in ways she never would have chosen. A couple of times her only recourse was to give way and step down another step, too fast, unconsidered, desperate. Every time that happened her feet stuck in some new crack, an unexpected salvation. Fuck! she said, time after time. Fucking hell. It couldn't last like that. Some step was bound to go wrong. Her heart was pounding, she was sweating, nothing else existed but this steep broken staircase of cracked black rock.

After an endless spell of this work, they got to the ice terrace she had seen from the pass above, which now sloped down as far as they could see, then ended in mid-air. Obviously some kind of cliff out of sight down there, ice or rock, it didn't matter, it was going to be bad, wickedly bad.

We wait here, Priska told her.

We wait? Mary repeated. For what?

Priska said, They are coming here for us. We're going to catch a ride.

Thank God, Mary said.

Far, far below, where they could see a different glacier's surface, on the far side of the firn, she spotted a teeny line of flags. Now there was enough light to see they were orange. After fifteen or twenty minutes, they heard a thwacking from down the valley. Up to them rose a helicopter, with the Swiss white cross on red painted on its side. It took a while to get up to them; it really was a long way down. And the air was thin, she could feel that right in her bones, a weakness she didn't like to feel.

The helicopter landed loudly and with a great rushing of air, some fifty meters away, on an almost flat stretch of ice. With its rotors still blurring over it in a roar, a helmeted and flight-suited person opened a door in its side and got down and gestured them over to the craft. They ducked and cramponed over to the helo, sat down on the ice and unstrapped their crampons (the others unhooked Mary's for her), then climbed up metal steps into the cabin of the thing. Here it was just as loud as outside, but after they sat down in the big central chamber, and were strapped into their webbing seats by a crew member, and got their climbing helmets off, they were given earmuff headsets. With those on, the world became very much quieter. And there were voices in her ears.

The discussion over the headset was in Swiss German, so Mary just relaxed and flexed her calves and feet. Right on the verge of cramping, they were, in several different places. Thighs too. She was thrashed and no denying it.

They had put her next to a little window, and as the helicopter rose she looked out: steep black mountains, vast swathes of white snow. Then out over an even deeper valley, immensely bigger, mostly green-walled, and floored by a river, and freeways, train tracks, miniature villages with church steeples and square towers and rooftops, and vineyards rising in rows contouring the sides of the great valley walls, especially the south-facing side to their right. The Rhone River, Mary assumed. If she was right, this was the Valais, one of the biggest valleys in Switzerland.

Down this immense canyon they flew, lower than the mountains to left and right. Then they turned left, southward, and flew up a narrow valley. Mary knew that the Matterhorn stood at the top of one of these southerly valleys, but she couldn't imagine they would go to Zermatt to hide; and it seemed to her they had flown farther west than that anyway. In the brilliant horizontal morning light, these side valleys were still dark in shadow. She lost her sense of where they were.

Eventually the narrow canyon they were ascending closed on them, and the helicopter descended onto a concrete pad underneath a concrete dam, a dam very tall and very narrow, incurved like dams often were. A surreal sight, a comic-book dam it seemed, exaggerated to caricature.

They got out of the helo, went into a building by the landing pad. Here they sat and ate a quick meal, went to the bathroom, changed out of the snow boots and into more conventional hiking boots.

Not done walking yet, Priska explained. Another short walk.

How short? Mary said, annoyed and fearful; she was tapped out, she could feel that in her legs.

Six kilometers, Priska said. Not so far.

And two hundred meters up, Thomas added, as if correcting Priska. Full disclosure kind of thing.

Mary bit her tongue and said nothing. She was wasted. This was going to be bad.

They left the building. By now it was late morning. The sun stood just over the top of the dam hanging over them. It was the tallest dam in Switzerland, they told her, the fifth tallest on Earth. Nearly three hundred meters tall.

Happily a cable car was strung up the slope to the west of the dam. They got in a car and rose swiftly, getting an excellent view of the concrete incurve of the dam. Ear-poppingly tall. Hard to imagine someone saying Let's put a dam here, let's fill the air of this deep narrow valley with a concrete wall a thousand feet high.

At the upper station of the cable car they got out and walked through a long tunnel, a tunnel with open gallery sections on its left side that gave views down onto the reservoir behind the dam. Water the color of radiator antifreeze.

Then they were out of the tunnel and in a high shallow valley, headed west. Another narrow glacial valley, with a trail that ran up its floor next to a chuckling brook. Mary tromped up this trail wearily. It was a lot easier than hiking in crampons and snow boots, she had to admit, but she was out of gas. Knackered. It was a little embarrassing that her guards didn't even seem to register it as effortful. Fucking Swiss and their fucking mountains. She had been outskied by three-year-olds who weren't even using ski poles, zipping by her not even seeing her; why should she be

surprised now, they were like the ibexes up here. Home ground. Rocky
dusty trail. Green grassy slopes bordering the creekbed, rocky walls above
the grass, high on the left, low on the right. Up and up, and up again.
Grinding. She felt weak in her bones.

Then she noticed that the creek to her right was a dark brown color. Its
clear water riffled quickly over a reddish-brown streambed. She looked
closer; the creek was floored almost entirely by rusty nails. Nails, bolts,
washers, nuts, L joints, other bits of small metal hardware, all of it a dark
rich rusty brown, carpeting the creek bottom in a dense tangle, like mus-
sels or sea urchins. It was like a vision in a dream, surreal such that she
could scarce believe her eyes. She actually wondered if in her exhaustion
she was seeing things.

What the hell's that? she asked her minders, pointing.

They shrugged.

From the old days, Priska said.

What old days?

You will see.

Then they topped a rise and found themselves in a round high basin,
a mountain wall to their left, green alp to their right. Grass covered the
floor of the basin, along with rocks of all size, from small to house-sized,
the big ones like dolmens in Ireland—and then she saw that some of
these really were buildings, blank-walled cubes of concrete, doorless and
featureless.

Then, as she looked around again, Mary saw that the cliff that walled
off the basin to the south had inlaid into it three massive concrete doors,
giant ovals perhaps fifteen or twenty meters high, and almost as wide.
Like the blockhouses on the basin floor, these walls in the cliff were as
blank as concrete could be.

What is this? Mary asked.

Military, Sibilla told her curtly.

Air force, Priska added, gesturing at the giant doors in the cliff as if that
explained what she meant.

Civilians not actually allowed in this basin, they told her. No one
comes here.

But we did? Mary said.

And then a small concrete door under the giant ones, so small Mary had not noticed it, opened from the inside. Her minders led her up and into it.

She followed them right into the mountain. One of the Swiss air force's secret bases, she gathered; Priska had gone uncharacteristically silent. Mary had heard rumors about these places, everyone had. A rocket-launched jet facility, remnant of the Cold War, designed to repel Soviet invasion. If the Soviet tanks had rolled, these big doors in the mountainside would have opened and Swiss jets would have shot out like bolts from Wilhelm Tell's crossbow. Swiss defense craziness: it wouldn't be the first time.

And in fact the jets were still there, racked on their launch slides. Little things like cruise missiles with stubby wings and bubble cockpits. Obviously outdated and old-fashioned, like weapons from the set of a James Bond movie. New cruise missiles made these things look like rowboats in a marina. Like shillelaghs in a museum.

It was a newer kind of warfare, she thought wearily, that had gotten her invited into this hidden fortress. They weren't just protecting her now, she guessed, nor her ministry; it was Switzerland itself under attack. This she gathered from the people there to greet her. The assault on the ministry had been part of a larger attack, one of them confirmed to her. Viruses had sabotaged not just the ministry's computers but other UN agencies based in Switzerland, and more importantly, their banks. So it was an emergency; they were on a war footing now, but for this new kind of war, invisible and online for the most part, but also including the possibility of drones, and of fast targeted missiles.

So now defending her ministry was part of defending Switzerland. And as this museum fortress served to show, the Swiss were very intent to defend themselves. Small country in a big world, as was explained to her by a Swiss military man as he escorted her along a big tunnel to a conference room deeper inside the mountain. Unusual things became necessary. He introduced himself; turned out he was the country's defense minister.

She sat down at a long table, suppressing a groan of relief, and was joined by a circle of officials. Such a relief to sit down, her legs were throbbing. She glanced around the room; it was broad and low-ceilinged, its long back wall made of the green-black gneiss of the mountain itself, cut and polished like an immense facet of semi-precious stone. Overhead a ceiling that seemed to be white ceramic glowed everywhere with a powerful diffuse luminosity.

Mary felt her face burning, knew by that feeling that she was sunburnt and trail-dusted. Wiped out: *Alpenverbraucht*, Priska had called it. Alp-wasted. She looked around, saw that everyone at the table recognized how she felt, and knew the feeling well. They had been there, they understood her state.

One of them shuffled his papers, glanced at his phone, waiting for something. Then seven people walked into the room together. The executive council, Mary understood suddenly. All seven of their presidents!

The seven-headed president of Switzerland sat down across the table from her. Five women, two men. She didn't know their names.

They spoke in English. Of course this was for her sake, but a question distracted her as she tried to attend to them; she wondered what language they would speak if they were just among themselves. She dismissed the thought and tried to concentrate on what they were saying, too exhausted to reply, almost to understand. Some were French speakers and some German, she thought, although with the Swiss it wasn't as easy to tell which was which as it would have been with actual French and Germans. And especially not now. It felt like she was reeling in her seat.

One of the presidential women told her that they had come here to meet her because they were now confronted with a crisis that seemed to have something to do with her. The recent attack on her ministry was part of a larger attack; also under assault were the UN offices in the country, Interpol, the World Bank offices in Geneva, and Switzerland itself. The international order, in effect, was now under attack.

Attack by whom? Mary asked.

A long pause as the seven presidents looked at each other.

We don't know, admitted one of the women. Suisse Romande—Marie Langoise, it came to Mary. A Credit Suisse veteran. She went on, There's been an attack on our banking regulators that appears to have come from the same source as the attack on the Ministry for the Future.

I see, Mary said, though she didn't.

Did your ministry plan the hostage taking of Davos? Langoise asked.

Don't know, Mary said sharply. Then she added, But maybe those people had it coming, right? Did anyone really regret that?

We did, one of the others said.

Unfriendly silence. Mary let it stretch out. Their move, she felt. Although none of them seemed to agree.

What did they do to your banks? she asked at last.

They looked around at each other.

We are not bankers, Langoise said (though she was), so we can't go into the details. But the attack apparently compromised many secret Swiss bank accounts.

Revealed them? Mary asked.

No. Private accounts are encrypted in multiple ways, they could not be revealed. But now the banks themselves are having trouble accessing files that decode owners' ID, in order to contact them and so on. So the danger is not so much exposure of clients, as loss of fundamental information.

Mary said, Your banks can't figure out who owns what?

Somewhat the case, one of the others admitted. Another banker, Mary thought. Out of seven Swiss presidents, how many came from a banking background? Four? Five?

Of course it would eventually get sorted out, one of the presidents said. Information all on paper and in cloud, as of course it should be. Time machine storage; it took Mary a second to understand this meant a computer back-up. But still, in the immediate aftermath of the attack, there was depositor fear. Even panic. Not good for stability.

Mary nodded. Silence as they watched her. She saw they were here to listen to her.

She began to talk to them, almost as if thinking aloud to herself. Why not? She was too tired to find and apply her usual filters.

It's the mystery of money, she said. Numbers that people trust; unlikely from the get-go. But then, if that trust was lost, boom, it was gone. Meanwhile they were all part of a global financial system that had become so complex that even the people running it didn't understand it. She looked around at them as she said this: yes, she meant them. An accidental megastructure, she went on, enjoying the sound of J-A's phrase, right at the heart of society. Right in this secret Swiss mountain fortress, which ultimately protects not just your countryside and your society, but your banks. Which means also people's trust in civilization. Their faith in a system that no one really understands.

The seven parts of the presidency regarded her.

Mary felt a fog pass through her; then she came to, it seemed, and became aware of them again. What do you want from this situation? she asked curiously.

They wanted the Ministry for the Future defended, they told her. Even strengthened. Just as part of Switzerland's own defense. They wanted better ways to make a better future, as part of making a safer Switzerland. It wasn't as if the country's eight million people could live off what could be manufactured and grown in Switzerland alone. Country half the size of Ireland to start with, and 65 percent of that mountains, useless to humans. The remaining 35 percent an agglomeration, satisfying human needs as best they could. They did what they could, but were part of a larger world. Not self-sufficient. Self-sufficiency was a dream, a fantasy, sometimes of xenophobic nationalists, other times just a decent wish to be safer. Swiss people mostly realistic, which meant being honest about what is possible. Thus engagement with the world.

So they wanted her ministry to succeed, because they wanted Switzerland to succeed, which meant the world had to succeed. The future had to succeed. That would take planning, it would have to be engineered.

All this is well enough, Mary told them. It's our project too. But you can do more than you are. Right now you're not doing enough.

She almost laughed as she heard herself doing a version of what Frank had done to her that night. But not a good idea to laugh at them for no obvious reason, and she suppressed it, recalling suddenly that vivid night, the way she had been transfixed by Frank's scorn. What had made his accusations so compelling? Because here they were not so convinced, she could see that. They thought they were doing all they could. As had she, before Frank caught her.

She asked if their banks really knew who their depositors were, even when their records were not damaged.

They looked puzzled at this.

I ask, Mary said, getting irritated (Frank had been more than irritated), because your banks are often regarded as tax havens, because of their secret accounts. Other countries lose tax money which gets put in secret accounts here. So you're rich in part because you're the bagman for criminals worldwide. A kind of organized theft. People are supposed to trust money, but then a lot of it gets stolen, by the very structure of money itself.

Very unfriendly silence at this.

Mary saw that and pressed harder. She might even have stood up if she could have mustered the energy. She might have shouted. It was time for redemption for Switzerland, she told them flatly, keeping right inside the line of civility. Or maybe on the edge of it. All that stuff you want to forget as if it never happened, the Nazi gold, the Jewish gold, the tax havens for oligarchs and kleptocrats, the secret bank accounts for criminals of all kinds. It's time to end all that. End the secrecy in your banks. Blockchain all your money, and put all your ill-gotten gains to good use. Leverage it for good. Forge an alliance with all the other small prosperous countries that can't save the world by themselves. All of you rich little countries join together, and then join up with India, follow India's lead. Create more carbon coins by way of investments in carbon drawdown. That's the safest currency there is now. Far safer than the Swiss franc, for instance. Stabler. More stable. Your best choice at this point.

Some of them were shaking their heads at this.

You need to join the world! Mary insisted. You've always been Switzerland alone, the neutral one.

We joined the Paris Agreement, one of them objected.

And Interpol, said another.

And the United Nations, said a third.

We've always been engaged, another clarified.

Okay, Mary conceded. But now, join the carbon coin. Gather the rich small nations into a working group. Help get us to the next world system. New metrics, new kinds of value creation. Make the next political economy. Invent post-capitalism! The world needs it, it really has to happen. And you've got to change your banks now anyway, to recover from this attack. So change them for good. Make them better.

Silence.

Mary looked at them. Alp-wasted, yes indeed. A feeling everyone in the room had felt: descend back into the world, after an ascent to that higher realm one encountered in the Alps, an encounter with the sublime— otherworldly, visionary—then afterward exhausted, sun-blasted, clarified. Transparent to the world, lofted into a higher realm. Mary knew she had an intense look she could fix on people, her laser, Martin had called it. She had known it all her life, even in childhood when she had been able to freeze people in place, even her mother. Now she leveled it on these people facing her, and they too went still. Something had set her off—her exploded office, her cramping legs. The Alps. She lasered them.

The Swiss presidents shivered collectively, shaking her off. They looked around the table at each other. Not happy. Not angry. Not panicked. Not dismissive.

They were thinking.

64

Once John Maynard Keynes wrote of "the euthanasia of the rentier class." This is a very provocative, not to say ominous, phrase. Euthanasia was a 1930s euphemism, one of many phrases used in that period to refer to state-sponsored execution of any perceived political rivals. A century later it still sounds deadly.

But it appears that Keynes used the word only to mean something like putting some poor creature out of its misery by a relatively painless procedure. This is true to the original Greek: in literal translation, euthanasia means something like "a good death." Dictionary: "The painless killing of a patient suffering from an incurable and painful disease or in an irreversible coma. The practice is illegal in most countries." Mercy killing is one synonym.

First use in written language is in Suetonius, describing the emperor Augustus's "happy death." First use in describing a medical practice is by Francis Bacon. Relief from suffering was crucial to the early connotation of the word.

Now, it could be argued that the rentier class is not suffering, and in fact is happily engaged in eating up everything. A parasite killing its host by overindulgence is not suffering. In which case, really the rentier class needs to be executed.

But perhaps it is overstating the case to compare the alteration of certain tax and inheritance laws to execution. Although possibly more than changes in tax and inheritance laws would be required to bring an end to the rentier class, often called "the ruling class." Still, the concepts get blurred on a regular basis: a shift in social structures is often regarded as a kind of killing. On the other hand, the name for a certain kind of fiscal decapitation is called taking a haircut, which clarifies just how minor and even trivial are most of the financial limitations on wealth that get considered in the neoliberal hegemony.

Euthanasia: "For the good of the person killed."

Now this is interesting, because capitalism is not a person, and the rentier class as such, though made up of people, is not a person. And as a class it is suffering, one could argue, from guilt, anxiety, depression, shame, a surfeit of everything, a sense of irredeemable criminal culpability, and so on. So to put this class out of its misery would be to relieve the individuals in that class from that horrible psychic burden, and possibly release them to a fuller happier life as guilt-free humans on a planet of equally guilt-free humans.

Capitalism: after a long and vigorous life, now incurable, living in pain. In a coma; become a zombie; without a plan; without any hope of returning to health. So you put it out of its misery.

But what about banishment, what about exile? What about a really short haircut?

Criminals in earlier times could simply be banished, and not allowed to return to their home territory. A punishment which did not match a mortal crime with another mortal crime. A judgment, and a harsh punishment sometimes, but also it could be simply a chance to start over somewhere else, being the same person you were before. It all depended on circumstances.

"The rentier class." Keynes meant by this the people who made money simply by owning something that others needed, and charging for the use of it: this is rent in its economic meaning. Rent goes to people who are not creators of value, but predators on the creation and exchange of value.

So "the euthanasia of the rentier class" was Keynes's way of trying to describe a revolution without revolution, a reform of capitalism in his time, toward whatever subsequent post-capitalist system might follow. It was his evaluation of the parts of the already-existing system, for their possible use value in a future civilization. He did not suggest ending capitalism; just end rent, and rentiers. Although that very well might come to the same thing in the end. He might have been using a euphemism to conceal the shock of his suggestion.

A just civilization of eight billion, in balance with the biosphere's production of the things we need; how would that look? What laws would create it? And how can we get there fast enough to avoid a mass extinction event?

The rentier class will not help in that project. They are not interested in that project. Indeed that project will be forwarded in the face of their vigorous resistance. Over their dead bodies, some of them will say. In which case, euthanasia may be just the thing.

65

We were slaves in that mine. Of course they told us we could leave if we wanted, but we were in the desert backside of Namibia and no way to get away, nowhere to go. We would have had to walk hundreds of kilometers without food or even hats to cover our heads. On the other hand if we stayed we got fed. Two meals a day, ten-hour work day, Sundays off. Hurt bad enough and you could go to the clinic and nurse would look at you, maybe a doctor if it bad enough. Broken bones were set. Dysentery pilled and IVed.

There were about five hundred of us. All men except for some of the nurses and cooks in the mess hall. Most from Namibia, some from Angola and Mozambique and SA and Zimbabwe. Most of us operated machinery or worked on it, but there was some digging too. Digging out machinery after collapses. Bodies too sometimes.

It was a pit mine. Open hole in the earth. Made in the shape of an oval that widened a valley that might have been there before. Roads spiraling down toward hell. Red rock of iron ore, and there were some yellow and greenish patches we were supposed to look for and dig out into separate trucks. We didn't even know what was in those colored rocks. Gold? Uranium? Rare earth, some called it. Not so rare there, but mostly it was red rock. Iron ore, common as dirt, and yet we were slaves to its taking.

Then a bad time came in the kitchens. Less food every week, and the water tasted of iron and made people sick. Finally one of the dorms got up one morning and sat down outside the kitchens. Feed us right or we won't work, they chanted together in a chant. Looking at them sitting there you could see they were desperate. They were scared men. We all saw it and one by one we went and sat down beside them, until every single miner in that mine was sitting there in the morning sun, expecting to get killed. Drones buzzing overhead like flies. It could have just as easy been these flies killed us as anything. The guards with their machine guns just watched us, like we were all waiting for something. Which we were, be it death or whatever. No matter what it was it couldn't be worse than what we were living. So it felt good to sit there that morning in the sun, scared and sweating. We were brothers in that moment in a way we had never been while working.

Finally a man came out with a bullhorn. We knew he was just the voice for a higher power. The mine was owned by Boers from SA, or China or somewhere far away, we heard all kinds of thing. This voice was speaking for them, whoever they were. It said, Get back to work and we'll feed you.

We sat there. Someone yelled, Feed us and we'll get back to work!

So there it stood. We weren't going to budge unless they fed us. They weren't going to feed us unless we budged. We talked it over with the men near us. Everyone agreed; might as well die now and get it over with. We encouraged each other to stick to that. It was that bad. We were scared.

Meanwhile the cloud of drones had been growing overhead, like vultures flocking over some dead body on the veldt. There were more drones up there than there were people on the ground. They hung there more like mosquitoes than vultures, with the same sort of whine as mosquitoes, but bigger. Most about as big as dinner plates, some bigger. Their whine cut at your head and itched in your belly.

Then all the drones or almost all came swooping down fast like hawks and we rose to our feet shouting our dismay and throwing our arms overhead and ducking down and the like. But the drones all went at the guards. They surrounded them dozens to each man, packed around them like coffins made of stacked black buzzing plates. One guard shot his gun and his cloud of drones collapsed on him and felled him to the ground somehow, we couldn't see what they did to him, but he didn't move, and the other guards saw that and no more shooting.

Then the drones spoke together, first in Oshiwambo, then Afrikaans, Swahili, English, Chinese, other languages I couldn't name.

"We are from the African Union Peace and Security Council. This mine has been nationalized by the new Namibian government, and will be protected from now on by AFRIPOL security forces. All countries of the African Union are now united in support of the Africa for Africans program. Representatives from the Namibian government and the AU will arrive shortly to help you with this transition. Please stay seated, or feel free to move into the dining halls or dormitories while the armed personnel here are escorted off the premises."

Which we were happy to do. The guards left on foot down the road. We cheered, we hugged our brothers, we cried for joy. The cooks broke into the pantries and freezers and made us a proper meal, trusting that more food would come in time to make up for the shortage they were creating. Which eventually it did. Troops from the AU arrived that night and declared us liberated. Nationalized, they said. Told us that now we were worker-owners of the mine, if we wanted to stay. If not, free to get on buses and ride away.

Some of us left as soon as the buses showed. Most of us stayed. We figured we could leave later if we wanted. But being an owner of the mine sounded interesting. We wanted to know what that meant. Like sweat equity, some said. Sweat equity! Hell, we had blood equity in that mine.

66

You think your birth was hard—my mom exploded! Literally, yes, in that when she went supernova the heat of the detonation exceeded a hundred megakelvins and in that pressure three helium nuclei stripped of their two electrons were crushed together and there I was, as elegant as anything in the universe: carbon, the king of the elements, sweetly six-sided and tetravalent, able to bond with the atoms of my kind in several different ways, and to compound with other atoms in almost countless ways, being so friendly. So, boom! and there I was, flying across the universe. My particular neighborhood was your Milky Way, and I flew right into the knot of dust that was swirling down into Sol, where I could very easily have been roasted or crushed into something else entirely, but happily for me I fell into a swirl of dust that was coming together around ninety million miles from mighty Sol, and not too much time later I was part of a rocky planetesimal.

Earth, you're probably guessing, since we're here now, but actually I first joined the Mars-sized rock coalescing at the Lagrange 5 point to Earth, a rock which now gets called Theia. So I was there for that big collision when Theia hit Gaia at speed, and they merged and tossed out a spray that quickly became the moon. A big bang! Although not compared to the real Big Bang, of course. And with that I found myself inside the hot new Earth, but in the mantle very near the surface, luckily for me or I wouldn't be talking to you now. That was it for my catastrophic childhood and youth, everything since has been fairly sedate and what you might call adult.

Well, but I forget my escape to the surface. That was pretty dramatic too. I came out in a volcanic eruption at a mid-ocean ridge between Pangea and I forget what land mass, they go away pretty fast. Hot lava sprayed into the sky, cooling almost instantly. A few million years exposed to the photon rain of sunlight softened me up, it was like getting a sunburn and I was part of the dead skin about to get sloughed off. Fine, I was ready, a million years is a long time, not to mention fifty million years; but the question was, which atoms would I join to effect my escape? I wanted to be eaten by a dinosaur, Jurassic era, and in those days it wasn't that hard—photons banged me, my four exposed connector electrons were all quivering tetravalently, hoping for a pick-up, and as it so often happens, I got interest from two suitors at the same time! and wham bang, I had been stuck simultaneously to two oxygen atoms, and I was in a marriage very convenient indeed, as carbon dioxide.

We made a good team. Life got busy. By flying low we kept getting picked up by plants. They would suck us in, and zoop, gurgle, I was part of a leaf, a twig, a tree trunk. I joined a proto-sequoia, that was a long date, then a fern, got eaten by an allosaur, pooped by same allosaur, yes I was a piece of shit then, and have been many times since, but bacteria love to eat shit, and quickly enough I ran into another pair of oxygen atoms and off again. But then, disaster: I was caught underwater in a muddy clutch of my fellow carbon atoms, and down we went back into the Earth, crushed there to graphite, in this case a seam of coal, where I spent many millions of years. Could even have been sucked down tectonically and crushed to diamond, and thus stuck forever in one small town of everyone-the-same, latticed in a veritable jail for all time, meaning really till the burning up of the Earth when the sun goes large, that would be my welcome release from that fate, but in this case I got lucky; my seam was mined by humans and burnt in a furnace, around the year 1634. Freedom! Back into the sky, and how I loved that. I like variety. So back to the sky, and hurray for organic chemistry, I was this and that, pangolin and rice stalk, mosquito and frog, frog poop and bacteria, then back to the sky yet again, hurrah!

There's that moment that comes when the water molecules drifting around in the air constellate on a tiny speck of dust and become a rain drop and begin their dive to Earth, and you can latch onto that, just get smacked by a downward droplet and join those happy people, your oxygen mates singing hi to the oxygen atoms in their hydrogen-twin marriages, trios are the best, everyone partying for the time of the dive. You lose the sense of gravity pulling on you when you fall at terminal velocity for your droplet, in fact sometimes you get hung up in a cloud or a mist or a fog and it's just delightful, a delicious no-g sensation, I would suppose that it might be like what you would call orgasm. Bonding, sure, that can be good or bad, but floating in the sky in an orgasmic cloud, wow.

But eventually the droplets are likely to coalesce until you are pelting down again. Snow is fun, sleet even more so. And then you crash onto Earth and things start again. Who would I join this time?

Well, shit—not this time! Turns out that people in Canada had begun to deal with asbestos mine tailings by feeding the toxic rubble into the tailing pools that form in and next to mine pits, then adding some local cyanobacteria. These cyanobacteria grabbed me and then bonded with the asbestos dust, and together we clumped into hydromagnesite, a form of magnesium carbonate. These local kidnappers were all happy to have locked me and many of my mates down again, and the asbestos too, but when you've floated in fog, and body-surfed through an alimentary tract, sitting there in a rock is boring as hell. My only hope is that I'll get ground up and used as a rock-climber's hand powder, that's about all magnesium carbonate is good for. Maybe I'll end up in the powder pouch of some awesome cliff climber, that would be exciting, but for now I'm stuck. Oh well, time for a nap.

67

Taxes are interesting. They are one way governments guide a society and fund governmental activities, more the former than the latter. They are as old as civilization. An ancient manifestation of the power of the state. It's possible that both debt and money were invented in the earliest cities, specifically in order to enable and regularize taxation. Both of them being forms of IOU.

Progressive taxation refers to the idea that the more citizens own, the higher their rate of taxation. A regressive tax takes more, proportionally to individual wealth, from the poorest.

Income taxes tax individual or corporate annual income, so these incomes are often manipulated by those earning them to appear lower than they really are. Various deferments and reinvestments and other methods slip money through tax loopholes, and tax havens are places where money, if it can be moved there before the annual accounting takes place, will not be taxed by the haven's host, or will be taxed much less. So a progressive income tax can become quite ineffective as such. Vigilance in application is required.

At certain moments in history excess personal wealth was frowned on, and the scale of progressive taxation grew quite steep. In the early 1950s, a time when many people felt that wealthy individuals had helped to cause and then profit from World War Two, the top tax bracket in the United States had earners paying in income tax 91 percent of all earnings over $400,000 (current value, four million dollars). This rate was approved by a Republican Congress and a Republican president, Dwight D. Eisenhower, a man who had commanded the Allied forces in the war, and had seen the death and destruction first hand, including the concentration camps. Later these top rates were lowered, over and over, until in the neoliberal period top rates were more like 20 or 30 percent. In those decades the tax loopholes and dodges and deferments and havens also grew hugely, so these already low percentages are actually inflated compared to the real amounts collected. Income taxes thus were made much less progressive; this was a feature of the neoliberal period, part of the larger campaign favoring private over public, rich over poor.

Capital asset taxes, sometimes called Piketty taxes, tax the assessed value of whomever or whatever is being taxed. Usually these have been applied to corporations, but the same kind of tax can be applied to individuals. France taxed its corporations one percent of their assessed value per year, and if applied globally, the effects of such a tax could be very significant. These asset taxes too could be made progressive, such that the larger the corporation, or the assessed value of any asset, such as property, the higher the annual tax taken from it. If set steeply enough, a progressive tax of this sort would quickly cause big corporations to break up into smaller companies, to decrease their tax rate.

Land taxes, sometimes called Georgist taxes, after an economist named Henry George, are taxes on property, meaning in this case specifically land itself as an asset. Again, these land taxes could be set progressively such that larger properties, or more valuable properties by way of location, or land not lived on by its owner, got taxed at a higher rate. As a great deal of profit and liquid assets more generally get turned into real estate as soon as possible, usually to own something tangible, the value of which is likely to rise over time, or at least not disappear entirely in a bubble's burst, a land tax properly designed could again swiftly redistribute land ownership more widely, while quickly swelling government coffers in order to pay for public work, thus reducing economic inequality.

A tax on burning fossil carbon, which could be called not a tax but rather *paying the true cost*, could be set progressively, or offset by feebates, to avoid harming the poorest who burn less carbon but also need to burn what they burn to live. A fossil carbon tax set high enough would create a strong incentive to quit burning it. It could be set quite high, and on a schedule to go even higher over time, which would increase the incentive to quit burning it. Tax rates on the largest uses could be made prohibitive, in the sense of blocking all chance of profits being made from any derivative effects of these burns.

If all fiat money everywhere went digital and got recorded in block-chains, so that its location and transaction history could be traced and seen by all, then illegal tax dodges could be driven into non-existence by sanction, embargo, seizure, and erasure.

Thus it will be seen that a fully considered and vigorous tax regime, using digital trackable currencies and instituted by all the nations on Earth by way of an international treaty brokered by the UN or the World Bank or some other international organization, could quickly stimulate rapid change in behavior and in wealth distribution. Some might even call it revolutionary change. And of course taxes are a legal instrument with a pedigree as long as civilization itself, its rates decided by legislatures and backed by the full force of the state, meaning ultimately the judiciary, police, and military. Taxes are legal, in other words, and accepted in principle and used by all modern societies. So, targeted changes to the tax laws—would that really be a revolution, if it were to happen?

It would be interesting to try it and see.

68

Mary was flown back to Zurich in a military helicopter. They landed at Kloten and she was taken into town in a black van like the one she had left Zurich in. She sat next to Priska, watched their driver take the usual route into the city. But then where?

Home, as it turned out. Hochstrasse, stopping curbside in front of her apartment building. "Here?" she asked.

"Just to get some of your stuff," Priska said. "They don't think it's a good idea for you to live here anymore, I'm sorry."

"Where, then?"

"We have a new safe house up the hill," Priska said. "We would like you to stay there. Once the situation becomes a little more clear, you can move back here. If that's all right with you."

Mary didn't reply. She wanted to be at home in her place, but also the idea made her nervous. Who was watching, if anyone? And why?

She went in and packed a couple of big suitcases they provided. As she did she glanced around the place. She had lived here fourteen years. The Bonnard prints on the walls, the white kitchen; they looked like a museum recreation. That stage of her life was over, this was like walking around in a dream. Her legs were still throbbing. She needed to sleep. Shower and bed, please. But not here.

They carried the suitcases for her, down the stairs and into the street, into the back of the van. Then off east, past the little trattoria she had sat in on so many nights, reading as she ate. Farther up the Zuriberg, into the stolid residential neighborhood on the side of the hill. These big old urban houses were worth millions of francs each, they gleamed with the finish of all that money, unremarkable boxes though they were. The van turned into the gated driveway of one of them, the driveway just a concrete pad the size of one vehicle, in the middle of a garden behind a tall white plastered wall topped with broken green glass shards, an unexpected touch of evil in all this bourgeois conformity. A gate closed off the driveway and made it a compound. Her new home. She stifled a groan, kept her eyes from rolling. She could still walk to work from here, if they would let her.

Which they did. She could call and within minutes a little club of them would be gathered at the walk-in gate to escort her down the hill to Hochstrasse and the ministry offices, their blown-up building being rebuilt, the rest already re-occupied. She was surprised that the Swiss security people felt it was safe to go back, but she was assured that the area was now surveiled in ways that made it safer there than anywhere else. They couldn't function from hiding, and it was important to show the world that Switzerland and the UN considered the ministry to be a crucial agency. Also that terrorism couldn't change the momentum of history. They were going to defend that principle, and she was one of the living avatars of history in their time.

Or just bait, Mary thought. Bait in their trap, perhaps. But then again her team was reassembled and back in their offices or jammed into replacement offices, doing their familiar work. Possibly the Swiss had caught the people who had attacked them and thus eliminated the danger. Their banks were said to be back online and functioning as before; whether there were structural changes included in the reboot wasn't clear yet. So if those assailants had been caught or rendered inoperative somehow, possibly they were safe now. There couldn't be that many people in the world who felt a toothless UN agency was worth attacking. Although the Paris Agreement had enemies, sure. It could be that the entire military apparatus of some vicious petro-state was now aimed right at her, as the symbol of all that was going wrong for them. It would be great to take some of those petro-states down, somehow. Jail their leaders or the like.

But thinking of prison reminded her of Frank. Did she want to see him? Alas, she did. Possibly something in her wanted to make sure he was still locked up; maybe she was still afraid of the idea of him at large. But also, given that he was certainly going to be there, seeing him had to be more than just that. It felt like some kind of duty. Which feeling also had its interests. It was impossible to deny that he had caught her interest.

*   *   *

Downtown on a tram like any ordinary person. This was all right with her minders, as long as one or more of them accompanied her. She glanced at the people in her tram car, wondering who they were. None of them looked likely. She recalled a line from a children's book she had loved, something like, *If you want to claim to be our queen, while yet always invisible and unknown to us, you are welcome to the task.* It was the same now; if you're going to guard me but I don't see you, fine, do it.

Down at Hauptbahnhof she got off and walked the narrow downtown pedestrian streets to the Gefängnis. So characteristic of the Swiss to keep the old jail downtown. Why proclaim one part of the city to be more valuable than any other? The whole point of a city was to smoosh the whole society together and watch it function anyway, daily life some kind of flaneur's bricolage. An agglomeration, as their urban designers called it, unembarrassed by the ugliness of the word in English.

She checked into the prison without fuss and went to Frank's dorm. He was in the living area there, reading a book. He looked up and his eyebrows rose.

"I thought you were run out of town."

"I was. They let me come back." She sat down on a couch across from him. "What are you reading?"

He showed her the cover; an Inspector Maigret omnibus. In a dark world, she thought, a place of safety. Diagnose the evil. Everyone should have a Mrs. Maigret.

"How's it going?" she asked, wondering as always why she had come, what she could say.

"It's okay," he said. "They let me out during the day. I work at the same place I was working before."

"The refugee center?"

"Yeah. They're expanding again. I've been there so long I've become a fount of institutional wisdom."

"I doubt that."

He laughed, surprised. "How come?"

"We're in Switzerland. The institutional wisdom all gets written down."

"You would think. Anyway I'm there."

"Feeding people?"

"Most of the time I'm in processing."

"What does that mean?"

"People arrive and we try to figure out where they got Dublined, if anywhere."

"It must have been somewhere, right? No coastlines in this Bohemia."

He shook his head. "Smugglers. They get to Greece or the Balkans, they don't want to get registered there. Switzerland has a reputation for quality, in this as in everything."

"Despite the kind of attack that got you arrested."

"But it's worse everywhere else. So they want to come here and then get Dublined. A lot of them have mangled their fingerprints so you can't ID them that way."

"Which means they probably got Dublined somewhere else."

"Sure."

"So what do you do?"

"If we find out they were tagged elsewhere we have to send them back there. So we don't try very hard. Most of them we can register here, fingerprints or not. They use retinal scans here. Then we try to find room for them in camps that already have people from their country."

"Where are they from?"

"Everywhere."

"Are they climate, political, economic?"

"You can't tell the difference anymore. If you ever could."

"So you think you're getting real refugees."

He gave her a look. "No one would leave home if they didn't have to."

"Okay, so you get them registered here, then you send them to a camp where they'll have people from their country?"

"We try."

"But you don't visit the camps?"

"No. I have to be back here by eight every night."

"Well, but you can get almost anywhere in Switzerland and back by eight."

"That's true. But I'm not supposed to leave the canton."

"Doesn't Zurich have any camps?"

"Yes, and I've gone to them. There's a big one out beyond Winterthur, in an old airport or something. Twenty thousand people there. I see how they're doing. Help in the kitchen. That's what I like. Although I can help in the kitchen here too."

"Does that get you time off?"

"I think so. I'm not so worried about that anymore. I don't have anywhere to go."

She regarded him for a while.

"Did you ever spend any time in the Alps?" she asked at last.

"A little bit."

"They're amazing."

He nodded. "They look steep."

"They definitely are." She told him about her crossing of the Fründenjoch. He seemed interested in her tale, which to her was about the Alps, but when she was done he said, "So who do you think was after you?"

"I don't know."

"Who loses the most when your ministry does well?"

She shrugged. "Oil companies? Billionaires? Petro-states?"

"That's not a giant list of suspects."

"I don't know," she said. "It could be anyone, I suppose. Maybe it was some individual or small group that they've caught. It would make sense if it was just some nut who thought we were the important ones. When actually we're just a cog in a giant machine."

"But they might have thought you were the clutch."

"What does a clutch do again?"

He almost smiled, which was his smile. "It clutches. It's where the engine connects to the wheels."

"Ah. Well, I don't know."

"Wouldn't your guards tell you if you asked them?"

"I'm not sure. We're not that close. I mean, their job is to protect me. They might think I'm safer not knowing."

"Why would that be?"

"I don't know. The Swiss banks were attacked too. So they're keeping pretty quiet about their counterattacks."

He was smiling his little almost-smile. He gestured at his book. "You need an Inspector Maigret. He liked to explain things to the people he saved."

"Or if he thought it would make them reveal themselves to be the actual criminal."

"True. You've read them?"

"A few. They're a bit too dark for me. The crimes are too real."

"People are twisty."

"They are."

"So you need Inspector Maigret."

"And you need the Alps."

A woman and girl entered the room, and Frank looked startled. "Oh hi," he said, then looked at Mary, and back at the two who had just entered. He didn't know what to say, Mary saw. Nonplussed; confused.

He stood up. "This is Mary Murphy," he told the two. Then to her: "These are my family."

Mary squinted. The woman's mouth had tightened at the corners.

The girl, about ten or eleven years old, broke to Frank and lifted the tension. "Jake!"

"Hey Hiba. How are you."

"Good." The girl gave Frank a hug. Frank leaned awkwardly into it, looked over the girl's shoulder at the woman.

"How did you get here?"

"We took the train."

"Where do you live?" Mary asked her.

"We stay at the refugee camp outside Bern."

"Ah. How is that?"

She shrugged. She was looking at Frank.

"Listen, I'll let you all catch up," Mary said, standing and holding a hand out to block Frank's objection. "It's okay, I should be going anyway. And I'll come back soon."

"All right," he said, distracted still. "Thanks for dropping by."

69

In Saudi Arabia, during the height of the hajj, what appeared to be a coup by the military resulted in the deaths of an unknown number of Saudi princes. Reports ranged from twenty to fifty, but no one knew for sure. The king was in New York at the time, and was said to be in hiding and not planning on going home. He called on the world to support his legitimate government, and a few governments did, but none of them offered active help. The United States offered asylum. The new government had the backing of most of the people in the country, as far as anyone could tell; with the hajj in disarray and two million Muslims either trying to complete their pilgrimage or get home, confusion on the Arabian peninsula was general. The only thing that was obvious in that first month was that no one outside the Arabian peninsula knew very much about what had really been happening in Saudi Arabia. Which was now to be called simply Arabia, the new government told the world. The Sauds were done.

The other Sunni national governments were cautiously approving or disapproving of this removal, reserving their sharpest criticism for the disruption of the hajj. No one had liked the Saud family, it now appeared, but the ramifications of this were unknown, and potentially volatile throughout the region. The Shiite nations openly applauded the coup. Other governments around the world stayed reticent. They seemed to be trying to calculate what the change meant and what the new government would do, especially with its immense reserves of oil. The formerly implicit was now uncomfortably obvious; no one had cared about these people, only about their oil.

Then word came from Riyadh that Arabians respected the pressing need to decarbonize the world's economy, and intended to use their oil only for plastics manufacture and other non-combustible uses. The new

Arabian government therefore made an immediate claim to the CCCB, the Climate Coalition of Central Banks, which recently had been established specifically to administer the carbon coin, saying that their full conversion to solar power, to begin immediately, and their refusal to sell their oil reserves for burning, deserved compensation in the form of the CCCB's newly created carbon coins, sometimes called *carboni*. At the rate of one coin per ton of secured carbon, the Arabic claim was estimated at about a trillion carbon coins; at current exchange rates this came to several trillion US dollars, which would make Arabia instantly one of the richest countries on Earth, at least in terms of national bank assets. If the present currency exchange rates held, they would be wealthier than if they had sold their oil for burning.

After a period of delay the CCCB agreed to this exchange, but stipulated it was to be paid out on a schedule pegged to how fast the Arabian oil would have been produced and burned in that now deactivated alternate history, only a bit front-loaded and accelerated to reward Arabia for doing the right thing for the planet and human civilization. Meanwhile they could leverage this assured income stream, which they did. They accepted the deal and went to work.

This sudden loss of supply sent oil prices and oil futures sharply up. Oil was rarer now, therefore more expensive, which meant that clean renewable energy was now cheaper than oil by an even larger margin than before. And as the new carbon taxes being levied in every country in compliance with the latest commitments to the Paris Agreement, made at the COP43 meeting, were also scheduled to rise year by year by an increasing percentage, price signals were now all pointing toward clean renewables as the cheapest way to power the world. The social cost of carbon was finally getting injected into the price of fossil fuels, and that old saying, ridiculed by the fossil fuels industry for decades, was suddenly becoming the obvious thing, as being the most profitable or least unprofitable thing:

Keep it in the ground.

Soon after this, Brazil's government entered another paroxysm of corruption charges, leading to the resignation of the right-wing president and then his arrest. Quickly there followed the triumphant return of the so-called Lula Left, now also called Clean Brazil, with a promise of clean government representing the entire populace, also an end to oil sales, clearly modeled on Arabia's move; also the full protection and caretaking of the Amazon basin's rainforest. They claimed compensation for this last policy also, to be paid in more of the CCCB's carbon coins. The CCCB agreed to that, and by way of Rebecca Tallhorse's negotiations, a lot of carbon coins were also given immediately to the indigenous groups of the Amazon, who had been keeping the rainforest's carbon sequestered for centuries. That act of climate justice along with newly scheduled payments to the Brazilian federal government meant a few trillion more carbon coins were added to the general circulation, and now mainstream economists everywhere were fearful that this sudden flood of new currency was going to cause massive deflation. Or perhaps inflation: macroeconomics was no longer so very clear on the ultimate effects of quantitative easing, given that the evidence from the past half century could be interpreted either way. That this debate was a clear sign that macroeconomics as a field was ideological to the point of astrology was often asserted by people in all the other social sciences, but economists were still very skilled at ignoring outside criticisms of their field, and now they forged on contradicting themselves as confidently as ever. Some of them were asserting that the carbon coins were merely replacements for petrodollars, which had always been pulled out of the ground like rabbits out of a hat, having not existed before the oil was pumped up and sold. Pulled out of the ground, pulled out of a hat; was there any real difference between petrodollars and carbon coins, these economists asked?

There was, other economists insisted. Petrodollars had first been real pre-existing money, paid for a commodity turned into electricity or physical movement, thus turned into economic activity; carbon coins, on the other hand, were created by the actual *removal* of that same electricity and transport potential from the world, and thus from the Gross World Product. Petrodollars thus fueled GWP; carbon coins depleted GWP. They were functionally opposite.

Then again, still others argued, that absence of carbon burn, and even the resulting lowering of GWP itself, would save some difficult-to-calculate but real amounts of damage to the biosphere, also the necessary mitigation and remediation and ecological restoration work and insurance pay-outs that would have inevitably followed the carbon burn; and these costs could be calculated; and when they were, ultimately it seemed to amount to almost a wash, petrodollars or carbon coins, and thus the whole thing was a tempest in a teapot, a nothing in economic terms.

So: either a huge boon, a complete calamity, or a non-event. Thus the economists, faced with explaining the biggest economic event of their time. What a science! They worked all over the world (including in the Ministry for the Future's offices) trying to calculate the gains and losses of this event in some way that could be entered into a single balance sheet and defended. But it couldn't be done, except in ways so filled with assumptions that each estimate was revealed to be an ideological statement of the viewer's priorities and values. A speculative fiction.

Some pointed out that this had always been true of any economic analysis or forecast ever made. In this case, these people insisted, please go back to the basics. Here's the true economy, these people said: since the Earth's biosphere was the only one available to humanity, and its healthy function absolutely necessary to humanity's existence, its worth to people was a kind of existential infinity. Gauging the price of saving the biosphere's functions against the cost of losing them would therefore always be impossible. Macroeconomics had thus long ago entered a zone of confusion, either early in the century or perhaps from the moment of its birth, and now was revealed for the pseudoscience it had always been.

The upshot was that they had no real way of knowing what the global economy was doing now, or what would happen if the central banks continued to fulfill their pledge to create and underwrite a massive infusion of new money into the world. Carbon quantitative easing, CQE, was a huge multi-variant experiment in social engineering.

This was volatility indeed! To use not just the financial term, but simply common human language. It was without doubt a volatile situation. But recall that the financial markets of that time loved stock price volatility, as it made money for financiers no matter what happened, their having gone both long and short on everything. These were not economists, but speculators. Finance in that late moment of capitalism's exhaustion meant gamblers, sure, but more than that, the casino in which people gambled. And the house always won. And carbon coins were the best opportunity to go long ever created. Almost a kind of sure thing. So in certain respects, the craziness of the time was simply good for investors. Those who had shorted fossil fuels and gone long on clean renewables were now making fortunes; and fortunes require reinvestment to actually be fortunes. Growth! Growth!

This was the world's current reigning religion, it had to be admitted: growth. It was a kind of existential assumption, as if civilization were a kind of cancer and them all therefore committed to growth as their particular deadly form of life.

But this time, growth might be reconfiguring itself as the growth of some kind of safety. Call it involution, or sophistication; improvement; degrowth; growth of some kind of goodness. A sane response to danger—now understood as a very high-return investment strategy! Who knew?

Really, no one knew. The remaining big petro-states each regarded the new situation uneasily, or even in a panic. Together they sat on fossil carbon reserves that at current market prices ranged into the hundreds of trillions. These reserves could easily become stranded assets in the very near future, in fact it looked a bit like a financial bubble starting to burst. In that context it made sense to sell as much of the product as possible before prices collapsed completely. But if everyone holds a fire sale at once, who's going to buy? The small prosperous countries had clean renewable energy already. The shipping industries, under the duress of their ships being sunk if they didn't shift, had shifted already to wind and electrics and hydrogen. Aviation, under the same annihilating pressure, was shifting to electric planes, and mainly, airships. Ground transport was going entirely electric, and where it still used liquid fuels, was completely committed to renewable biofuels that bypassed fossil sources.

Power plants were therefore the last interested customers, but even there, solar was cheaper and batteries getting better, and non-battery energy storage by way of water levels or salt temperatures or flywheels or air pressure were all becoming more and more robust. The developed petro-nations therefore tried to sell their oil to the developing non-petro nations, at a big discount. The developing petro-nations decided to power themselves with their own oil assets, and sold to their fellow developing nations-without-petro at even bigger discounts. These non-petro developing nations therefore were the last big carbon burners of note, and they were significant. But India had already shown what could be done by making their pivot to clean renewables after the heat wave. And China was leading the world in solar panel production. They were still selling and shipping their coal, and Japan was still importing it and burning it, among others. Russia and Australia were still exporting coal where they could. Despite all the sudden shifts, carbon was still being burned. Oh yes; to the tune of twenty gigatons a year. Did you think that because Arabia was virtuous, there would be no more cakes and ale? In fact, there was no quick end in sight. The image of a bubble bursting was apt in some respects, illusory in others. People still needed electricity and transport. Money still ruled, fire sales still sold carbon. The bubble that was bursting might still be the biosphere itself. The battle for the fate of the Earth continued.

Russia kept selling its oil and gas, which was a good thing for Europe, as much of Europe was heated in winter by Russian gas. As was made clear when the pipelines were bombed in the coldest part of that winter.

They were also selling their pebble-mob missiles. Either that or else they had sold, or given, or lost to theft or espionage, the plans for building pebble mobs. And in fact these missiles were not so hard to reverse engineer that they weren't quickly becoming a part of all major militaries. Even part of some private armed forces' arsenals, it appeared, which had to have purchased such high-tech devices from some major country. It was in fact somewhat terrifying that anyone at all owned such missiles.

They had been introduced by the Russians in the 2020s, and spread rapidly after that. They spread faster than the spread of the implications of their existence. Nation-states kept spending billions that could have been used elsewhere on navies and air forces and military bases, none of which could be defended against these new weapons. They were more powerful than the atomic bomb, in this very particular sense: you could use them. And they couldn't be stopped. That was the main thing that was not being understood, either accidentally or on purpose, so that people wouldn't have to change their armaments or their purchase orders: these missiles couldn't be stopped. They were small, they launched from mobile launchers, they came from all directions in a coordinated attack in which they only congregated at their target in the last few seconds of their flights.

They did not give off radioactive signals, and thus could be hidden until the moment of launch. And they were relatively cheap.

After you launched them, they flew at about a thousand miles an hour. That and the fact they only coalesced to their mob moment in the last seconds before impact was enough to forestall any realistic defense against them. They were lethally explosive versions of the drones that had brought down all the planes on Crash Day.

Aircraft carriers? Sunk. Bombers? Blown out of the sky. An oil tanker, boom, sunk in ten minutes. One of America's eight hundred military bases around the world, shattered. Death and chaos, and no one findable to blame.

The war on terror? It lost.

Either everyone's happy or no one is safe. But we're never happy. So we'll never be safe.

Or put it this way: Either every culture is respected, or no one is safe. Either everyone has dignity or no one has it.

Because why? Because this:

A private jet owned by a rich man—boom.

A coal-fired power plant in China—boom.

A cement factory in Turkey, boom. A mine in Angola, boom. A yacht in the Aegean, boom. A police station in Egypt, boom. The Hotel Belvedere in Davos, boom. An oil executive walking down the street, boom. The Ministry for the Future's offices—boom.

What people then had to consider was that this list of targets could be greatly extended. The US Capitol, various houses of parliament, the Kaaba in Mecca, the Forbidden Palace in Beijing, the Taj Mahal—and so on.

No place on Earth was safe.

Meetings at Interpol and many other agencies concerned with global security were inconclusive as to the source or sources of these attacks when they happened. Although, as with the nerve gas made by the Russians and used to kill Russian dissidents in Britain, these pebble mobs were definitely complex military devices, not something you could cook up in your garage. They were nation-state devices, in effect, made for fairly big and sophisticated militaries by fairly advanced aerospace and computer companies, then sold or given to smaller actors.

So, after the Interpol meetings, a rumor began to circulate that it was in Russia where the Arabian coup against the Saudi royal family had been planned. But wait, why would the Russians be party to that? Because with Arabian oil off the table, and then Brazilian oil too, Russian oil was that much more valuable. But this was speculation only. The Russian government denied all these rumors and identified them as part of the ongoing anti-Russian campaign common in the West and elsewhere. Nothing to it. Each nation responsible for its own security. Russia was a keeper of all its treaty obligations and a force for stability in the world. Pebble mobs might even be a force for good, because now war was rendered impossible. It was mutual assured destruction, not of civilian populations, but of war machinery. An end to the twentieth-century concept of total war, a return to the focus on military-against-military that had characterized armed conflict before the breakdown of civilized norms established at Westphalia in 1648, then forgotten in the twentieth century. Now back again.

But not really. Because anything could be targeted. So it was not really military-against-military, as Russia claimed, but anyone-against-anyone. If they could get hold of one of these assemblies.

A heat wave hit Arizona, then New Mexico and west Texas, then east Texas, then Mississippi and Alabama and Georgia and the Florida panhandle. For a week the temperature/humidity index hovered around wet-bulb 35, with temperatures around 110 F and humidity 60 percent. For the most part electrical power remained functioning, and people stayed inside air-conditioned buildings; if they didn't have air conditioning themselves, or were nervous about losing it, they congregated in public buildings where it existed. All fine, but then the heat wave was met by a high-pressure cell coming up from the Caribbean, and the so-called double wave created wet-bulb 38s, seriously fatal. Demand for power grew until there were power failures and brownouts, and even though some of these were planned to avoid blackouts, and all of them eventually contained, and relatively brief, they still exposed millions to fatal levels of heat. Somewhere between two and three hundred thousand people died in a single day during that heat wave.

Later that figure was revised, as it was discovered over time that the

previous decadal census had significantly undercounted the at-risk population. But no matter the exact number, it was huge. Not as large as
the great Indian heat wave had been, but this time it was Americans, in
America. That fact made a difference, especially to Americans.

Although still, in the months that followed, people's biases emerged. It
was the South where it had happened. It was mostly poor people, in particular poor people of color. It couldn't happen in the North. It couldn't
happen to prosperous white people. And so on. Arizona's part in it was
forgotten, except in Arizona.

This was yet another manifestation of racism and contempt for the
South, yes, but also of a universal cognitive disability, in that people had a
very hard time imagining that catastrophe could happen to them, until it
did. So until the climate was actually killing them, people had a tendency
to deny it could happen. To others, yes; to them, no. This was a cognitive error that, like most cognitive errors, kept happening even when you
knew of its existence and prevalence. It was some kind of evolutionary
survival mechanism, some speculated, a way to help people carry on even
when it was pointless to carry on. People living just twenty miles from a
town flattened by a tornado in Ohio would claim that the flattened town
was in the tornado track and they were not, so it would never happen to
them. The following week they might get killed, in the event itself surprised and feeling that this was an unprecedented freak occurrence, but
meanwhile, until then, they swore it couldn't happen. That's how people
were, and even the torching of the South didn't change it.

CO_2 levels that year were around 470 parts per million.

They had gone from driving a car to hanging on to a tiger's tail. Hanging on for dear life.

70

COP meetings of the Paris Climate Agreement kept happening every year, despite the increasing sense of irrelevance everyone in attendance felt in the face of the world's ever-widening disasters. It was clear that the worsening situation vis-à-vis carbon in the atmosphere meant that many of the developing countries, those which could least afford to cope with climate catastrophes, were now being struck by weather disasters on a regular basis, and these were perhaps the prime drivers in the human conflicts now breaking out everywhere. For those who held to that perception of ultimate causes, the Paris Agreement remained something to hold on to, however weak it was beginning to look relative to the crisis.

The UN's climate negotiations had always made a strong distinction between developed and developing nations, with lists of each specified, and repeated injunctions made that developed nations were to do more to mitigate climate problems than developing nations could. Much of this call for "climate equity" was spelled out in Article 2 of the Paris Agreement. Clause 2 of Article 2 states, "This Agreement will be implemented to reflect equity and the principle of common but differentiated responsibilities and respective capabilities, in light of different national circumstances." Article 9's clause 1 repeats this principle: developed nations are to assist developing nations, they can and should do more than developing nations.

These were crucial clauses in the Agreement. The text of these articles and their clauses had been fought over sentence by sentence, phrase by phrase, word by word. The delegates who had pushed hardest for the inclusion of these articles had given their all, they had spent years of their lives working for them. On the subway rides during summit meetings they compared notes on divorces, bankruptcies, broken career paths, stress-related illnesses, and all the other personal costs accrued by throwing themselves so hard into this cause.

Were they fools to have tried so hard for words, in a world careening toward catastrophe? Were they fools to keep on trying? Words are gossamer in a world of granite. There weren't even any mechanisms for enforcement of these so carefully worded injunctions; they were notional only, the international order of governance being a matter of nations volunteering to do things. And then when they didn't do them, ignoring the existence of their own promises. There was no judge, no sheriff, no jail. No sanctions at all.

But what else did they have? The world runs by laws and treaties, or so it sometimes seems; so one can hope; the granite of the careening world, held in gossamer nets. And if one were to argue that the world actually runs by way of guns in your face, as Mao so trenchantly pointed out, still, the guns often get aimed by way of laws and treaties. If you give up on sentences you end up in a world of gangsters and thieves and naked force, hauled into the street at night to be clubbed or shot or jailed.

So the people who fought for sentences, for the precise wording to be included in treaties, were doing the best they could think of to avoid that world of bare force and murder in the night. They were doing the best they could with what they had.

Now, as the situation continued to deteriorate, there were delegates at every annual Conference of the Parties who kept on focusing on words and phrases. Many of them were now arguing that all the young people on Earth, and all the generations of humans in the centuries to come, and all their cousin creatures on the planet who could never speak for themselves, especially in court—all these living beings added up to something like a poor and vulnerable developing nation, a huge one, appearing inexorably over the horizon of time. These new citizens were young and weak, in many cases utterly helpless. And yet they had rights too, or should have rights; and under the Paris Agreement's equity clauses, which every nation had signed, one could argue that they had rights equivalent to those of a developing nation. And without quick and massive efforts from the Annex One nations, meaning the developed world, the "old rich" countries, that giant new developing nation's development, even its very existence, looked less and less likely. So the COP meetings had to keep insisting on equity as a fundamental value and policy. Which meant that support for the Subsidiary Body popularly known as the Ministry for the Future should continue to be supported in full.

71

Notes taken for Badim again, regular executive meeting with Mary and leadership group. The usual summary of points, plus impressions (B wants more). Will clean this up later. Typing as they speak. I keep forgetting to go to the bathroom first.

Badim next to Mary, silent, distracted.

Tatiana Voznesenskaya next to him, also distracted. She's being sued back home, can't go to Russia safely, here no longer living in same place. Her report: Lawsuits in over a hundred countries. Also joined defense of nearly four hundred groups being sued for doing good things. T not happy.

Imbeni: MftF being attacked might have caused others to step up. African Union backing all nationalizations in Africa, means a united front toward China, World Bank, all outside forces. Africa for Africans the fastest growing party in every country there. Pressure on Nigeria in particular to claim the carbon coin like Arabia and Brazil. Doing so could fund a lot of other things. Basic infrastructure and education, then more. Possession of oil now seen as a curse to be exorcised. Chance here to help leverage good change. Africa led by Africans.

Bob, Adele, Estevan in a team report. Mostly Antarctica. Test projects pumping water out from under glaciers getting positive results. Pine Island Glacier slowed from hundreds of meters a year to tens of meters. We should help scale this up, support it fully. 60 Antarctic glaciers, 15 Greenland glaciers. Big push, but amazing cost-benefit. Bang for buck. Let's do it. M nodding.

Kaming not so cheery. Rate of extinctions still rising. Sumatran tiger, northern white rhinoceros, more river dolphins, these just the latest charismatics confirmed to be gone in wild. Orangutans next, along with 350 other mammal species in red zone danger of extinction.

Indra: Direct air capture now more powerful and less expensive, need to scale up appropriately, then find places to sequester billions of tons of dry ice. Progress being made in this technology means more should be invested in it. Like most of the MftF budget, and a huge dose of carbon coins.

Elena: The 4 per 1000 movement has made an accurate, uncomplicated, inexpensive test kit for year-by-year changes in carbon in the soil. Measurement now possible, need a whole army of certifiers to certify, then good to go.

Mary: Anyone paying farmers for losses incurred in transition to new ag methods?

Elena: No. No payers found.

Mary: But this is perfect for carbon coins! Why shouldn't they be getting carbon coins?

Dick: The banks have to make it clear the carboni can be issued in fragments of a coin. Like carbon pennies.

Elena: We would like to see that. But we're only just now able to quantify what's being sequestered. Also, defining sequestration becoming an issue.

Dick: The standard definition from CCCB is being held in storage for a century.

Mary: Is that long enough?

Dick: Long enough for now. Kick the can down the road a century, not bad compared to not kicking it at all. Emergency definition, in effect.

Indra: This is part of why geoengineering no longer a useful word or concept. Everything people do at scale is geoengineering. Glacier slowdown, direct air capture, soil projects like 4 per 1000, they're all geoengineering.

Mary: But solar radiation management is definitely geoengineering.

Indra: Sure but so what? The American heat wave has brought that one back again for sure. Indian results still debated. Claim that a double Pinatubo lowered global temperatures by three degrees for the five years following the event, a degree in the decade after that. Now we're back to pre-intervention levels. But so many confounding factors, all these figures contested.

Mary: Not in India. They're going to do it again.

Indra: Generally agreed there that their intervention worked. So you can see why they might want to. Talk of it now also in US Congress. Controversial. Meanwhile, we're working with CCCB to list all ways carbon drawdown could be quantified and confirmed, in ways that would allow for carbon coins to be created and paid to individuals. All geoengineering, all good. The word itself needs to be rehabilitated.

Mary: Good luck with that. Dick, what's going on with finance? World still in Super Depression but finance sailing along just fine, what's up?

Dick: Effect of carbon coins on global finance a nice stimulus for them! (joking as usual). Finance unfazed by anything, carbon coins just another tradeable commodity, listed in currency exchanges just like any other currency. Betting on the spread between the stick and the carrot. Thus you see people shorting carbon coins, meaning the worse the climate does the more money they make. They've hedged the apocalypse.

Mary: Can we stop them?

Dick: A falling price on carbon coin is a sign that the incentives to not burn carbon are not strong enough yet. Also, measurement of anything always leads to financialization of the measurement method itself. So, just the usual thing.

Mary: What can we do to up the pressure on carbon burning?

Dick: Get WTO to change rules in ways that penalize carbon burn of any kind. Up carbon taxes progressively. Publicize sabotage of petro now happening more and more frequently, this will encourage more sabotage copy-cat style. In short: arbitrage and sabotage.

Mary: Ha ha.

But some of us laughed for real.

Dick: The market needs the state healthy, to back money itself. The state needs the market healthy, to keep the economy liquid. But state and market aren't working hand in hand. Or they are hand in hand, but only because they're arm wrestling! Struggling for control of the situation they comprise together.

Mary: And we want the state to win.

Dick: States make laws, laws run the system. So yes, state is the crucial actor. But we can't just banish market. Not now. It's too big, it's the way of the world. We just have to force it to invest in the things we want.

Badim nods at this.

Janus Athena: Something Dick said before bombing hiatus struck those here in the AI group. (Meaning JA themself, but they never use any pronouns referring to themself.) That ministry should form a sort of shadow government, Dick said, so that when the system breaks down, people have a workable Plan B to turn to. So AI group has been trying to sketch a workable shadow government and put it up on websites. Plan B open source. We're seeing an increasing rate of uptake on YourLock. Already a new internet; now its users may be turning into a new kind of citizen of the world. Gaia citizenship, or what have you. Earth citizen, commons member, world citizen. One Planet. Mother Earth. All these terms used by people who are coming to think of themselves as part of a planetary civilization. Main sense of patriotism now directed to the planet itself.

Matriotism, Dick jokes.

JA nods. Support growing fast. Could cross a tipping point and become what everyone thinks. A new structure of feeling, underlying politics as such. Global civilization transcending local differences. A different hegemony for sure. Shadow government plans are just one part of that larger movement. Like a software for a feeling.

Mary: The global village.

JA: Sort of. That's an old name. Not really a village. Planetary consciousness, biospheric governance, citizen of Gaia, One Planet, Mother Earth, etc. More like that. Village not really the right word.

Badim: It should be an explicit religion, like I've been saying. A call for devotion or worship.

Mary, not hearing Badim: What about AI? Are your machines learning?

JA squints, looking pained. Embarrassing question, somehow. Naïve. Says: Names here keep deceiving people. Data mining tells us things we wouldn't have known unless we did it. That could be called artificial intelligence, but it's what we used to call science. What we have really is computer-assisted science. Best to call it that. It's getting stronger. But we still have to figure out what to do with it. Main potential for advancement here is in human understanding—

Mary, chop chop, cuts off JA's vocabulary worries. Brings meeting to a close. She says, We probably got attacked because of something specific we were doing. The Swiss got attacked too, so now they're very much on our side, and looking into it. That means it was probably financial, somehow. We may be able to see things they can't, you never know. Keep on the lookout for something of that nature. Because we're in some kind of crux now. Push has come to shove.

72

The habitat corridor idea was just one early move in the larger Half
Earth project, but first things first. With wild animals critically
endangered everywhere, it was necessary to do what we could right now,
without delay. And habitat corridors had an already-existing tradition
of methodologies and legal instruments already worked up. The famous
Y2Y corridor, Yukon to Yellowstone, was going great. Not that this was
the hardest corridor on Earth to establish, the Canadian Rockies being
almost empty down their spine, and including a lot of federal and tribal
land protected as parks and the like, on both sides of the border. Not
to mention including two of the great remaining ecosystem populations,
with the Greater Yellowstone Ecosystem animals prospering at the south
end, and the Arctic ecosystem animals doing fairly well in the north, if
you didn't count climate change melting their land or ice out from under
them. Both populations might start migrating to the middle, but the corridor allowed them to do that. And both ecozones had relatively healthy
populations of mammals and birds, in that they were not too poisoned,
and big in number and geographical extent. As proof of concept, Y2Y has
done wonders for showing people how it all could work. Animals had free
passage up and down it, and protection from hunting. Roads still active
in the region had been given under- and over-passes. Fences had breaks
in them or were torn down entirely. Millions of animals were tagged,
and thus now participating in the so-called Internet of Animals, which is
basically a gigantic suite of scientific studies. It was not far from true that
we had a better census going with the mammals and birds in the Y2Y corridor than we did for the humans in it.

All good! And tacking on a Y2Y-Cal, an east-west corridor from Yellowstone to Yosemite, had not been very difficult either. California was

already in the vanguard of animal protection, going back to the days when they had saved the Sierra Nevada bighorn sheep from extinction by an effort now being studied and copied all over the world, and around that same time they had forbidden outright the killing of mountain lions. Connecting that great bio-island to Yellowstone meant getting Wyoming, Idaho, Utah, Oregon, and Nevada on board, and though there were political complications in all these states, the truth was they were mostly emptying of humans across huge swathes of their rural land, and if public and private entities were offered a good enough deal, which mainly just meant offering them enough money for conservation easementlike arrangements, they usually went for it. The corridors could be patched together, the interstates finessed by under- and over-passes, the hunting laws changed, and the nights in particular given over to animals, as they mostly always had been. People diurnal, animals nocturnal, this wasn't entirely just them trying to avoid us. A lot of them lived that way even in the wilderness. So, lots of people were paid for motion-sensitive night photos of animals they sent to us, like bounties in the old days but reversed, people paid for keeping them alive rather than killing them. Local governments were often enthusiastic, sensing tourism, and various federal agencies liked the plan, especially BLM and the Forest Service, which for the purposes of this project were the two agencies that counted most. If we could keep under the radar in Washington, where assholes congregate and bray loudly about the God-given right to kill everything in creation without restraint, we could usually make corridor creation work quite well.

So, fine. Corridors were extended from the south end of Y2Y in multiple directions, southwest to California, then right down the Sierra spine of California to its big deserts, then also west and north to the Cascades and Olympic Range, and then also down the continental divide, down Colorado and New Mexico right to the border. People were talking about a Y2T, meaning Yukon to Tierra del Fuego, just following the great line that forms the spine of both Americas. Most of the Latin American countries involved were already doing things like it, and Ecuador and Costa Rica had been leaders all along. It could happen.

So, but what about going east? Meaning east into the eastern half of the United States, and the eastern half of Canada?

Well, most of Canada is empty, really. Of course they have their wheat belt, and the Highway One Corridor, and their big cities, sure. But most of that is near the US border. And Canada is big. Really, they could lead the Half Earth movement without even changing much; shift two percent of their human population, and over half of their country would be left to the animals. Of course a lot of the world turns out to be like that. But Canada, wow.

The United States, not so much. The farm states have a well-distributed population, and farms occupy every square inch of land that can be culti-vated, and they've killed off all the wild animals they can, in particular the top predators. So naturally they have a deer infestation, thus a tick infesta-tion, thus a human plague of Lyme disease and so on. Oops! Ecology in action! And it's true they still have the usual super-competent omnivore scavengers, the coyotes and raccoons and possums and such. But other than that, no. The Midwest has been treated like a continent-sized factory floor for assembling grocery store commodities, and anything that got in the way of that was designated a pest or vermin and killed off. Just the way it was. Part of a long-standing culture. It had been the same in California's central valley, and it was still in the ag regions of the South.

So when you advocated for wild animals in these old-fashioned parts of the country, it was like advocating for locusts or your favorite plant disease. Even though they were getting it backwards in terms of source of disease. And living in pools of pesticides that were chewing on their hor-mones and their DNA in ways sure to kill them. But that was a case that had to be made, and for sure there were headwinds. There were people screaming at you at meetings. There were men with guns foaming at the mouth, sick with the anticipation of shooting and killing some animal, any animal, such that their target might easily shift from wolf to man, if they thought they could get away with it. The situation had to be handled with a touch of delicacy.

First, money. Significant applications of money. Then persuasion. Hedgerows often saved soil, they built soil, they were considered worth the land they took. Native plant strips, the same. No-till ag, the same. Habi-tat corridors had to be seen first as extensions of that kind of agriculture, done to increase soil building and soil resilience. Wide hedgerows were the wedge for this topic, the least objectionable innovation. Then the idea of wild animals had to be brought in as kind of pest control devices. Of course those who grazed domestic animals were not pleased, but since the mad cow disease scare in the previous decade, with its subsequent collapse of beef demand, there were simply far fewer domestic beasts out there to worry about. Hogs were enclosed, chickens were enclosed; those supposedly terrible wolves would now mostly be eating tick-infested crop-eating deer; it was the deer who were the pests, deer who devastated crops! It was a matter of crop protection to have wild predators on the land! And you could even hunt them later on, if some culling was found necessary. Although making this argument was a bit disingenuous, as some of the more hotheaded among my colleagues were all for doing their culling by hunting the hunters. But we who were friendly Midwestern spokesperson types emphasized the pest control aspect of re-introducing wild animals, without going into detail concerning which pests we were talking about.

And to tell the truth, the upper Midwest, and the states west of it all the way to Seattle, were hurting bad. They were emptying out anyway. People could make more money ranching buffalo and tending wildlife sanctuaries than they could by farming. Those upper plains were never meant to be farmed, and people had learned that the hard way right from the start. Now all the young people were taking off and never coming back.

What would make them stay? Wildlife protection! Especially when you could make a good living at it, better than the debt-ridden drought-stricken winter-blasted poisonous hardscrabble farming that people had been attempting for the previous two centuries. All that effort had gotten them nothing but a dust bowl and mounting debt, and kids moving away, and early death. A category error from the start, an ecological illiteracy. Time for a change.

So, we would go to county supervisors, and town council meetings, and church meetings, and state legislature meetings, and county fairs, and trade shows, and school assemblies, and every kind of meeting, all the meetings no one ever thinks to go to and deeply regrets going to the moment they do, and we would make our case and show the photos and the figures, and see what we could do. Offer them woolly mammoths and saber-toothed tigers if that's what it took, although to tell the truth people in the upper Midwest seldom go for that. Their idea of charismatic megafauna is their chocolate Lab.

It was working pretty well, when we ran into a snag. A militia group with a burr up its ass declared that even though we had the right to cross the state border from Montana into North Dakota, leading a herd of buffalo as the vanguard of a suite of animals that included, yes, lions and panthers and bears, oh my!—they were going to stop us. It was their God-given right as Americans to stand their ground and kill trespassers of any kind, animal or human. And some private property owners along the state border were willing to let these guys congregate and block our passage when we crossed the state line. They organized a mob who all drove to these properties in their fat pick-up trucks and prepared to meet us with their guns blazing. It was a flashpoint, a media event.

Well, fine. Media events can be good. The trick is to handle them right. Which of course includes not getting killed.

So, we could send in a herd of ten thousand buffalo and crush them underfoot, like in that movie. Very satisfying, but not really serving the larger purpose. Not a good hearts and minds kind of move. Or we could walk in front of the animals with our families, put the kids in the vanguard holding their pet ducks and raccoons, and overwhelm them with love and kindness and puppies. Uncertain that, and dangerous. A great media image, and yet not a good idea.

We ended up going with cowboys wrangling a herd of wild horses. Also some sheep and sheepdogs, as if to reconcile the old Western fight between those two competing ways of wrecking the land. And these were wild horses, and wild sheep and wild mountain goats. The sheepdogs were domestic of course, they were just there to get dogs involved. Not that they aren't great at sheepherding, because they are. But they were useful for more than that. They were the Midwest's emotional support animals, their link between man and beast.

Then, behind that vanguard of cowboys and sheepdogs, the whole parade. Buffalo, elk, moose, bears—it was like an old-fashioned circus come to town. We kind of wished we did have a woolly mammoth or two. And we lowballed the wolves and the mountain lions; they didn't like to keep company with us anyway, and they could sneak in later at night, in their usual way. It wasn't natural to even see a mountain lion, they like to lie low and are extremely nocturnal. In my whole life I've only seen one once in the wild, and in truth it was terrifying. I thought I was done for.

So the day came, and we alerted the press, and people showed up from all over the world. So many people wanted to march with us that there were more people than animals, and the animals were getting spooked, naturally. But we started up anyway, right after dawn, and hit the state line across a ten-mile front, like a World War One over-the-top assault. And it did look kind of over the top.

Those poor animal murderers never had a chance. Actually quite a few of them stood their ground and shot a bunch of animals. Mostly deer, it turned out, as we had sent them out as scouts, poor guys. The first wave. Deer are the sad sacks of American wildlife, so beautiful, so defenseless, so numerous, so dim. Their chief predator is cars. They never seem to get it, about cars or anything else; or maybe they do, but they don't have a good way to transmit what they learn to their kids, if they happen to get lucky and live past their youth. It's important always to remember that even if they're as common as rats, some kind of mammal weed, they are still beautiful wild animals, getting by on their own in a dangerous world. I always say hi when I see them, and try to remember to get the same thrill I would if I were to see someone unusual, like a wolverine. It's hard, but it's a habit you can build. Love your deer! Just fence your veggie garden really well.

So, the day went about as well as we hoped. A few deer were killed, a few animal murderers and all-round jerks were embarrassed by the lame ways they had to stand down or slip away and pretend it had never happened. We even got good film of pick-up trucks driving away at speed, also of a duck hunter's blind demolished in a charge of buffalo who didn't even notice they had run something over. Cowboys did figure-eights and stood in their saddles and twirled their lassoes, sheepdogs nimbly nipped sheep through gates, and the images and stories went out worldwide. It was just one moment of the storm, but after that, habitat corridors were more of a thing. E. O. Wilson's great books shot to the top of the non-fiction bestseller lists, and we could continue the work with more understanding and public support.

On to the Half Earth!

73

Modern Monetary Theory was in some ways a re-introduction of Keynesian economics into the climate crisis. Its foundational axiom was that the economy works for humans, not humans for the economy; this implied that full employment should be the policy goal of the governments that made and enforced the economic laws. So a job guarantee (JG) was central to MMT's ideas of good governance. Anyone who wanted a job could get one from the government, "the employer of last resort," and all these public workers were to be paid a living wage, which would have the effect of raising the private wage floor also to that level, in order to remain competitive for workers.

MMT also reiterated Keynes's point that governments did not experience debt like individuals did, because governments made money in the first place, and could create new money without automatically causing inflation; the quantitative easing (QE) after the 2008 crash demonstrated this price stability despite major infusions of new money. So MMT recommended robust stimulus spending in the form of carbon quantitative easing (CQE) as well as a job guarantee. Both were to be directed to the effort to decarbonize civilization and to get in a sustainable balance with the biosphere, humanity's one and only support system.

Critics of MMT, who sometimes called it "Magic Money Tree," pointed out that Keynes had advocated deficit spending during economic contractions, but also the reverse in times of expansion, governments gathering in enough in taxes to fund things through the next crisis. To ignore this counter-cyclical necessity and regard money as infinitely expandable was a mistake, these critics said, because there was a real relationship between price and value, no matter how distorted that got by various historical forces. Also, if governments offered full employment then they were in effect setting the wage floor, and if that caused inflation and governments then stemmed that inflation with price controls, then government would be in effect setting both wages and prices, thus taking complete control of the economy, and at that point they might as well dispense with money entirely and go to the Red Plenty solution of computer-assisted production of everything needed; in other words, to communism. Why not just admit that and go there?

There were some who responded to that question with: yes, why not?

The MMT advocates replied that they hoped to retain what was empirically useful in conventional economics as a social science, for purposes of policy analysis, while re-orienting economics' ultimate goals to human and biosphere welfare, thereby changing its policy perspectives and monetary theories, leading to recommendations for actions that would help get civilization through the narrowing gate of their crisis. Economics was a tool for optimizing actions to reach goals; the goals could be adjusted, and should be. So the MMT crowd admitted they were proposing a move to a new political economy, rather than merely adjusting capitalism. It was not just Keynes Plus, nor just the ad hoc theory or rather praxis that had gotten them through the 2020 crash, nor just the theory or praxis that had bolstered and ultimately paid for the Green New Deal, that early shot in the War for the Earth. It was more than that: it was trying to think through how to do the needful in the biosphere's time of crisis, while orthodox economics failed to rise to the occasion, and stayed focused in its old analysis of capitalism, as if capitalism were the only possible political economy, thus freezing economics as a discipline like a deer in the headlights of an onrushing car.

Enough governments were convinced by MMT to try it. That it influenced so much policy through the late thirties was regarded as a sign either of progress or of desperate fantasy solutions. Similarly split responses had of course greeted Keynesian policies exactly a century before, so for some observers the interesting thing became to watch the next steps, and see whether this time around, having reiterated in this realm the twentieth century's thirties, they would manage to avoid reiterating its forties.

74

Days passed for Frank one like the next. He didn't mark a calendar, or keep track of the day of the week. Every once in a while Syrine and her younger girl would come by to visit; seemed like once a month or so. He got the impression the older girl was too mad at him to come. Mary Murphy came by more like every week or two. She worked nearby, he thought.

The meals in the prison dining hall were solid Swiss food. He was gaining a little weight on it. He read over his bowl or plate, books from the library. Occasionally an English-language newspaper, published weekly in Paris. The prison library had a lot of books in English, and he made his way through them unmethodically. John le Carré, George Eliot, Dickens, Joyce Cary, Simenon, Daniel Defoe. Robinson Crusoe was funny. Lucky to have been able to ransack the wreck of his ship like that. All that stuff he saved had given him a good life. Not unlike Frank's in some ways. He too was stranded on an island, getting by.

Most days he got on the 8 AM van that ran prisoners around the city. He kept getting off at the refugee camps. This was always a little disturbing, but he did it anyway. Maybe it was what one of the therapists had called habituation. Go right at what bothers you, face up to it. One of the books in the prison library had been written by an African man who had traveled up the coast of Greenland in the early twentieth century, staying in Inuit villages. Eskimaux, he called them. He wrote that they had a saying in their cold little villages, to deal with the times when fishermen went out and never came back, or when children died. Hunger, disease, drowning, freezing, death by polar bear and so on; they had a lot of traumas. Nevertheless the Eskimaux were cheerful, the man wrote. Their storm god was called Nartsuk. So their saying was, You have to face up to Nartsuk. This meant staying cheerful despite all. No matter how bad things got, the Inuit felt it was inappropriate to be sad or express grief. They laughed at misfortunes, made jokes about things that went wrong. They were facing up to Nartsuk.

Which they all had to do. One day, working a camp food line and seeing a distraught refugee's face out of the corner of his eye, he understood that eventually everyone was post-traumatic, or even still mid-trauma. These people he served had been variously beaten, shot at, bombed, driven out of their homes, seen people killed; all had made desperate journeys to get here, sleeping on the ground, hungry. Now they were in a new place where possibly new things could happen, different things, good things. It was a matter of being patient, of focusing on the people right in front of your face. Possibly they could get past their traumas, eventually. You had to talk to people.

Frank seldom talked to anyone, but sometimes he did, and then he found himself babbling a little. But asking questions too, and listening to what people said to him. No matter how bad their English was, it was always better than his attempts at their language. They used English like a hammer to get their meaning across, they banged in nails of meaning. Strangely articulate and expressive sentences often emerged from them. Sometimes they sounded like Defoe's characters. The situation has become urgently urgent, someone said to him one day. I blue the sky! one little girl exclaimed.

The news often disturbed him. Heat waves, terrorist attacks. All the militaries of the world were focused on counter-terrorism. There weren't any state-on-state clashes serious enough to distract the militaries from trying to discover and root out terrorists. But with limited success, it seemed. A hydra-headed foe, someone called it. And to Frank it seemed different than it had when he was a child, when terrorists were universally abhorred. Now it felt different. Many attacks now were on carbon burners, especially those rich enough to burn it conspicuously. Car races and private jets. Yachts and container ships. So now the terrorists involved were perhaps saboteurs, or even resistance warriors, fighting for the Earth itself. Gaia's Shock Troops, Children of Kali, Defenders of Mother Earth, Earth First, and so on. People read about their violent acts and the frequent resulting deaths, and shrugged. What did people expect? Who owned private jets anymore? There were blimps now that flew carbon negative, as the solar panels on their top sides collected more electricity than needed for the flight, so that they could microwave it down to receivers they passed over. Air travel could now also be power generation—so, a jet? No. If a few people got killed for flying, no one felt much sympathy. Fools conspicuously burning carbon, killed from out of the sky somehow? So what. Death from the sky had been the American way ever since Clinton and Bush and Obama, which was to say ever since it became technologically feasible. People were angry, people were scared. People were not fastidious. The world was trembling on the brink, something had to be done. The state monopoly on violence had probably been a good idea while it lasted, but no one could believe it would ever come back. Only in some better time. Meanwhile hunker down. Try to stay lucky. Don't fly on private jets, or maybe any kind of jet. It was like eating beef; some things were just too dangerous to continue doing. When your veggie burger tasted just as good, while your beef package proclaimed *Guaranteed Safe!* with a liability waiver in small print at the bottom, you knew a different time had come.

One afternoon after he returned from the big camp in Winterthur, Mary Murphy came by. They crossed the street and sat down at a table in a café. Nice afternoon, still in the sun. Kafi fertig, its attractive little clash of bitternesses, clash of effects. This strange woman watching him. Life in prison not so bad. Indeed at the next table another pair of prisoners were sharing a spliff. The prison wardens approved of prisoners using cannabis, it kept them calmer. Wardens looked the other way even when it was smoked in the prison's smoking yard, much less across the street. They were right about the calming effect, so it was just being sensible. And the Swiss were all about being sensible.

Frank said to Mary, "These attacks on carbon burners. There's a lot of them now."

"Yes." She looked at her glass.

"Do you think some of them are done by your people?" he asked.

"No. We don't do that kind of thing."

She was never going to admit anything to him. She had no reason to. They had once or twice passed through certain elusive moments of closeness, starting even that very first night in her apartment, but now they seemed to have drifted apart and come to rest in something more distant, more formal. He didn't know why she visited him anymore.

"Why do you visit me?" he asked.

"I like to see how you're doing." She paused, sipped her drink. "Also I like knowing where you are."

"Ah yeah."

"You seem calmer, but..."

"But what?"

"Not quite here. Not happy."

He blew out a breath, a poof of dismissal. "I'm not."

"It's been a long time," she said.

"Since when?"

"Since the bad things happened."

He shrugged. "They're still happening."

She seemed to lose patience. "You can't take on the whole world's troubles. No one should try to do that."

"It happens without trying."

"Maybe you should stop reading the news. Stop watching the screens."

"I'm reading *Moll Flanders*. It was the same for her."

"Who's she again?"

"A character in Defoe. Sister to Robinson Crusoe."

"Oh yeah. I sort of remember." She smiled briefly. "A survivor."

"For sure. They didn't worry like we do. They faced up to Nartsuk. There was no such thing as post–traumatic stress disorder."

"Or else it was everywhere. Just the water they swam in."

"That's still a difference."

"Maybe. But Moll Flanders didn't try to take on the whole world's trauma, as I recall. People weren't so worried about other people."

"But we should be, right? For them, people dropped dead all the time. You had to move on. Could be your partner, your kid. Now it feels different. You and yours probably won't drop dead, not today."

"Mine did," Mary said shortly.

That startled him, and he looked at her more closely. She watched her drink. He recalled a moment that first night, when he had asked her about her life, how she had flared up at him. Angrier at a personal question than being kidnapped, almost. "Okay," he said, "but still somehow it's different. Maybe what we know now. We know we all live in a village of eight billion neighbors. That's our now. It's all of us succeed or none of us is safe. So we take an interest in how the others are doing."

"If that's all true."

"Isn't it?"

"I think a lot of people don't do the global village part. Janus Athena says village is the wrong idea. And nationalism has come back big time. Your language is your family. Pull in the perimeter like that and it gets easier. You still get to have your us and them."

"But it's wrong."

"Maybe."

He felt a little jolt of irritation. "Of course it's wrong. Why do you say that, are you trying to tweak me?"

"Maybe."

He glared at her.

She relented a bit. "What is it they say about us only really knowing a few score people? Like back in the ice ages?"

"It's different now," he insisted. "We know more now. Those people in the caves, they only knew there were a few hundred people alive. Now we know better, and we feel it."

She nodded. "I suppose it could be. Eight billion people, all stuffed in here." She tapped her chest. "No wonder if feels so crowded. All smashed into one big mass. The everything feeling."

Frank nodded, trying that on. That feeling of pressure in his chest. The headaches. Call it the everything feeling. A new feeling, or a new blend of feelings, bitter and dark. Caffeine and alcohol. Uppers and downers. Lots of everything. The everything feeling. Made sense that it resembled being somewhat stunned. Not unlike despair.

"Maybe," he said, mimicking her.

She grimaced, acknowledging that she had been annoying. "Oceans of clouds in my chest. Some poet said that. So, say we feel the global village, but in a mixed-up way. Is that what you're saying you are, mixed up? Mashed together?"

"No. Yes." He glanced at her, looked down again. "Maybe."

She was regarding him with a very curious look. "You should go up to the Alps, have a walk around. I found it very clarifying, even though I was up there for bad reasons. It could be a day trip from here, you could be home by curfew."

"Maybe."

Later he considered what she had said. That he was not all there: true. That he was mashed together into a thing he couldn't grasp: true. The everything feeling. But the project was to face up to Nartsuk. That wasn't just acceptance, but defiance. You had to laugh at whatever the world threw at you, that was good Inuit style.

He took the train to Luzern, a bus to the forest under Pilatus. Hiked up one of the trails through a strange parklike forest, then up onto the big clean grassy alp above the forest and below the gray peak. Cable car high above, swinging up and down across a giant gap of air. He ignored it and contoured around the peak until he couldn't see it anymore. He only had a couple hours before he would have to head back, so it was kind of an exercise in getting as high as possible on this trail and then turning back.

Still mid-alp, crossing a tilted rumpled lawn of immense size, he came over a small vertical ridge in the trail and there was an animal standing there. Ah—four of them. Chamois or ibex, he didn't know, he was just guessing. He had heard they were up here. This group was maybe a male and female, and two youngsters, but not too young; he couldn't really tell.

They didn't seem disturbed by his presence. They were aware of him, alert, heads up, sniffing; but they were chewing their cuds, it looked like. Slow and regular chewing, a lump inside their cheeks, had to be a cud being chewed, or so it seemed.

Their bodies were rounded and full, they looked well-fed. If they ate grass, he could see why that would be so. Their heads looked like goats' heads. They had short horns, slightly curved back but mainly straight. Horizontal ridges ringed the horns, possibly annual growth; looked like that made for strong horns, could really stick you if they tilted their head down. Although they'd have to be looking back between their forelegs to have their horns pointing forward, so that was a mystery. Short brown hair over most of their bodies, but finer beige hair, like fur, on their bellies, with a dark band separating brown from beige.

The biggest one was looking at him. Then Frank saw it: the creature's irises were rectangular. Like a goat, then? It gave him a little shock to see it. Rectangular irises, how could that be? Why? Was it really looking at him?

Seemed like it was. Steady regard of another animal, chewing as it watched him. What would this person do? Was he a problem?

The creature did not seem to think so. It was more a case of interest. Frank, standing there in his windbreaker. They looked at each other. Frank mimed chewing a cud of his own. The creature tilted its head to the side, interested at this, perhaps. It blinked from time to time. A gust of wind ruffled the hair on its back, then Frank's beard. Frank smiled to feel it.

Something reminded him to look at his watch. It was late! Apparently he had been locked in a gaze with this chamois or ibex or mountain goat, this person of the alps, for around twenty minutes! It had felt like two or three.

He stirred, raised his hand uncertainly, waved weakly at the beast. Turned and headed back.

75

In the United States, the National Students' Union website showed that thirty percent of the union members had now responded YES to the union website's standing poll asking them if they were in so much financial distress caused by their student debt that they would like to see the union initiate a fiscal non-compliance strike, by not paying their next debt payment. On joining the union, members had agreed to join any strike requested by thirty percent of the membership, so now the union coordinators called for a strike vote to be sure, and got an eighty percent yes vote, with ninety percent participation. None of this was surprising; student food insecurity, meaning student hunger, was widespread, also student homelessness. So the strike began.

Student debt was a trillion-dollar annual income stream for the banks, so this coordinated default meant that the banks were suddenly in cash-flow hell. And they were so over-leveraged, and thus dependent on all incoming payments being made to them on time to be able to keep paying their own debts, that this fiscal strike threw them immediately into a liquidity crisis reminiscent of the 2008 and 2020 and 2034 crashes, except this time people had defaulted on purpose, and precisely to bring the banks down. The banks all rushed to the Federal Reserve, which went to Congress to explain the situation and ask for another giant bail-out to keep liquidity and thus confidence in the financial system itself. There were calls from many in Congress to bail out the banks, as being essential to the economy, and too connected for any of the big ones to be allowed to fail. But this time the Fed asked Congress to authorize their bailing out the banks in exchange for ownership shares in every bank that took the offer. This was either nationalizing finance or financializing the nation, in that now it was clearer than ever that the country was in effect run by the Fed. And since Congress ran the Fed, and people voted in members of Congress, maybe it was all beginning to work, somehow, because of this strike. Definancialization of a sort. End of neoliberalism.

This possibility was shocking enough, but in that same month, the African Union informed the World Bank and the Chinese government that they were declaring all African debts to these organizations to be odious debts. All the national governments forming the African Union were together backing complete debt forgiveness, the haircut of haircuts, to be followed by a new set of agreements, negotiated by the African Union in collaboration with all the African nation-states. They called this the end of neoliberal neocolonialism, and the definitive start of Africa for Africans. Even Egypt and the rest of North Africa joined in on this, plumping for people over capital, continent over history, and in Egypt's case, maybe all the Muslim north, splitting with Arabia.

And in the very next month, in China the workers known informally as the billion took over Tiananmen Square, by the simple expedient of walking into it by the millions, on the way causing all the rest of the traffic in the city to grind to a halt. By taking over Beijing's highways by way of a crush of pedestrians, they were able to bypass the barriers and checkpoints around Tiananmen Square and fill it and the surrounding district so full that the police and army couldn't react, confronted as they were by about five million people jammed into the city center. Similar demonstrations appeared in all the largest Chinese cities; there were literally hundreds of such demonstrations, far more than the Chinese army could deal with in any typical crowd control way. Coordinated demands for the end to the hukou system were supplemented by other demands that together insisted that the Chinese Communist Party be more responsive to all its citizens' needs. As the protests had been timed to the five-year Party congress meeting, it proved possible to pressure the Party into appointing an entirely new standing committee, one including women and younger people, and devoted to reforms as demanded.

Seeing an opportunity in all this turmoil, the Kurds declared Kurdistan in the part of Iraq they already controlled, with significant chunks of Syria, Turkey, and Iran included within the new boundary, to underline the point that the Kurds were asserting themselves here, and there was no one strong enough to stop them. All the surrounding countries were outraged and antagonistic, but they were also already antagonistic among themselves, so they could not form an effective response that would not also constitute an attack on a neighboring sovereign state, either ally or enemy, but either way a nation dangerous to attack.

All these events occurred at once; and these were not all. There was so much going on, such a spasm of revolts occurring spontaneously (if it was spontaneous!) all over the world, that some historians said it was another 1848, some kind of reemergence of the spirit of 1848 around the zone of its bicentennial. And just as it had been during that period of enormous unrest and revolutionary upheaval, no one could explain why it was happening in so many different places at the same time. Coincidence? Conspiracy? World spirit, Zeitgeist in action? Who knew? All they knew for sure was that it was happening, things were falling apart.

In all this turmoil and uncertainty, the market sought something to be sure of. Volatility was good for traders, sure; but at the end of the day you had to have somewhere to hang your hat. Short the dollar—really? Short everything? Maybe, but where then did you hide your emergency fund in case everything went pear-shaped? The cash-in-the-mattress move wasn't the same as going long, which actually at this point was looking almost impossible. This meant things were getting more and more existential, as it was a question of ultimate value, of trust in the act of exchange itself. And when definitions of value shifted from talking about interest rates to talking about social trust—when finance and theories of money fell through a trapdoor in daily normality, down into the free fall of philosophy's bottomless pit—when people began to wonder why money worked at all—wonder why some people were as gods walking this Earth while other people couldn't find a place to lay their head at night—it turned out there was no very good answer. Certainly no answer at all when it came to investment strategies you could count on.

Money was made of social trust. Which meant, in this spasmophilic moment, with everything changing and the ground falling under one's feet in immense tectonic jolts, that money itself was therefore in limbo. And that was scary.

Vast amounts of paper turned to vapor. The banks of the developed Western world were too connected to fail; if one or two of the big ones went down, the rest would shrink in on themselves and wait for the state to reestablish trust before either lending money or even paying what they owed. Why pay a creditor that might be non-existent next week? Best wait and see if they survived to press that debt in court.

In other words, liquidity freeze. The various forms of paper that are in effect IOUs between banks all became worthless; the only money was cash. But that couldn't work, because every day there were trillions of dollars' worth of exchanges on the various markets, including the dark pools where people working in unregulated data space were trusting each other to pay despite the lack of regulatory oversight; and honor among thieves is kind of a feudal notion, more appropriate to Robin Hood and his merry men than to the world of contemporary finance. No. Since money is an idea, a system based on social trust, when things go south, and trust disappears in a poof, then there simply isn't as much money as there used to be.

This was not shocking news to some; which was why much of the wealth on the planet was invested in property. Real estate values may drop, but ownership of that capital asset still remains, and will be there later no matter what happens. But property is not liquid. So the money problem remains even if the latent wealth problem has been solved in advance by buying up land, houses, apartments in Manhattan towers, and so on.

So with groans and clanks and huge ripping sounds, the world's economy ground to a halt. A great depression was back at last and in full, after almost a decade of recession. The depression they had been living in up until now, now got called the Little Depression, or the Super-stagnation, and so on. This new one was the Super Depression, here at last. Very little money out there; and without money, people can't be paid. No loans, no purchases. Unemployment quickly surpassed the 1930s high-water mark of twenty-five percent. Indeed, this time it looked like unemployment could rise to—to what? Fifty percent? Seventy percent? No one had any idea.

People spoke of barter coming back, especially in rural areas, where one could almost believe in it. But not really; barter was always mostly an idea in the minds of economists, a fantasy history. And in the cities it didn't work at all. The pawnshop is the bartering site in cities. But that's just money for stuff, stuff for money. It only works if money works. Same with the internet, only more so. The internet as a market didn't work when no one trusted money.

Local currencies were proposed and introduced, backed by the town one lived in; but the town needed local banks; and the local banks needed central banks. A lively set of exchanges nevertheless began in many local regions, often watersheds big enough to support their populations; and people began to use their YourLock accounts as sites for digital micro-banking that was of some real use, and showed potential for some kind of post-capitalist crowd banking.

But all this was too new, too provisional; there were too many people, and all of them strangers. Despite all the interesting efforts, as the economy circled the drain, it became clearer: for this moment, it was the central banks or nothing. They were the Dutch boy's finger plugging the leak, the last stitch that might stop the hemorrhaging. The central banks spoke for the state. The states in question were there with their armies and police to back these central banks, all of them in theory owned by the public they reported to. If public banks held the line somehow—perhaps by creating more money, by keeping all the private banks afloat by way of even more quantitative easing—then all might be well.

Since some people had been pinning their hopes on the central banks anyway, this sudden onset of chaos and disorder was seen as an opportunity. Possibly the public could now insist on the right to be properly repaid for public money backing private banks, as they had been all along. Extract reparations from the profit takers; abolish profit if it was necessary to create the reparations. If the private banks objected, let them crash, then move to a fully nationalized financial system, owned by the public and used for the benefit of the people.

So strangely, in this new utmost financial crisis, people, ordinary people en masse, as the material manifestation of "the public," now seemed to hold the ultimate power. When push comes to shove, it's always humans looking at humans; and when a thousand people stand looking at one person, it's clear who has more power. So it was a matter of realizing that, then acting on that realization. Maybe that shouldn't have felt strange, but it did; it felt like free fall. Inventing the parachute after leaping off the cliff.

Which meant it had to be arranged fast.

Thus the shadow government devised by the Ministry for the Future in

Zurich, Switzerland, became one template for a new plan. Not completely new, of course. In fact it was a rearrangement of various elements of old plans, in many ways. Mondragón, Kerala, MMT, blockchain, Denmark, Cuba, and so on: all the elements had been out there working all along. Which made the new methods easier to implement. Not complete revolution, no ten-day weeks with new names and so on, no dive into that revolutionary euphoria that tries to change everything at once. Just ownership adjustments. Numbers. Representations. Reversals in some valences of value. Improvisations. The sun still rose, plants kept growing. But people lived in ideas, so despite the sun in the sky and so on, things felt crazy. It was a panic spring.

But as it became clear that the central banks were stepping in to keep things stable and liquid, certain markets calmed. Bakeries kept baking. The US Congress got very busy adopting new laws. Chinese people ended their demonstrations and went back to work, with a different standing committee in charge. Kurdistan secured its borders and signed treaties with every nation and organization that would co-sign. People began to look for ways to earn a carbon coin or two. Only a few of these would be a lot in the local currencies. Surely sequestering a hundred tons of carbon couldn't be so hard. DIY DAC became a vibrant side activity, like growing a truck garden for food; and sometimes the two were even the same thing.

It was quite a month, then quite a year; a year that became one of those years that people talk about later, a date used as shorthand for a whole period. A tectonic shift in history, an earthquake in the head.

76

I joined the US Navy out of high school. I wanted to get out of town. See the world kind of thing. From Kansas this seemed like a good way. My mom was worried but my dad was proud. The Navy needs more competent women, he said. You'll show them what for. I love my dad.

So I put in eight years. Parts of those years were messy, mostly relationship stuff, I'm not going to go into that, everyone is the same, we all fuck up until we get lucky, if we do. Then if you're smart enough to see your luck and act on it, things can work out. They have for me. But this is about the US Navy. When I joined I was paid about 25 grand a year. It doesn't sound like much, but I was also given free room and board, and educated in a number of different jobs, so I was free to bank most of my salary if I wanted to, and I did. Eventually it adds up. More on that later.

The main thing I want to say is that being Navy was something I was proud of. This was not at all uncommon among my fellow sailors. We had good esprit de corps. I don't know if it's the same in the other branches of the military, I wouldn't be surprised if it was, although we often made fun of them for being stupid compared to us. Not unusual I know. The thing is, the Navy is well-run. I have the impression it's one of the more widely respected and well-run institutions anywhere. Among other things, we've run 83 nuclear-powered ships for 5,700 reactor-years, and 134 million miles of travel, all without nuclear accidents of any kind. I lived within a few feet of a nuclear reactor for three years, no harm no foul. My dosimeter showed just the same as yours would, maybe better. How can that be? Because the system was engineered and built for safety, whatever the cost. No cutting corners to make a buck. Done that way, it can work. Probably the Navy should run the country's electricity system, I'm just saying.

A couple more things about Navy: now that pebble mob missiles exist, none of our ships could survive an attack by hostiles who have such systems. It just wouldn't work. Nothing can stop those. You don't have to have gone to Annapolis to figure this out. The upper brass don't talk about it, no one talks about it, because it would be too mind-boggling. Would you just throw in the towel, say Whoops, we are now like the cavalry, or flint-tipped spears or a sharp rock in the hand? No, not right away you wouldn't say that. So what it means is that the submarine fleet is all we've got in a real war. Since the threat from submarines is nuclear, as in atomic end of the world kind of stuff, hopefully they won't ever be used. So in practical terms, until the subs re-arm, maybe with pebble mobs of their own, I wouldn't doubt that that has even already happened, as a military force the Navy has been taken completely off the table. All navies have. In any real war, the whole surface fleet would soon be on the bottom. Not a happy thought.

But so think of the Navy in peaceful times, and given these pebble mobs, maybe peaceful times are what we've got now, very fucked-up peaceful times, low-intensity asymmetric insurgency terroristical climate-refugee peaceful times. And in that kind of world, the surface fleet of the US Navy can serve to deploy protective services, and also emergency relief. Like the Swiss military over the last few centuries, its main function will be to bring disaster relief in coastal areas. A force-for-good, US ambassador to the world kind of organization. You'll see, I'm not just making this up—if the Navy persists at all, it will be doing these kinds of things. It already has been for quite some time.

Then the last thing I want to say about the US Navy is this. Occasionally our ships would get visited by admirals. Even destroyers and minesweepers get visited, like for inspections or courtesy tours or whatever. And as they wander around, they see a female able seaman, they sometimes stop to chat and ask questions. These were usually older men, once a woman, that was fun. They all started at Annapolis and made Navy their life, and no matter how fast they rose through the ranks, they've lived on ships and know the drill. So they're well-informed, and interested in how things are for sailors now. Curious and friendly, I'd say, and surprisingly normal. Like a captain, but less pretentious.

Then later I looked it up and learned that admirals' salaries top out at $200,000 a year. No one in the Navy gets paid more than that per year. So they call this the pay differential, it's sometimes expressed as a ratio from lowest pay to highest. That ratio for the Navy is about one to eight. For one of the most respected and well-run organizations on Earth. Sometimes this gets called wage parity or economic democracy, but let's just call it fairness, effectiveness, esprit de corps. One to eight. No wonder those admirals seemed so normal—they were!

Whereas in the corporate world I've read the average wage ratio is like one to five hundred. Actually that was the median; one to 1,500 happens pretty often. The top executives in these companies earn in ten minutes what it takes their starting employees all year to earn.

Ponder that one for a while, fellow citizens. People talk about incentives, for instance. A word from business schools. Who is incentivized to do what in a wage ratio of one to a thousand? Those getting a thousand times more than starting wage earners, what's their incentive from out of that situation? To hide, I'd say. To hide the fact that they don't actually do a thousand times more than their employees. Hiding like that, they won't be normal. They'll be bullshitters. And for the lowest income folks, what's their incentive? I'm not coming up with one right off the bat, but the ones that do eventually come to mind sound cynical or beat down or completely delusional. Like, I hope I win the lottery, or, I'm going to shoot up now, or, The world is so fucked. You hear that kind of thing, right? Maybe incentive isn't the word here. Disincentive, to keep it in that lingo. When you get one pay amount, and someone doing something easier gets a thousand of that pay amount, that's a disincentive to care about anything. At that point you throw a rock through a window, or vote for some asshole who is going to break everything, which may give you a chance to start over, and if that doesn't work then at least you have said fuck you to the thousand-getters. And so on.

So, what if the whole world ran more like the US Navy? What if the standard, or even the legally mandated, maximum wage ratio was set at say one to ten, being so easy to calculate? With the lowest level set high enough for life adequacy or decency or however you want to call it. Enough for a decent life. Which then, ten times that? That's a lot! I mean think about it. Count it on your fingers and thumbs, seeing the enough amount on the tip of each digit, all ten stuck together at the end of your arms looking back at you. Enough times ten is fucking luxurious.

Works for the US Navy. Hooyah!

77

Everyone knows me but no one can tell me. No one knows me even though everyone has heard my name. Everyone talking together makes something that seems like me but is not me. Everyone doing things in the world makes me. I am blood in the streets, the catastrophe you can never forget. I am the tide running under the world that no one sees or feels. I happen in the present but am told only in the future, and then they think they speak of the past, but really they are always speaking about the present. I do not exist and yet I am everything.

You know what I am. I am History. Now make me good.

78

He flew into Lucknow and got on the train into the city and then took the subway and bus out to the branch of the City Montessori school that he had gone to as a child. Biggest school in the world, winner of a UNESCO Peace Prize, it had been a turning point in his life for sure. His father having married a Nepali woman, he could have ended up in Nepal forever, in a Rawang hill village where the police station had been blown up by Maoists and never rebuilt. His father had been stubborn and had not wanted to expose his wife to the pressures of Lucknow. The second-happiest city in India could be tough on hill folk. So he could have lived his whole life in the middle ages, trapped by parents who had met through a young man's desperate matrimonial answered by a girl who could read and write, and dream.

But a German had passed through with an aid group, and when he got caught stealing from them, by no means the worst of his boyhood crimes, it was Fritz who sat him down and interrogated him, as stern as a policeman, but cheerful too, skeptical of his badness, unconvinced by his tough demeanor. Fritz said to him, firmly but kindly, To get anywhere in this world you must hitch your tiger to your chariot. No more stealing, that only hurts you. You're clever and you have a burning desire, I can see that in you, so use your cleverness and get yourself a freeship to a school in a city. That way you'll get what you want without hurting people. Learn all you can in your school here, that won't be much of a challenge for you, I can see that. And then Fritz had spoken to his father, saying, Send this boy to the city. Give him a chance. And his father had done it and he had ended up in Lucknow, his father's home town.

The city amazed him, it stunned him. It was such an upgradation of his fortunes that he had not slept more than three hours a night for the rest of his childhood and youth, and all because of the violence of the spin of thoughts in his head, the day's inrush tumbling up there like clothes in a washer. Lucknow: now luck. The place had made him.

And now he was back. He wandered away from his old school, off into the tight dense neighborhood south of it, the crush of old buildings caught between the subway line and the river, between the present and his past. He had done a lot of stupid things here. Despite the city's many exhilarations he had not discontinued the truant ways of his Nepali childhood, he couldn't remember why. Chain-snatching, market theft; maybe all that had been somehow a way to stay connected to home. His parents would have beaten him had they known what he did in the city, so senselessly endangering everything, but maybe the danger of it had been part of the allure. He liked doing it, and he liked the toughs he did it with. He was one of them. They reminded him of himself. He was a hill beast, no city could tame him. He took what he wanted and no one could stop him. Only a close call entailing a broken arm had slowed him down. And then when he moved to Delhi he had changed again, given up all that kind of thing. Again he couldn't remember why he had done it, how he had justified it to himself. It was just the way he had been then. Things happened. Even though he had rarely slept in Lucknow, he had not really ever woken up until he moved to Delhi. At that point many things came clear, and he never looked back.

Now he was back. He wanted to look back. He walked across town to his school and spoke to the assembled students, and all the young faces were enough to slay him on the spot. Talk about burning desire. They wanted what he had, and he didn't know what to say to them. He said, To get anywhere in this world you must hitch your tiger to your chariot.

He went out with them to the fields outside the city where they were working in the India Regenerative Agriculture job guarantee program. There was full employment in India now, and the work was hard but it was scientifically based too, and drawing carbon into the soil year by year in ways making them all safer. He worked with them planting corn and then repairing a terrace wall, and ended the day feeling cooked. I'm still a hill boy, he told them, I can't hack the terai, it's too hot. But look you, this is good work you are doing, so you must persist. Gandhi made up this word, *satyagraha*, that's Sanskrit for peace force, you all know this word, right? But the Mahatma made it up himself, and I think he would be happy to imagine another word that puts the two parts in reverse order. *Grahasatya*. Force peace. It changes it from a noun to a verb, maybe. And you are exerting that force for peace. The work that you do here helps save the world, it forces peace on the world. Keep at it.

Then as he was preparing to leave, he got a note handed to him in the street requesting a meeting. And this was interesting enough to pursue. Indeed he had wanted to speak to some of these people and had not been able to figure out a safe way to reach them. So he went to the address on the note.

When he got there he was startled and amused to find it was just one street away from the very intersection where he had spent much of his truant youth, the same X of alleyways meeting at a big plus sign of crossing avenues. A very messy intersection, as messy as his young mind and life had been, the same tram wires overhead, same narrow wrought iron balconies on the buildings. It gave him a little smile to think the people he was meeting, no doubt some of this generation's young toughs, had accidentally called him back to his old neighborhood.

These were not the same kind of people he had been, however. They had a purpose; their burning desire was already directed, hitched to a chariot outfitted for war. They stepped out of a doorway and gestured to him to follow them into an empty tea stall. They were older than he had been, and a woman led them. His childhood gang had been happy and boisterous; these people were angry and cautious. Of course: lives were at stake, theirs included. And they had probably been in the heat wave. That would change you. Forged in the fire: yes, these were Children of Kali, staring at him as if calculating where to insert the knives.

We want you to stand up, the woman told him bluntly.

And I want you to stand down, he replied, as mildly as he could.

She frowned heavily, as did the four men with her. They looked like the demon faces on Kali's necklace.

You don't tell us what to do, she said. You're like firangi now.

I am not, he said. You don't know what I am. You know enough to ask to meet with me, I'll give you that, but that's all you know.

We see what's happening. We brought you here to tell you to do more.

And I came when you asked, to tell you the time has come to change tactics. That's a good thing, and it's partly because of what you did. You were doing the needful, I know that.

We are still doing the needful, she said.

It's a question of what's needful now, he said.

We will decide that, she said.

He looked at each of them in turn. He felt how it could be more intimidating than anything one might say. It was almost like touching them; like an electric spark jumping the gap from mind to mind. A hard look; but he let them see him, too.

Listen, he told them. I understand you. I've helped you, I've helped work like yours all over the world. That's why you asked me to meet with you. And it's why I agreed to meet with you. I am putting myself in your hands here, to make you understand I am your ally. And to tell you that conditions have changed. Together we helped to change them. So now, if you keep killing the wicked ones, the criminals, now that all the worst of them are dead, then you become one of them.

The worst criminals are not dead, there are many more of them, she said fiercely.

They always find replacements, he said.

We do too.

I know that. I know your sacrifice.

Do you?

He stared at her. Again he shifted his stare one by one to the men with her. Faces to fear, faces to love. That burning desire.

He said slowly, This is Lakshmi's city. I grew up here. I hope you know that. I grew up right here in this neighborhood, when it was far tougher than it is now.

You weren't here in the heat wave, the woman said.

He stared at her, feeling a strain inside him that might break him apart. His whole life was cracking inside him. Trying to control that, he unsteadily said, I've done more to stop the next heat wave than anyone you have ever met. You've done your part, I've done mine. I was working for this neighborhood long before the heat wave struck, and I'll keep doing that work for the rest of my life.

May you live ever so long, one of the men said.

That's not the point, he said. My point is, I see things you can't see from here, and I'm your ally, and I'm telling you, it's time to change. The big criminals are dead or in jail, or in hiding and rendered powerless. So now if you keep killing, it's just to kill. Even Kali didn't kill just to kill, and certainly no human should. Children of Kali should listen to their mother.

We listen to her, but not you.

He said, I am Kali.

Suddenly he felt the enormous weight of that, the truth of it. They stared at him and saw it crushing him. The War for the Earth had lasted years, his hands were bloody to the elbows. For a moment he couldn't speak; and there was nothing more to say.

79

The time approached when Frank was to be released from prison. Term served. It was hard to grasp, he didn't know what to think of it. Years had passed but he wasn't sure how. Part of him was still stuck outside himself, beyond life and its feelings. That was a relief in many ways—to be spared all the pain, the fear, the memories. Just cold sunlight on the corner terrace of a Zuri day. That he could spend four or five hours without a thought in his head—this was what the life here had given him. He didn't know if he wanted to give that up. Dissociation? Serenity? He didn't care what he called it. He wanted it.

Because something else had gone away too: he wasn't afraid. As far as he could tell. And surely he would be able to tell, if it weren't the case. He was a creature of habit now. Eat, walk, work, read, sleep. He was neither happy nor unhappy. He didn't want anything. Well, that wasn't quite true. He wanted to be free of fear. And he was interested to see more animals. And he wanted the people in the refugee camps to be released like he was going to be. These were all different kinds of wants, and some of them he could try to pursue, others were out of his power.

Every morning, he either took the prison van or the city trams and buses to refugee centers and helped clean the kitchens; or he walked the downtown, criss-crossing the Limmat on its many bridges and often ending up at one of the lakefront parks.

On this day he went into Grossmünster to have a look around. Say hello to the spirit of Zurich, so gray and austere. Like a big old concrete warehouse, immensely tall and almost completely empty. That this was their place of worship always struck him funny. Zwingli as some kind of zen monk, an advocate of nothingness. Purity of spirit. A devotional space reacting against the baroque church, the very idea of church. Did it reveal

as much about the Swiss as it seemed to? Wasn't the graceful Enlightenment church across the river a better image of what they were in modern
times? Possibly so. He walked out, recrossed the river, passed the "Goethe
slept here" sign, went into Peterskirche. No, this wasn't it either. Smooth,
tasteful, kitschy, alabaster; the Swiss weren't like that now. Bauhaus had
struck them harder than anyone, they were all about design now, leap-
frogging back toward Zwingli, or forward to some space-age clean line.
Function as form, yes that was Swiss style. Do it right, make it last. Clean,
sober, elegant, stylish. The old-fashioned Heidi gestures banished to the
touristic parts of the Alps where they belonged. Here in Zurich it was all
about function.

He passed by the women's club set right out over the Limmat, where bath-
ers lay in the sun. Across the river stood the Odeon, looking at the sunbath-
ing women like prurient Joyce himself. Then to the bridge at the river
outlet, Quaibrücke, the beginning of the Limmat. West along the lake to
the first lakeshore park. Sit at a bench above the tiny marina and watch the
statue of Ganymede hold his arms out to the big bird before him. A simple
gesture, enigmatic to the point of blankness. His kind of statue. Ganymede's *ta-da*—that was Frank, maybe, going forward. Offering something
to a great eagle. The sight gave him a shiver. In the sun, wan though it
was, he shouldn't have been cold, but he was. Then he felt a slight wave
of nausea, and a cold sweat burst out all over him at once. He sat there,
willing the sensation to leave him. To his relief it did. But now his clothes
were damp, and he sat there feeling weak and cold.

This had happened a few times recently. He hadn't told anyone, had
brushed it off. Somehow out here in the pale sunlight it felt worse. He
stood up, caught at the bench arm unsteadily. Walked down the broad
steps to where the lake lapped against the concrete abutment. The sight
reminded him of something he couldn't recall—couldn't afford to recall—
he knew what it was, but he ignored it in order to plunge his hands into
the water. Cold alpine water, clean and fresh. You could drink right out
of the lake, Mary Murphy had told him. She swam in it and knew. He
scooped a handful and lifted it to his mouth, sucked it down. Cool and
bland, a little organic. He could taste that it had been snow a week before.
He sucked down several handfuls, ignoring a couple of passersby who
thought it was strange of him to be drinking from the lake. It was James

Joyce who had said you could eat your breakfast off the Zurich streets. Now you could certainly drink their lake water. He had swum in it once or twice, he recalled now. It had been years. Jake in the lake. Strange it was so long ago.

He took a deep breath. Something wrong there, some light-headed chilled weakness he couldn't identify, couldn't put a word to. People said it was a shock to be released, that the days stretched out forever in the weeks before it happened, that you went crazy, that you got afraid of the freedom, wanted back inside. None of that was right when it came to him; it was not any of his old familiar reactions, which he had been told were mental states manifesting in his body. Post-traumatic stress disorder, yes, but this phrase always hid more than it revealed. What was the trauma, what was the stress, what was the disorder? No one knew. In the jungle of each mind a wandering went on ceaselessly, finding a clearing here, a pool there, all in the murky light of one's sputtering thoughts, half awake, half asleep. Why the helpers tried to put words to it he didn't know. Well, they were trying to help. People were wordy creatures, they felt their feelings as words. Sometimes. But sometimes it didn't work. No words fit.

A shaft of fear cut through him like a blade. Something was wrong.

He took the steps up from the lake carefully, looking down to make sure his feet were set right. Not a place to stumble, there were too many Zurchers out for a walk, and if he fell and they helped him and saw his ankle monitor they would think he was on drugs. No, he had to maintain.

He got up to street level, breathed deeply. He took stock, shook his limbs, felt them move as he expected them to, took heart. Across the busy street and into the little sidestreets between Bahnhofstrasse and the river. Up here was a candy shop that sold wedges of candied orange half-dipped in various types of chocolate. He liked the darkest chocolate. Best possible wedges of orange, bittersweet, not quite dried, half-coated with the best possible chocolates. He had made a habit of dropping by and buying just one, to nibble on while he walked. He went in now, and the saleswoman recognized him, plucked out a wedge with tongs and put it on a sheet of waxed paper without him having to ask, a nod from him was enough. Then back out into the narrow pedestrian streets, smooth flattened pavers almost like cobblestones but not, back toward Paradeplatz, across the tram tracks running down Bahnhofstrasse and up into the neighborhood around the prison, so well known to him.

Orange and chocolate, chocolate and orange. Bitter and sweet, dark and light. The complementary tastes forming a composite taste of its own, full and chewy. An infusion of sugars flowing into him, some fat, probably a little jolt of caffeine too. He turned the corner and saw the Gefängnis and felt better. He would shake off this malaise, await his release as stoically as he had endured it, get out and take an apartment in the neighborhood. There was a co-op apartment nearby, overlooking the bus garage, that he had been on the waiting list for since the beginning of his sentence. Now a little bedroom in it was available. He would take that and keep living just as he was now. Keep his head down and get through the days just as he had been.

Mary Murphy didn't come by for a while, until he began to wonder. Then when she did, he joined her at the table across the street and told her about seeing the chamois under Pilatus. He couldn't tell it right; he could see she wasn't getting it. You should go up there and see for yourself, he said finally, irritated.

I'm glad you went, was all she said. It sounds good.

Then he asked her about her refugee division and what they were doing. He said to her, There's something like 140 million of them now, and growing all the time. That's like the entire population of France and Germany combined. It's as bad as it's ever been.

I know.

You have to work up a plan all the governments will agree to, he told her. Have you looked at what happened at end of the world wars? There were millions of refugees wandering around starving. They put Fridtjof Nansen in charge of the problem after World War One, and he came up with a system they called Nansen passports, which gave refugees the right to go wherever they wanted to, free passage anywhere.

Is that true? she asked.

I think so. I've been reading around, not very systematically, but you've got a team to throw at this. There should be Nansen passports again.

She sighed. There's a lot of countries won't accept such a thing, I'm afraid.

Do like you did with the central banks. It's a plan or chaos. The camps,

I know you've visited them, but what I see there is that it's like this jail here, but worse, because they don't have any sense of how long they'll be held, and they never did anything in the first place. Europe is just punishing the victims. Sudan takes care of more refugees than all of Europe, and Sudan is a wreck. People come to Europe and they get called economic migrants, as if that wasn't just what their own citizens are supposed to do, try to make a better life, show some initiative. But if you come to Europe to do it you're criminalized. You've got to change that.

She shook her head. It isn't just Europe.

But you're in Europe, Frank said. He stared at her. She was looking at her kafi fertig. They were falling back again into the pattern of their first night, probably not a good idea: he hectoring her to do more, her resenting that.

On the other hand, here she was, so many years later. It was strange. He didn't know what to make of it. But suddenly he realized it was important to him. He wanted her to visit. That represented something he couldn't name. But he needed it, whatever it was. This Irish woman was kind of crazy, kind of ominously interested in him, really, and quite often a bit vengeful and harsh, pushing him around in ways that bothered him, very irritating; but he had gotten used to her visiting. He needed her.

You should go to the Alps, he said. Remember you told me that, and you were right. Now I'm telling you the same thing.

She nodded. Maybe so.

80

I've had to push him every step of the way. He's just like one of his oxen, that's why he likes them so much, also why he doesn't like me. I'm like one of the birds standing on the oxen poking them in the back. He'd be so much happier if I were an ox. Instead I'm his wife and it's a stupid fate but I have no one to blame but myself, and truthfully, I love him; but I don't want to starve for that.

So he inherited the butt end of his father's property, two hectares as far from the river as his family's land got, which meant it had been used as a dump for many years, and first we had to dig through a thick layer of various kinds of crap, even pay to get some of the worst of it carted off, at which point we had a triangle of dirt hard as a marble floor. First job was breaking up the hardpan surface, second was getting an irrigation channel cut over to us from the cousin's property upstream. I drove him to drive his brothers and nephews to help us, and eventually that all got done and it was time to amend the soil. Here his stupid oxen were of some help, as we could rent nearby pasturage for them and collect their manure and turf it into our land. Of course the water from the ditch just ran over our property at first, carrying everything loose to the river, so we had to deal with that, berm, terrace, channelize, polder, whatever. I did most of that, being the only one who could work an hour straight, also read a level. Progress was slow.

Then we heard the rumors that the district council would be giving out money for carbon retention. Given the state of our property, this would be getting paid for what we had to do anyway to keep from starving, so I told the ox to get registered right away. He dithered and mooed as always, why waste my time, he complained, those things never work. Quit it! I said. Get down there now or I'll divorce you and tell everyone why. He went and got us registered.

That meant a team that came through the village dropped by our place for about an hour, and took samples to get a benchmark figure. One of them was looking around at our place with an expression that made it clear we were obviously going to be setting a good low benchmark. Our daughter was pestering him as he worked, and he took some of our soil and put it in a glass of water, swirled the water, then stopped and showed her how at the moment he stopped moving the glass, the water in it cleared almost immediately. All the grit and mud floating around sank to the bottom. You need to add compost, he told her. Organic material will float in water, but you can see you don't have much of that. A good starting point. We might as well have been living on a linoleum floor, or a rock.

After that I pushed every day. Doing no-till agriculture is all very well, but first you need soil to not till. That takes first doing some serious turning over and plowing under, I'll tell you; years of backbreaking work, in our case, and always pinched for cash, as we used everything I could afford to set aside to pay for various neighbors' manure and crop waste.

But shit to gold, as they say; we did all that. I drove him and he drove his workers, and we got some trees and perennials planted and left them alone, and during the harvests we harvested their usufruct with gratitude. We suffered a drought and a flood, but saw our land do a little better through those catastrophes than some of our neighbors' properties did, because of what we were doing. And it was all without any tractors or fertilizers or pesticides, just the good old poisons that had always been there. All the right kinds of old ways, and all documented by me, as these were going to be factors in the eventual carbon reckoning. We grew most of what we ate, we grew some things to sell, and we put all we earned back into the land. My ox grumbled; who ever heard of growing a crop of dirt?

Finally came a time when the team from the district office was coming through again to check carbon levels. The moment I heard I went down to the district office to sign up for it. Soon after that, the day came when the team, a different one of course, visited to make its evaluation of our little farm's soil. They wandered the property taking samples, sometimes digging with a tool like a posthole digger, other times with a pole like a long corkscrew. Samples, then evaluations over at their truck, which held in its back some big metal machines.

When they were done with their evaluation they came over to us.

You've done well, they told us. We're authorized to pay you right now, but first you have to know, we subtract an eleven percent fee out of your pay-out, to pay for our expenses, and also your taxes. So if you'll sign here to agree to that, we'll get it done.

My big ox bristled. I've never heard of any such cut, he said. What's ours is ours, just pay us what we're owed, we'll deal with the rest of it ourselves.

The one talking sighed and looked at his colleagues. I can't do that, the procedure is set. You have to sign to get your part.

I won't do it, my husband declared. Let's go have it out at the district.

No! I said. I dragged him off to confer in private; I didn't want to embarrass him too much in front of these strangers. Around the corner of the house I wagged my finger under his nose. You take the deal or I'll divorce you, I told him. We've worked too hard. People like this always take a cut. We're lucky it's only eleven percent, they could have said fifty percent and we'd still have to take it! Don't be an idiot or I'll divorce you and then I'll kill you, and then I'll tell everyone why.

The ox thought it over and went back to the visitors. All right, he said, my wife insists. And she can be very insistent.

The men nodded. We signed their form, then looked at what they had given us.

Twenty-three? my man asked. That's nothing!

Twenty-three carbon coins, they said. Actually, twenty-three point two eight. One coin per ton of carbon captured. Which means, in your currency, if that's how you want to take it, about...He tapped on his wristpad. At the current exchange rate, it comes to about seventy thousand. Seventy-one thousand, six hundred and eighty.

My ox and I looked at each other. That was more than we spent per year on everything, by a long shot. Almost two years of expenses, in fact.

Is that before or after the eleven percent is taken out? my ox inquired.

I had to laugh. My husband is funny.

81

Transcript Mary S/Tatiana V, phone conversation, secure line. M in office, T in safe house, location undisclosed.

M: How are you doing?

T: Bored. How about you? Shouldn't you be in hiding too?

M: I don't think anyone wants to kill me. It wouldn't change anything.

T: Maybe.

M: So what are you doing?

T: Working. Telecommuting, like this. Advising our legal efforts.

M: Anything interesting?

T: Well, I think some parts are working.

M: What do you mean?

T: I think the bet that the super-rich will take a buy-out is turning out to be correct. For most of them, anyway.

M: How can you tell?

T: We've been trying it. Offer them fifty million they can count on, or endless prosecution and harassment, even a situation like mine, to stay safe. Many of them are taking the deal.

M: This is legal? It sounds like extortion.

T: There are legal forms for it. I'm just speaking plainly for your sake.

M: Thanks. So they don't just shift their money into tax havens?

T: We've killed those. That's maybe the best thing about blockchain for fiat money—we know where it is. There aren't any hiding places left. If you do manage to hide it, it isn't really money anymore. Only money on the books has any real value now. The older stuff is like, I don't know, doubloons. Real money, we know where it is and where it came from.

M: Wasn't that always true?

T: No. Remember cash?

M: I still use it! But it was numbered, right?

T: Sure. But once it got moved around a couple of times, it was just cash. There were lots of ways to launder it, and it couldn't be traced. Now it can be traced, in fact it has to be to stay real. So there's no place to hide, there are no tax havens. We blockaded the last ones, got the WTO to declare them a disqualifier, all that. No. For individuals, if you want to stay rich in the current moneyscape, it's best to take the haircut and accept your fifty million and walk.

M: I guess it makes sense.

T: Yes. Fifty million in the hand is worth a billion in the bush. Maybe it's the rich who are most like *homo economicus* was supposed to be. They have all the information, they pursue rational self-interest, they try to maximize their wealth. But if a maximum level of wealth gets mandated by society, fighting that isn't rational. Especially if you've got something like twenty times as much as you need to be secure.

M: I don't know about that. The kind of ambition that gets you a billion won't be brooked. It's a sociopathic thing. There's nothing rational about it. It's a man thing, most of the time. Although I've met women who feel just as entitled.

T: Of course.

M: So they'll lash out.

T: Some of them, sure. You're always going to have crazy people. It's the system I'm talking about. If crazy people lash out in a sane system, they do some damage but then they end up in jail, or someone kills them. So it's the system that matters. And that's where I'm seeing results.

M: Is the Russian government on board with all this?

T: Hard to say. Maybe so. Soviet nostalgia is getting stronger. And Siberia is melting, which turns out to be no joke. Some people thought it would be a good thing, that we would grow more wheat and so on, but turns out we just get a bunch of swamps, and you can't drive on the frozen rivers like they used to. It's a mess. Also it's releasing so much methane and CO_2 that we might make jungle planet. Nobody in Russia wants jungle planet. It's too messy, it's not Russian. So ideas there are changing.

M: So they're coming around.

T: Maybe so. It's still a battle inside the Kremlin, but the evidence is clear. And that Soviet regard for science still holds for a lot of Russians. It's a Russian value too. And they think it's funny that the Soviet way might save the world. It's a kind of vindication.

M: Every culture wants respect from all the rest.

T: Of course. Now the Chinese have it, and India too. The ones still hungry for it are Russia and Islam.

M: So how do you get that respect?

T: Not by money. The Saudis showed that. They were fools. Obvious fools get no respect. In Russia we worry about that. We think we are always seen as fools. The great bear, dangerous and uncouth. Provincial.

M: Best novels in the world, best music in the world?

T: That was all czarist stuff. Then the Soviet Union, it had some respect for standing up to the Americans, and getting out there in the sciences, and standing for solidarity. Or so it gets remembered. Now we are just the great losers to the Americans. And with the whole world speaking English, that impression can never go away. Not unless we use Soviet methods to save everyone from American stupidity.

M: Or go back to the czar.

T: Yes, that's the bad response. We saw that with Putin. But the Soviet dream is better. We assume our past, use it to save the world. Mother Russia saves the day.

M: I hope so. Someone's got to do it. I don't think America will.

T: America! They are the rich person who has to accept owning just fifty million rather than infinity. They'll be the last ones to come around.

M: I guess I should go to San Francisco again.

82

Stores are bottlenecks, being distribution centers and not that numerous. You kick them in the balls when you attack their distribution centers. Their stock price drops at news of such attacks, and they have no way to counter that. And their valuations are already at historic lows. Of course police might arrest and prosecute, but that doesn't bring the share price back up. A hundred thousand dollars of physical damage can leverage a hundred million dollars in lost asset value. Big pension funds notice there's a problem and move their monster assets elsewhere, then endowments and trusts and universities and non-profits and hedge funds all notice the big dogs moving, and they try to get out of the house before it falls on their head. And suddenly a big famous corporation, which is also of course a legal person, has suffered something like a stroke, and is now lying there paralyzed in a hospital bed, on life support, his heirs arguing over who gets the last of his stuff.

So stores were torched, sure. In the past that would have been the end of it. People like direct action because it's quick, and afterwards you don't have to face any real change. But around this time the Householders' Union backed the Student Debt Resistance in support of its payment strike. That was non-compliance in action, meaning stay-at-home for almost every job. It's a form of general strike.

Then on July 16th big parts of the internet, the online store of stores, stopped working. That felt freaky. And we had done it. Was it smart? Wasn't the internet like our nervous system now? It was like that guy who cut his arm off to get out of a canyon in Utah. A very desperate measure. We had cast ourselves out into an interregnum, the chaos between dynasties. The Crisis, Year Zero: oh my fucking god.

*   *   *

In a situation like this, there has to be a plan. You can't make it up on the fly in the middle of the breakdown. Not in the modern era of hyper-complexity. Say the internet stops working, your savings suddenly vanish and money doesn't work anymore: Jesus H. Christ in a bucket! Can you make up a new society from scratch at that point? No, you can't. Things just fall apart and next thing you know you're eating your cat. So take this in: there has to be a pre-existing Plan B. And it can't be a secret plan, popped on the world in the time of chaos. No conspiracy theories, please, so fucking tedious those people—as if things secretly made sense! No. Obvious bullshit. We're winging it here. Not that there aren't conspiracies, it's just that they're all well known. So it's in that spirit that Plan B has to be a known plan, an open conspiracy known to all in advance, like the shadow government of an opposition party, putting out all its plans for citizens to consider and hopefully vote for. All of the proposals on the table and argued for. Step-by-step assembly instructions. Yes, this sounds like politics—because it is. Very depressing.

A canonical example of how much the lack of a Plan B can hamstring a revolution is—well, pick any revolution you've ever heard of! They're almost always spasms and so you get the usual spastic result, history as fuck-up, as pinball machine, as nightmare. But consider this one example, which is actually an example of how the lack of a Plan B can stop a revolution from even starting in the first place, despite the crying need for one, so it's especially relevant for us now: meaning Greece and the failure to Grexit, back in the early years of the century. Greece had fallen into arrears in paying its debts to the European Union central bank, also the World Bank, and a raft of private bank creditors. These global financial powers then put the screws on. They told Greece to quit giving its citizens pensions and health care and so on, so the Greek government would be able to afford to pay back the international lenders who had so foolishly extended credit to such a bad risk. Syriza, the party in power in Greece at the time, refused to do this. The so-called Troika, representing international finance, insisted they do it. They couldn't give in on this one, or all the little PIIGS would run off, meaning Portugal Italy Ireland Greece and Spain.

So who was going to lose here, the Greek population or international finance? Syriza put the question to its people in the form of a poll. The Greek people voted by a large majority to defy the Troika and refuse austerity. Syriza then promptly accepted austerity and the EU's leash, going belly-up and begging for a bail-out.

Why did Syriza do that, why did they betray the wishes of the people who elected them? *Because they had no Plan B.* What they needed at that moment was a plan that would get them out of the EU and back to the drachma. They would have needed IOUs of some sort to stand in and do the job of money while they printed new drachmas and made all the other necessary changes as they transitioned back to a country in control of its own currency and sovereignty. And in fact there were people in Syriza working furiously to design that Plan B, which they called Plan X, but this turned out to be a case of too little too late, as they couldn't convince their colleagues in government to risk trying it.

So, in the absence of such a plan ready to be enacted, Syriza had to cave to the EU and global finance. It was that or chaos, which could have meant starvation. People in Greece were already hungry as it was; unemployment was 25 percent, 50 percent for the young, and the austerity regime already imposed earlier by the Troika meant there was no money for basic social services, for relief. The government of an advanced nation, a European nation, the cradle of civilization blah blah blah, was reduced to choosing hunger and unemployment rather than starvation and chaos—those were their only choices, because they hadn't made a Plan B.

This time, our time, when the whole thing broke all over the world, there had to be a Plan B.

What was it? Big parts of it have been there all along; it's called socialism. Or, for those who freak out at that word, like Americans or international capitalist success stories reacting allergically to that word, call it public utility districts. They are almost the same thing. Public ownership of the necessities, so that these are provided as human rights and as public goods, in a not-for-profit way. The necessities are food, water, shelter, clothing, electricity, health care, and education. All these are human rights, all are public goods, all are never to be subjected to appropriation, exploitation, and profit. It's as simple as that.

Democracy is also good, but again, for those who think this word is just a cover story for oligarchy and Western imperialism, let's call it real political representation. Do you feel you have real political representation? Probably not, but even if you feel you have some, it's probably feeling pretty compromised at best. So: public ownership of the necessities, and real political representation.

Details can be arranged on a case-by-case basis, and even though the devil is in the details, they are still details, a matter of making the pieces of the puzzle fit. These details can be worked out, and often they already have been. The Zurich plan, the Mondragón system, Albert and Hahnel's participatory economics, communism, the Public Trust plan, the What's Good Is What's Good for the Land plan, the various post-capitalisms, and so on and so forth; there are lots of versions of a Plan B, but they all share basic features. It's not rocket science. The necessities are not for sale and not for profit.

One scary thing, there has to still be money, or at least some exchange or allocation system that people trust, which means the already-existing central banks have to be part of it, which means the current nation-state system has to be part of it. Sorry but it's true, and maybe obvious. Even if you are a degrowth devolutionist, an anarchist or a communist or a fan of world government, we only do the global in the current world order by way of the nation-state system. Or call it by way of the family of languages, if it makes you feel better. Hundreds of different languages have to be mutually comprehensible. It is what we've got now, and in the crux, when things fall apart, something from the old system has to be used to hang the new system on, hopefully something big and solid. Without that it's castles in air time, and all will collapse into chaos. So yes: money, meaning central banks, meaning the nation-state system. It's a social agreement, nothing more. This is what makes it so creepy. It's like being hypnotized; you have to agree to it for it to work. So we are all hypnotized in a giant dream we hallucinate together, and that's social reality. Not a happy thought.

Especially since the current order is so unequal, so unfair. Old story, of course. Biblical; detailed in Genesis; it's the oldest story, inequality, and never much changed from the start of civilization. So how can we change that? What do we do now?

Now, everyone knows everything. No one on the planet is ignorant of the real conditions of our shared social existence. That's one real thing those stupid smartphones have done; you can be illiterate, many are, and still have an excellent idea of how the world works. You know the world is spinning toward catastrophe. You know it's time to act. Everyone knows everything. The invisible hand never picks up the check. The money is already here, it just isn't evenly distributed. Which is to say properly distributed. So now things have broken. We broke them; we broke them on purpose! Riot, occupation, non-compliance, general strike: breakdown. Now it's time for Plan B. Time to act—as in, act of parliament. It will be legislation that does it in the end, creating a new legal regime that is fair, just, sustainable, and secure. Public utility districts, state-owned (meaning citizen-owned) enterprises, cooperative enterprises, real political representation, and so on. We have to enact a Plan B as law, as soon as possible. The best Plan B will emerge from the multitudes.

83

Back in Pita, in the season of white nights, midnight and yet still not dark, on this night a dim eerie gray light, the seaside city under low clouds, as so often. On the Trinity Bridge, looking inland at the great wall of the Peter and Paul Fortress in its spotlights, glowing yellow in the dusk. Raw cold, as usual. She had her old coat on, its fur collar warm under her chin. Her winter cap down over her ears. Long underwear under her pants, her old lined boots. She was okay, she had come prepared. Though it was always a shock to get back into such a raw cold.

Ah Tatiana, her friend said from behind her, you were such a beauty before you left, and now you're just another babushka like the rest of us.

Fuck you, she said. You look even worse.

They hugged.

Actually you are looking pretty fucking good, Svetlana said, inspecting her as they held hands. I guess all that money in Switzerland has been good for you. Spas and diets and workouts, it looks like.

Not.

Well, you've always been lucky to have a figure like yours, you can pack it on and still look good, just more gush.

Lots of Swiss cheese, Tatiana agreed unhappily. But quit being stupid, tell me what you want.

I want out.

Tatiana sucked in on the cold air. Are you sure?

Svetlana gestured at the city. Who wouldn't?

I thought it was looking pretty.

Svetlana gestured at the Lakhta Center tower, puncturing the white night like a giant silver needle. Are you kidding me?

I kind of like it.

You would. Come on though. It isn't a case of how it looks, or how you look, or how you felt about it when you were a girl.

I hated it when I was a girl. It's now I like it.

Now that you're gone.

I'll be back someday.

I doubt it. Don't kid yourself. This place would kill you.

I thought things were getting better. Putin's gone, the communists are making a comeback, the oligarchs are dead or in jail.

Don't say that! Why do you say that? The first generation is gone, but their children are thugs, you know that. We'll never be rid of them.

But they don't care about political power. They're all in Monaco or New York, right?

Not all. Some, but not all. And while there's still a rump party around to wreak havoc, it's not over. Not even close.

The democracy movement has the momentum. Democratic communism—you have to love it. We'll end up the reverse of the Chinese. They became dictator capitalists, we'll become democratic communists.

In Zurich you can maybe believe that. Here it's not so simple.

You have to admit it's getting better.

Better, yes. Because there was so much room for getting better. But the more we win, the more resistance there is. The better we do the more dangerous it gets for us, do you understand that?

I do.

So, think about it. You've been having some success, and we have too. And so the danger rises. I want to work from the outside now, like you.

Do you think it matters where you are?

I do. This bridge has the highest murder rate of any bridge in the world.

And so you picked it as our meeting place?

I wanted to remind you. I wanted you to feel it.

Tatiana heaved in a cold breath, feeling depressed. Mother Russia, the unhappy bear. Not a surprise, really. She said, Is there something I can do from Zurich?

Of course. Get me out.

Beyond that?

Yes, of course. The new prosecutor general is on our side, at least to the extent of wanting better land protection. He's some kind of animal lover. I think with his help we can get the Council of Judges to approve Yevgeny's slate of reforms.

Do the judges matter?

More than they used to. Makarov is trying to be the anti-Putin, don't you see? Since no one could out-Putin Putin, he's trying the opposite.

Tatiana said, Are you sure no one could out-Putin Putin?

Svetlana gestured at the city. Maybe Peter the Great. Or Stalin. No. That whole line is done. And no one wants to be a second-rate version of the guy who came before. So to make his mark, Makarov's trying to be the opposite. The great reformer, the globalist, the democrat. Tilt more to India than China. Which means more rule of law. Which a lot of people definitely want. And so we have our chance.

Even Yevgeny's program?

Yes. Including legal standing for animals, which is just what you need to forward some of your cases here. With legal standing you can go crazy in your usual style.

I like that, Tatiana said, feeling buoyed at the thought. I want to sue some people here real bad.

Help me and I'll help you.

As always. So let's get off this fucking bridge and go find a drink.

As always. Time for *kiryat*.

Time for *kvasit*.

We will anoint ourselves with one hundred grams.

Or two hundred.

No wonder you're getting fat. Alcohol has calories you know.

Good. I'm hungry too. I'm cold and I'm hungry and I need a drink.

Welcome home.

84

After several years of container ships being sunk on a regular basis, taken out by drone torpedoes of ever-increasing speed and power, the shipping industry had finally begun adapting to the new situation. It was adapt or die; there were only about eleven thousand container ships afloat, only two hundred of them in the Very Large class, and after forty of those were sunk the verdict was clear, the writing on the wall. They weren't going to be able to stop the saboteurs, who still remained unidentified. Maersk and MSC (a Swiss company) both began to rebuild their fleets, and all the big shipyards followed. It was that or die. That one of the biggest shipping companies on Earth was a Swiss company says something about the Swiss, and the world too.

An ordinary container ship was massive but simple. They were very seaworthy, being so big and stable that even when caught in cyclones and hurricanes they could ride them out, as long as their hulls kept their integrity and their engines kept running. And of course their capacity for cargo was immense. They were well-suited to their task.

So the first attempts at transitioning to ships the saboteurs wouldn't sink involved altering the ones that already existed. Electric motors replaced diesel engines to spin the props, and these motors were powered by solar panels, mounted as giant roofs over the top of the cargo. This could work, though the speed of the behemoths was much reduced, there being not enough room on them for the number of solar panels needed to power higher speeds. But if the supply chain of commodities was kept constant, in terms of arrivals at destination ports, these reductions in ship speed, and thus in economic efficiency, were just part of the new cost of doing business. "Just in time"—but which time?

Because they were slow. Fairly quickly there emerged specialized

418 *Kim Stanley Robinson*

shipyards devoted to taking in container ships and cutting them up, each providing the raw material for five or ten or twenty smaller ships, all of which were propelled by clean power in ways that made them as fast as the diesel-burners had been, or even faster.

These changes included going back to sail. Turned out it was a really good clean tech. The current favored model for new ships looked somewhat like the big five-masted sailing ships that had briefly existed before steamships took over the seas. The new versions had sails made of photovoltaic fabrics that captured both wind and light, and the solar-generated electricity created by them transferred down the masts to motors that turned propellers. Clipper ships were back, in other words, and bigger and faster than ever.

Mary took a train to Lisbon and got on one of these new ships. The sails were not in the square-rigged style of the tall ships of yore, but rather schooner-rigged, each of the six masts supporting one big squarish sail that unfurled from out of its mast, with another triangular sail above that. There was also a set of jibs at the bow. The ship carrying Mary, the *Cutting Snark*, was 250 feet long, and when it got going fast enough and the ocean was calm, a set of hydrofoils deployed from its sides, and the ship then lifted up out of the water a bit, and hydrofoiled along at even greater speed.

They sailed southwest far enough to catch the trades south of the horse latitudes, and in that age-old pattern came to the Americas by way of the Antilles and then up the great chain of islands to Florida. The passage took eight days.

The whole experience struck Mary as marvelous. She had thought she would get seasick: she didn't. She had a cabin of her own, tiny, shipshape, with a comfortable bed. Every morning she woke at dawn and got breakfast and coffee in the galley, then took her coffee out to a deck chair in the shade and worked on her screen. Sometimes she talked to colleagues elsewhere in the world, sometimes she typed. When she talked to people on screen they sometimes saw the wind scatter her hair, and were surprised to learn she wasn't in her office in Zurich. Other than that it was a work morning like any other, taking breaks to walk around the main deck a few times and look at the blue sea. She stopped work for birds planing by, and dolphins leaping to keep up. The other passengers aboard had their own work and friends, and left her alone, although if she sat at one of the big round tables to eat, there were always people happy to talk. Her bodyguards left her alone. They too were enjoying the passage. If she wanted she could eat at a small table and read. She would look up and observe the faces talking around her for a minute or two, then go back to her book. Back out on deck. The air was salty and cool, the clouds tall and articulated, the sunsets big and gorgeous. The stars at night, fat and numerous—the salty air more than compensated for by the truly dark skies. Then the new moon fattened, night by night, until it threw a bouncing silver path out to the twilight horizon, sky over water, indigo on cobalt, split by a silver road.

It was beautiful! And she was getting her work done. So—where had this obsession with speed come from, why had everyone caved to it so completely?

Because people did what everyone else did. Because first no one could fly, then everyone could fly, if they could afford it; and flying was sublime. But also now a crowded bus ride, a hassle. And now, on most of the planes Mary flew on, people closed their window shutters and flew as if in a subway car, never looking down at all. Incurious about the planet floating ten kilometers below.

On the eighth day her ship sailed into New York harbor, a dream of a harbor, Cosmopolis itself, and she debarked on a Hudson dock and took a cab to Penn Station and got on a train headed west.

Okay, this was less interesting. But still she could work, sleep, look out the window. And the Americans had finally gotten some high-speed rail built, including this cross-continent line, so it was only another day and then she was in San Francisco, coming out of the ground and walking over to the Big Tower and taking an elevator up to the top, where she had met with the central bankers years before. Her trip had taken nine days, and she had worked every day as if home in Zurich, except that she got more done. And the carbon burn, as calculated by Bob Wharton's own personalized calculator, had been the same as it would have been if she had stayed at home. And the ocean crossing had been beautiful. She had sailed across the Atlantic! And now stood before a picture window in the Big Tower, looking across a huge wedge of the Pacific. Amazing!

"We've been so stupid," she said to Jane Yablonski, still chair of the

Federal Reserve; she had been reappointed by a new president who had been terrified by all the changes. Yablonski looked mystified as she tried to parse which stupidity Mary might be referring to. Mary did not elucidate.

Not immediately, anyway. She needed to hear how they thought things were going first.

The group talked over events they often named—the Heat Wave, Crash Day, the Little Depression, the Transition, the Intervention, the Strange Times, the Super Depression—and she saw it again: these were the rulers of the world, if indeed anyone could be called that. Maybe not, maybe that was the change they were trying to catch with all their glib names: the system as such had escaped them. Change itself was changing.

Still, they undoubtedly held a lot of whatever power remained. No doubt about that. Because money mattered.

Now they were gathered to discuss the first years of their big experiment, what Bob Wharton called their Hail Mary: the carbon coin. It was very like meetings convened to discuss the Indians' Pinatubo interventions: climatologists in those meetings, bankers in this one. Here too they had made an experimental casting into the air, in this case of gold dust. What had happened?

They listened to their assistants give reports on that for a couple of hours. Abstract after abstract, information crushed to crystalline density. Then it was back to them, looking at each other around the big table. Time for reckoning: had it done what they hoped it would? Had it worked?

Yes and no and maybe. The usual answer to any question these days. But in many senses, yes, it had worked. This was what Mary heard them saying to each other, cautiously and indirectly. Around this table set so high above the beautiful city, looking at each other, she saw that their faces were pleased, even when expressing nervous concern.

They agreed on these points:

The new carbon coin had stimulated many short-term investments in carbon sequestration projects, and many longer-term investments in the coin itself. It had caused some of the biggest carbon owners to cash out and keep fossil carbon in the ground, or use it for plastics if they could. Coal had become just a black rock you could turn into money by leaving it alone. They had created and paid out trillions of carbon coins, and yet had seen no signs of inflation, or deflation, for those who held that theory; no noticeable price change.

There were currency exchanges where they had seen efforts to manipulate the value of the carbon coin against the values of other currencies, including people trying to drive it down in value, in hopes of buying low and selling high later on. Combatting those efforts, if they needed to do that, had been best accomplished by attacking the manipulators in ways that would cause them to desist, while also warning others away from such actions. Tools existed to strongly sanction those who tried it. One report on this had used the phrase *financial decapitation*.

As part of altering the investment climate further, their staffs had written up draft legislation to propose to their governments several related reforms, including strong currency controls; attacking and eliminating tax havens; shifting all money from cash to digital forms tracked by blockchain technologies; and mounting the pressure on carbon by way of increased taxation and regulation. A lot of momentum on those fronts now.

This all sounds good, Mary said at the end of the day. So now, what else can you do?

They regarded her. You tell us, they said. What kind of thing did you have in mind? And why?

Mary sighed. They were never going to be the source of change. It just wasn't in their DNA, either institutionally or individually.

She ran down a list that Dick and Janus Athena had made for her. Outlawing dark pools and killing them off. Putting significant delays in high-frequency trading. Creating high-frequency trading taxes big enough to get trading back to human speed. Calculating basic necessities needed and providing these gratis to under-served communities.

Oh dear oh my! Not what central banks do!

Their first response, as always. But they had already strayed far out of their lane, as one of them put it.

More like a bull over the pasture fence, she thought. A herd of big bulls off into the wild, looking around, didn't see any more fences—this then so scary a sight that at first they tried to conjure back some fences, enclosure being so much more comfy than freedom. Of course trust in money must be protected, as their prime directive; and that directive itself was a fence, hurray!

But wouldn't killing the dark pools create more trust in money, Mary inquired, rather than less?

They had to agree. And in general, less autonomy for loose money might increase trust in fiat money generally.

And high-frequency trading, Mary asked, wasn't it simply rent, parasiting on productive exchanges?

She found it funny to be speaking of rent as if it weren't the fundamental action of the economic system these people upheld, but they liked to think otherwise. Anything to get them to screw their courage to the sticking place. And courage was always necessary when contemplating regicide. Could anyone kill King Capital? Mary doubted it, but if anyone could do it, it was surely these insiders, the people running the system. These thirteen people were close enough to the royal body to get the knives in before anyone could stop them. Et tu, Brute? Yes, Caesar, me too. Die. Into the ash can of history.

So now they nodded as they discussed things, and translated Mary's proposals into more respectable policy formulations. They avoided any large conclusions that would make them aware of their temerity. Just numbers, juggling some numbers, and all for the sake of stability. And since this was partly true, she could let them get on with it, encourage them more.

At the end of the meeting, the Chinese finance minister, Madame Chan, joined her at the west window, where the late sun was making the Farallons again look like a sea serpent's spine. What was that rising, there in the west?

These are just trims, she said to Mary with a smile. Buttoning up the coat you already made. Surely you don't mean to stop there?

What do you have in mind?

We've been looking at what India is doing. They're leading the way now in all kinds of things.

They were radicalized, Mary said.

Yes, and who wouldn't be? We don't want what happened to them to happen in China, or anywhere else. So now they're teaching us regenerative agriculture, and we need it. But of course it keeps coming back to how we pay for these good things.

I suppose, Mary groused.

Chan smiled. Of course. Think of it as land reform. That's a financial arrangement too. So, land taxes, which in China means a tenure tax. Creation of a commons for every necessity. Also, simply the legal requirement that private businesses be employee-owned.

Mary shook her head skeptically as she listened, but she was smiling too, thinking that now the baton had passed to this woman. A woman with real power, huge power. We'll back you, she said happily. Take the lead and we'll back you. And Madame Chan nodded, pleased with her.

Afterward Mary had been planning to fly home, but she cancelled that, and got her team to book a train sleeper and then a clipper ship, New York to Marseille. She worked all the way home.

85

Hi, I am here to tell you about Argentina's Shamballa Permaculture Project. We are representatives of Armenia's ARK Armenia, happy to be here. Down in Australia we've connected up our Aboriginal Wetland Burning, Shoalwater Culture, Gawula, Greening Australia, How Aboriginals Made Australia, Kachana Land Restoration, One Acre Small Permaculture Project, Permaculture Research Institute, Purple Pear Farm, Rehydrating the Landscape, Regenerate Australia, and the Yarra Yarra Biodiversity Corridor. We're a busy crowd!

We are from Belize's Coral Reef Restoration. I represent Bolivia's Food Security. I, Borneo's How to Restore a Rainforest. We are from Brazil's Centro de Experimentos Forestaisis, also Restoration Through Agroforestry. We come to you from Burkina Faso's The Forest of Lilengo, and the group Reforesting with Ancient African Farming Practices. Cameroon here: the Bafut Ecovillage.

Canada's delegation represents Bkejwanong Traditional Knowledge, Jardins d'Ambroisie, the Great Bear Rainforest, Miracle Farms, Taking Root, and Water for the Future. China's delegation is happy to speak for A Man Plants a Forest, Eco-Civilization, Greening China's Desert, Horqin Desert Reforestation, Karamay Ecological District, Kbuqi Desert Greening the Silk Road, Loess Plateau Watershed Restoration, Transforming Deserts to Cropland, World's Largest Man-made Forest, and Zhejiang Green Rural Revival Program.

From the Congo we are Bonobo Conservation, Participatory Mapping, and Virunga National Park. From Costa Rica, we are the Orange Peel Experiment, Punta Mona, and Vida Verde Water Retention. I am here to proudly speak for Cuba's Sustainable Agriculture Revolution, on its eighty-fifth birthday. We are from Denmark's Vitsohus Permakultur.

I have been sent to you by Ecuador's Cloudforest Agroforestry. I from Egypt's Creating a Forest in the Desert. From England we represent Agroforestry Research Trust, the Eden Project, Knepp Estate Rewilding, Rewilding Britain, and River Restoration.

From Eritrea we are Manazares Mango Regeneration, from Ethiopia, Ethiopia Rising, the Miracle of Merere, Regreening Ethiopia's Highlands, Restoring Ecosystems, and the Watershed Movement. For France we happily represent Pur Projet Agroforestry. From Guatemala we are Reserva de Biosfera Maya and the Asociación de Comunidades Forestales. We speak for Reforesting Haiti. From Holland we represent the Land Life Company and Tiny Forest.

Honduras sends you the Roots of Migration. The Hopi Nation is here with Hopi Raincatchers. Iceland sends you Afforesting Iceland, and Regenerating Forests. From India we come to tell you of Agroforestry in Arau, Aravali Institute of Management, Barren Land into Luscious Forest, Cooperative Sustainable Agriculture, Creating a Forest in a Cold Desert, Crowd Foresting, the Farmer Scientist, Fishing Cat Conservancy, Food Sovereignty, Foundation for Ecological Security, Hand-Planting a 300-Acre Forest, Mangrove Restoration in the Sundarbans, Miracle Water Village, Miyawaki Afforestation Program, Natural Farming, Navdanya Biodiversity Farm, Planting 50 Million Trees in One Day, Planting 66 Million Trees in 12 Hours, Protecting Rice Diversity, Rejuvenated Lakes in Bangalore, Sadhana Forest, Sai Sanctuary, Seeds of Life, Sikkim the First Organic State, Water Fields, Water Harvesting, and River Restoration.

Indonesia greets you with its Biorock Coral Reef Restoration, Mangrove Action Project, Mangrove Restoration, and Manta Reef Restoration Project. Israel present its Growing Forests in the Desert, and Kibbutz. We are from Japan and bring news of Creating Forests to Reduce Tsunamis. I come from Jordan to tell you of Greening the Desert. We Kenyans are proud to represent East African Hydrologic Corridor, Green Belt Movement, Kaikipia Permaculture Center, Northern Rangelands Trust, and Rainwater Harvesting. Madagascar celebrates its Great Forest Restoration by the example of Project Moringa, and Using Trees to Save Lives.

Mexico presents Greening the Chihuahuan Desert, Intensive Silvopasture, Hidden Rivers, Reforestation and Water Protection Group, and Via Organica. Morocco brings you its organization Making the Desert Bloom.

We come from Nepal to describe the Beyul Project and the Anthropocene Wilderness Group. New Zealand here: Hinewai Reserve, Mangarara Eco Farm, and The Regenerators. As with all the other countries gathered here, tip of the iceberg really.

I come from Niger to tell you of Farmer Managed Natural Regeneration, and Re-greening in Niger. From Norway we speak of Polar Permaculture Solutions. The Oglala Lakota Nation sends a representative from the Oglala Lakota Culture Economic Revitalization. We are from Panama's Mamoni Valley Preserve. Peru wants to tell you of its Biocorredor Martin Sagrado. We from the Philippines have news of Forest Regeneration, National Greening Program, and Saving a Fishery and Coral Reef.

We from Portugal celebrate the Fazenda Tomati Permaculture, Sown Biodiverse Pastures, Tamera Water Retention Landscape, and Wildlings. I am from Qatar's Sahara Forest Project. From Russia with love: Pleistocene Park, Wet Carbon Storage, and Siberian Wildlife Support. I am from Rwanda's Forests of Hope. We Scots are here for Britain's Oldest Forest, Dendreggan Forest Restoration, Isle Martin Regeneration, On the Deep Wealth of This Nation, and Peat Bog Restoration. I speak to you for Senegal's Great Green Wall Initiative, also Rolling Back the Desert. From South Africa we speak about EcoPlanet Bamboo, Joe's Garden, Making the Desert Bloom, and Restoring the Baviaanskloof.

I am from the South Korean project Reforestation with 350+ Million Trees. I am from the Spanish project Camp Altiplano. I am from the Syrian project The Art of Regeneration, please help us. I come from Tajikistan to tell you of Reforesting Tajikistan. From Tanzania we speak for Reforesting Kokota Island and Reforesting Gombe's Surround. Tasmania here, we've got a great Giant Kelp Restoration Project, also Restoring Pine Plantations. I am from Thailand to tell you about Indigenous Knowledge and Forestry, Planting Mangroves, and Sahainan Permaculture Farm. Uganda sent me to speak of Permagardens and Uganda's New Forests.

We come from the United States as representatives of Accelerating Appalachia, American Prairie Reserve, Broken Ground Permaculture, Food Forest Farm, Holistically Managed Bison, Homegrown Revolution, Institute of Permaculture, Kiss the Ground, Klamath River Basin Restoration, Lake Erie Bill of Rights, the Land Institute, the Leopold

Institute, Los Angeles Green Regeneration, Mirroring Nature's Management, Permaculture for the People, Planetary Healing, Planting Justice Farms, Regenerative Ranching, Restoring the Colorado River, Restoring the Redwoods, Restoring the Snake River, Rodale Institute, Saving the West, the Sierra Club, Singing Frogs Farm, Soil Health Institute, Stewards of the Wild Sea, Tabula Rasa Farms, Tending the Wild, Weaving Earth, Wild Idea Buffalo Company, and Wild Oyster Project.

I am from Zambia to tell you of the Betterworld Mine Regeneration. I am from Zimbabwe to speak for the Africa Centre for Holistic Management, and the Chikukwa Ecological Land Use Community Trust.

We are all here together to share what we are doing, to see each other, and to tell you our stories. We are already out there working hard, everywhere around this Earth. Healing the Earth is our sacred work, our duty to the seven generations. There are many more projects like ours already in existence, look us up on your YourLock account and see, maybe support us, maybe join us. You will find us out there already, now, and then you must also realize we are only about one percent of all the projects out there doing good things. And more still are waiting to be born. Come in, talk to us. Listen to our stories. See where you can help. Build your own project. You will love it as we do. There is no other world.

86

Back in Zurich, Mary emerged from a couple of weeks of intense work, every waking moment right at the face of the seam, so to speak. She was ready for a break. Frank May was now out of jail, living in co-op housing near the jail; what was he doing? How was he doing? She had lost track of him now that he was out, and was almost afraid to check in.

"I'm doing okay," he told her when she called. "Hey listen, I'm going to the Alps tomorrow to look for chamois. *Rupicapra*. It's some kind of goat-antelope, and the maps show they're all over up there. Do you want to come?"

"Look at animals?" Mary said dubiously. "Like at the zoo?"

An impatient snick. "Except no zoo."

One could use tracking apps to see where the creatures were hanging out, then go there and probably see some. He had done it before and liked it. There were some above Flims, it was a nice region.

"I guess so," Mary said.

So they met at Hauptbahnhof just before the 6 AM departure to Chur, and an hour later, after eating breakfast on the train, sitting silently next to each other, uncomfortably aware perhaps that they had never done anything like this before, they got off and switched to the narrow-gauge train that headed upvalley to the Vorderrhein. This was a much slower train, but they didn't stay on it long, getting off and taking a waiting bus up to Flims. A cable car from the station there lofted them high up into a big south-facing basin, elevation about 2,000 meters where they got left off. It was 8:30 AM and they were in the Alps. Mary had told her bodyguards in advance to leave them once they got on the trail, and they did; as so often, there was a little restaurant at the cable car's upper station, and they would wait there.

She and Frank found themselves hiking up into a big indentation in the Alpine range that formed the northern sidewall of the Rhine River's uppermost headwaters. Near the end of the last ice age, after the ice had melted out of this part of the valley, a massive landslide, one of the biggest ever known to have occurred, had slumped down this south-facing wall. The entire village of Flims rested on the flat top of the remains of this landslide. Above it the green alps filled overlapping stacked bowls of rock, rising more and more steeply to the Tschingelhörner ridge, a wall of steep gray crags with a horizontal crack running through it. This crack was so deep that it had created a gap in the range, a giant window of sorts through which one could see a big patch of sky, well under the gray crags above. Yet another strange Alpine feature resulting from millions of years of ice on rock.

Trails ran up the green alps under the crags. Mary and Frank ascended one that led them westward, away from ski lifts and farms and other human sites, toward wilder territory, what in the Alps passed for wilderness. The wild creatures of the Alps couldn't afford to be too picky, Frank told her as they hiked, when it came to hanging out near people; if you weren't on sheer vertical rock, people were going to be passing by pretty often.

Though it looked like a gentle upward slope, this was partly because the gray cliffs ahead and above were so steep. In fact their ascent was quite a slog of a climb. By the time they got up into one of the highest bowl meadows, floored with a rumpled carpet of short grass, studded with rocks and spangled with alpine flowers, they were tired and hungry, and the sun was well overhead. They sat on a low boulder and ate.

The meadow was littered with big fallen chunks of the ridge above, gray boulders that had detached and crashed down and rolled onto the meadow; or perhaps they were erratics, conveyed by the ice of a long-departed glacier and dropped there when the ice melted. As they sat on their low rock, eating their bahnhof sandwiches and drinking from water bottles, they were rewarded for their silence by a first sighting of alpine creatures: in this case, marmots.

These were fat gray things, like groundhogs or maybe badgers; Mary didn't have much basis for comparison. The color of their fur no doubt made predators take them for rocks, including the hawks soaring overhead. Perhaps because of these overseers, the marmots seemed to have a tendency to stay still; except when moving from one spot to another, they were as motionless as the rocks they were on. They spoke to each other by way of high staccato whistles. As Mary and Frank listened, it became clear that this must be a language much like any other.

"Down in Flims they speak Romantsch," Frank mentioned.

The marmots did not mind them talking, Mary saw.

Frank saw this too, and went on. They had heard a little Romantsch in the bus on their ride up from the train station to Flims, he said. It was like Italian and German put in a blender, they agreed. They shared their pleasure in the story of how Romantsch had become one of Switzerland's four national languages, by way of a rebuke to Hitler. Thus the national myth, and they were both inclined to believe it.

The sun beat on the meadow, causing it to shimmer. It was warm in the sun. The marmots got comfortable with them sitting nearby, even though they had apparently occupied a marmot outpost; piles of little dry turds were clustered in the cracks on the top of their boulder. Herbivores, by the looks of it.

Frank and Mary sat there like big marmots. They didn't say much. Mary thought it a little dull. Then one of the younger marmots, judging by size and fineness of fur, ambled their way, unconcerned by their proximity. It stopped, reached out a forepaw, pulled all the grass stalks within its reach toward its face. This created a small cluster of tiny grass seeds clutched inside its forearm, which it then munched off the top of the stalks. It only took a few bites. When that gathering and eating was done, it let the topless grass blades spring back into position and moved on. Did it again. Then again.

Seeing this, then looking around the meadow at the level of the grasstops, Mary suddenly realized that the little beast's source of food was almost infinite. At least now, when the grass was seeding. Probably it was the same for all the alpine herbivores.

Frank agreed when she mentioned this. The marmots would eat all day every day, until they were fat enough to get through the coming winter. They hibernated through the winter like little bears, tucked in holes under the snow and living off stored fat, their metabolisms slowed to a crawl. In the spring they would emerge to another summer of eating.

Then a clutch of larger animals appeared over a low ridge. Ah ha! Chamois!

Probably these weren't the animals Frank had seen online, as none wore a radio collar. If that was in fact how GPS got attached to animals these days; she didn't know. Frank was watching them closely.

They were odd-looking beasts. Round-bodied, short-necked, short-legged, snub-faced. Short curving horns. They had the rectangularly pupilled eyes of a goat. Devil's eyes. After Frank had proposed this trip, she had read they were "goat-antelopes," whatever that meant. Obviously they were their own beast, neither goat nor antelope, and not even much like the other species in their same family. The youngsters were slender and hornless, and stayed near their moms. They nibbled, looked around, walked calmly over rocks from one patch of grass to the next. They regarded Frank and Mary curiously. She was surprised that they seemed so unconcerned to see people; there were Swiss hunters, or so she had heard, and these beasts were among the most commonly hunted. Why were they so unafraid?

She muttered this question to Frank.

"Why should they be afraid?" he replied.

"We might shoot them."

"We don't have any guns."

"Do they know that?"

"They've got eyes."

"But have they seen guns before? Wouldn't that put them off people entirely?"

"Probably. So maybe they haven't seen any guns."

"I find that surprising."

"The Alps are wild."

"I thought you said they weren't."

He thought about it. "They are and they aren't. Lots of people up here, yes. But the Alps can kill you quick. They're savage, really. Didn't you say you went over one of the high passes?"

"I did."

"That should have taught you."

She nodded, thinking it over. "That was definitely wild. Even savage, yeah—if the weather had turned, sure. Nothing but rock up there."

"And that was a pass. There are lots of places up here where people don't ever go. They're really hard to get to, and they don't lead anywhere. If you look on the maps you see them all over."

"Not like this, then," Mary said, gesturing around. On the grass under their boulder lay scattered about twenty varieties of alpine flower, either tucked into the grass or waving over it, flowers at different heights for different plants, so that the air was layered by color: at the top yellow, waving over a white layer; lower still a blue layer; then the grass, spangled with a variety of ground flowers.

"Not like this." Frank smiled.

Mary saw that with a start; it seemed to her that she had never before seen him smile. Flower-filled meadow, wild beasts grazing all careless of them, the young ones literally gamboling, defining the word as they popped into the air and staggered around on landing, then did it again. Gray wall above, with a window in it to make it Alpine-strange. Blue sky. It was definitely a cheerful sight. Even a little hallucinogenic. Breeze flowing over the flowers like a tide, so that they bobbed in place. The young marmot still there near them continued to draw grass stalks to its mouth. The oily sheen of the bunched seeds it had caught in its paw gleamed in the sun. Quick little fans of food. The demon eyes of the chamois just a bit farther away, placidly chewing their cuds, unafraid of anything.

"I like this," she said.

"Seeing the animals?"

"Yes."

The slight smile returned. "Me too."

They watched for a while.

"Have you seen many?" she asked him. "Animals in the wild, I mean?"

"Not many. I'd like to see more. So far almost all the ones I've seen have been up here in this basin, or basins like it. These guys, mainly. Marmots and chamois. Once an ibex, I think it was. Another time something like a marten. I looked it up later and it seemed like it must have been a marten. It was by a creek up here, with some trees on the other bank. It was running around like a crazy person, back and forth. I couldn't understand what it was doing. Really fast, but erratic. Didn't seem to be hunting or building a den, or whatever. Just dashing. Slinky thing with dark fur. Very intent on its own business. I wanted to pull my phone out of my pocket and take a picture, but I didn't want to spook it. I didn't want it to notice me. So I stayed still."

"How did that end?"

"I had to leave to make the last chairlift."

She laughed. "That's the Swiss for you."

"So true." He plucked a grass stem from under him and began using it as a toothpick. "What about you?"

She shook her head. "Galway and Dublin aren't really places for wild-life. I liked going to the zoo when I was a girl. That wasn't quite the same."

"Yes. What you would want is animals around you where you live."

"Probably so. That's what they're working on in California. They've got so much land there, and they've been rewilding for a long time. My contacts there called it the Serengeti of North America, but they were referring to before Europeans arrived. It's something they're trying to get back."

"Nice for them. But we live in Zurich."

"Right. I don't know. I walk the paths on the Zuriberg pretty often, but I've not seen any animals up there. I'm a bit surprised, now that I think of it. It's quite a big forest."

He shook his head. "Surrounded by city. That doesn't work."

"So they need habitat corridors, you mean. Do those work?"

"I think so. If they're wide enough, and connect up big enough areas."

"That's what they're doing in California."

"In lots of places, I gather. I think it might work in Switzerland too. Although these beasties might not like it if wolves come back."

"Is that who used to eat them?"

"Yes."

"I can't imagine the Swiss getting on with wolves either."

"You never know. I read that with their glaciers going away, they're thinking of reforesting some lower alps, and helping start plants on the exposed glacial basins. That would make more wolf zone, I think. Wolves need forests, but they're good in open country too, and there's going to be more rocky empty areas exposed higher up, with marmots and moles and squirrels. And those areas will be pretty high, and as far from people as they can get in this country."

She shook her head. "Hard to believe wolves could come back."

"No more so than these guys. At this point, every wild creature is unlikely. It's going to be tough to come back."

"Not if the Half Earthers have their way."

He nodded. "I like that plan."

They sat there. No reason to leave. Nowhere else they could get to that afternoon would be better. So they sat. Frank chewed a grass stalk. Mary watched the animals, glanced at him. His body was relaxed. He sat there like a cat. Even the marmots and chamois were not as relaxed as he was at that moment. They were busy eating. And indeed that was what would get Mary and Frank off their rock; hunger, and the need to pee.

And darkness. The sun hit the ridge to the west and immediately the air felt cooler. Shadows fell across the meadow.

Frank glanced up at the ridge, at her.

"What say?"

"We should get back down I guess. There's a last cable car here too."

"True."

They stood, stretched. The chamois looked up at them, wandered off. Without appearing to hurry, they were soon across the meadow; and when they had gotten among the rocks bordering the meadow, they disappeared. It was like a magic trick. Even trying, even knowing they were there, Mary couldn't see them.

The marmots didn't seem to care that they were moving. Then one whistled, and the young one who had been feeding near them galumped away and ducked under a boulder. Frank looked up, pointed. A bird far overhead, soaring. Hawk, maybe.

They started toward the trail that led down to the cable car. Then Frank lurched forward and fell on his face.

Mary cried out, rushed to him, crouched by his side. He grunted something, looking stunned. Put his hands to the ground and pushed himself up, rolled into a sitting position, sat there with his head in his hands. Felt his face, his jaw.

"Are you all right?" Mary exclaimed.

He shook his head. "I don't know."

"What happened!"

"I don't know. I fell."

87

The town meeting included pretty much every single person left in the area; that meant about four hundred people. Looking around the old high school gym, where most of us had gone to school back in the day, we could see each other. Everyone we knew. We knew each other by name.

The deal had come down from some UN agency called the Ministry for the Future, by way of the Feds and the Montana state government. Everyone was offered a buy-out that pretty much covered the rest of your life; housing costs in expensive places, enrollment in the school of your choice, and options that if taken, might allow most of us to move to the same city. Probably Bozeman. Some argued for Minneapolis.

Everyone already knew the plan. The night before, the one movie theater still running had screened *Local Hero*, a Scottish film in which an international oil company based in Texas offers to buy a Scottish coastal village from the inhabitants, so they can rip it out and build a tanker port. The pay-out will make everyone rich, and the townspeople all cheerfully and unsentimentally vote in favor of it. They have a final ceilidh to say goodbye to the town and celebrate everyone becoming millionaires. Then the owner of the oil company arrives by helicopter and declares the town and its beach need to be saved for an astronomical observatory, astronomy being his personal hobby. Burt Lancaster. A funny sly movie. We watched it in silence. It was too close to home.

Our situation was not so different, although they weren't going to knock the town down. It would be left to serve as some kind of emergency shelter, and headquarters for local animal stewards, who could be any of our kids, if they cared to do it, or even us, if the idea of coming back to the town empty appealed. And we could come back once a year to visit the place. The movie theater had screened *Brigadoon* a few nights before, probably to show how stupid that would be. No one had laughed at that one either. Obviously Jeff, the owner of the theater, who had kept the business going at a loss, didn't want us to close the town. He was whipping on us a little. As we came in he was playing Simon and Garfunkel's "Homeward Bound" over the speakers, really piling it on. But the vote had been decisive.

The main thing was, it was going to happen anyway. Or it already had happened. Jeff could have screened a zombie movie to show that aspect of it. Because all the kids were gone. They graduated high school, having bused to the next town over for it, and went off to college or to find work, and they never came back. Not all of them of course. I myself came back for instance. But most of them didn't, and the fewer that came back, the fewer came back. Positive feedback loop with a negative result; happens all the time, it's the story of our time. The town's population had peaked in 1911 at 12,235 people. Every decade after that it had gone down, and now it was officially at 831 people, but really it was less than that, especially if you didn't count the poor meth addicts, who were zombies indeed. One store, one café, one movie theater, courtesy of Jeff; a post office, a gas station, a school for K through 8, a high school the next town over with not enough students and teachers. That was it.

And of course we weren't the only one. I don't know if that fact made it worse or better. It was happening all over the upper Midwest, all over the West, the South, New England, the Great Lakes. Everywhere on Earth, we were told. You could buy an entire Spanish village for a few thousand euros, we were told. Central Spain, central Poland, lots of eastern Europe, eastern Portugal, lots of Russia—on and on it went. Of course there were countries where villages were turning into cities right before their eyes, cardboard shacks melting in every rainstorm, but no. No one thought that was a good way, and anyway it wasn't our way. We were in Montana, and our little town's numbers had dipped below the level of function and habitability. Our town had died, and so here we were looking around at each other.

The buy-out was generous. And if enough of us agreed to move to the same city, we would still have each other. We would constitute a neighborhood, or part of a neighborhood. We would have enough to live on.

We could come back here once a year and see the place, see the land. After a while, those of us who had actually lived here would die, and then there would be no reason to come. Then they would let the buildings fall down, presumably, or salvage them for building materials. The land would become part of the Greater Yellowstone Ecosystem, one of the greatest ecosystems on Earth. Buffalo, wolves, grizzly bears, elk, deer, wolverines, muskrats, beaver. Fish in the rivers, birds in the air. The animals would migrate, and maybe if the climate kept getting hotter they would move north, but in any case it would be their land, to live on as they liked. The people still here, or still visiting here, would be like park rangers or field scientists, or some kind of wildlife wrangler, or even I suppose buffalo cowboys. Buffaloboys. The authorities were vague on that. They admitted it was a work in progress.

They planned to pull out a lot of the roads in the region. A few railroads and the interstate would remain, with big animal over- or under-passes added every few miles. Some regional roads too, but not many. Most would be pulled up, their concrete and asphalt chewed to gravel and carried off to serve as construction materials elsewhere as needed. Making concrete was bad in carbon terms, so the price on new concrete was astronomical now, taxed to the point where anything else was cheaper, almost. Recycled concrete from decommissioned roads and old foundations of deconstructed buildings was a way to get rich, or at least do very well. We were given shares in the roads that came through town, also the town's foundations and so on. A kind of trust fund going forward. Later, with the roads gone, the animals and plants would have it better than ever. Fish in the creeks, birds in the air. The Half Earth plan, right here in the USA.

So we talked it out. Some people broke down as they spoke. They told stories of their parents and grandparents, as far back up their family tree as they could climb, or at least back to the ancestors who had first come here. We all cried with them. It would come on you by surprise, some chance remark, or a face remembered, or a good thing someone did for someone. This was our town.

All over the world this was happening, they kept saying. All these sad little towns, the backbone of rural civilization, tossed into the trash bin of history. What a sad moment for humanity to come to. City life—come on. All the fine talk about it—only people who never lived in a small town could say those things. Well, maybe it only suited some of us. And not the majority, obviously. People voted with their feet. The kids were leaving and not coming back, plain as that. So we would move too. Seemed like about half wanted Bozeman, half Minneapolis. That would be like us. Maybe all little towns had that kind of either-or going on, a nearly fifty-fifty split on everything; the mayor, the high school principal, the quarterback, the best gas station, best café, whatever. Always that either-or. So we would split up and go on. Become city folk. Well, it wouldn't last forever. This was degrowth growth, as the facilitator pointed out. The facilitator was really good, I have to say. She encouraged us to tell our stories. She said in towns like ours it's always the same. She had done this for a lot of them. It was her work. Like a hospice preacher, she said, looking troubled at that. Everybody cries. At least in towns like this. She said that when they do this same thing in suburbs, no one cares. People don't even come to the meetings, except to find out what the compensation will be. Sometimes, she said, people are told their suburb is going to be torn down and replaced by habitat, and they cheer. We laughed at that, although it was painful to think of people that alienated. But that was the suburbs. For us—we had a town.

But it was happening everywhere, she said. All over the world. And after that, in the years and centuries that followed, there would come a time when the world's population drifted back down to a sane level, and then people would move back out of the cities into the countryside, and the villages would come back. Villages used to be just part of the animals' habitat, she said. Animals would walk right down the streets. They do here already! someone shouted. Yes, and it will happen again, she replied. People who like knowing their teachers, their repair people, their store clerks and so on. The mayor. Everyone in your town. All that is too basic to go away forever. This is just one stage in a larger story. People lived like this for a long time. But now it's some kind of emergency.

And we would still have each other. And we would be rich, or at least well off—secure—with an annuity for life, and a trust fund for the kids. We would still have something to do on a Saturday night. So it would be all right.

But after a while your eyes began to hurt. People broke down and couldn't finish what they were saying. Their friends helped them off the stage. This was our town. This was who we were.

Finally there was nothing left to say. It was midnight and we closed up the town like in a fairy tale. Nothing left to do but go home, feeling hollow, stumbling a little. Go in your house and look around at it. Pack your bags.

88

We eat shit and we shit food. A time comes when we need to move together. We follow the sun widdershins, moving all together. We are a herd made of individuals. We move in lines one after another. The land we walk over is mostly water. When we walk on water we grow frightened and hurry to return to land. Some of us lead astray the stupid, others urge fools to rash adventures. If we follow the wrong leader we die. If we panic we die. If we stay calm we are killed. You could eat us but we are of more use to you in other ways, so you rarely do. By our passing we render the land in ways you need more than ever before. We are caribou, we are reindeer, we are antelope, we are elephants, we are all the great herd animals of Earth, among whom you should count yourselves. Therefore let us pass.

89

Mary went to work the next day feeling uneasy.

"How was your day in the Alps?" Badim asked her.

"It was grand," she said. "We sat in a meadow and looked at marmots and chamois. And some birds."

He regarded her. "And that was interesting?"

"It was! It was very peaceful. I mean, they're just up there living their lives. Just wandering around eating. It looked like that's what they do all day."

"I think that's right," Badim said, looking unconvinced that this would be interesting to watch. "I'm glad you enjoyed it."

Then Bob Wharton and Adele burst in, excited; word had come in that the latest CO_2 figures showed a global drop, a real global drop, which had nothing to do with the season, or the economy tanking—all that had been factored in, and still there was a drop: it was now at 454 parts per million, having reached a high of 475 just four years before. Thus 5 ppm per year down: this was significant enough that it had been tested and confirmed in multiple ways, and all converged to show the figure was real. CO_2 was going down at last; not just growing more slowly, or leveling off, which itself had been a hugely celebrated achievement seven years before, but actually dropping, and even dropping fast. That had to be the result of sequestration. It could only be anthropogenic. Meaning they had done it, and on purpose.

Of course it was bound to happen eventually, they told each other, given everything that had happened. The Super Depression had helped, of course, but that impact had been factored in, and besides that would only have caused things to level off; for a real drop like this one, drawdown efforts were the only explanation. Bob said that reforestation and the greening of the ocean shallows with kelp were probably the major factors. "Next stop three-fifty!" he cried, giddy with joy. He had been fighting for this his whole career, his whole life. As had so many.

The rest of that hour was a celebration, mainly. They toasted the news with coffee. No one had ever seen Bob so exuberant before; he was usually a model of the scientist as calm person.

But when everyone left her offices, Mary realized that she was still uneasy. She texted Frank to see how he was doing. Fine, he replied, and nothing more. As if nothing had happened.

So at the end of her day she trammed down and walked to his co-op, just to see for herself. There he was in the dining hall, sitting on the piano bench with his back to the piano. He looked only mildly surprised to see her. Or like he knew why she was there.

"What?" he said defensively, when he saw her looking at him.

"You know what," she said. "Did you go to the doctor yet?"

"No."

She regarded him. There was something about him more pinched than usual. "That was no ordinary fall," she told him.

"I know. I felt faint."

"So you need to see a doctor. Get checked out."

He pursed his lips unhappily. She could see he wasn't going to do it. It was like looking at a child.

She sighed. "If you don't, I'll tell the people here. As it is, you're a danger to them. You'll fall and hurt yourself, and maybe someone else, and their insurance costs will go up."

He gave her a bitter look.

"Come on," she said. "I'll go with you to the clinic. After that we can go get a drink."

He grimaced. After a long pause, looking down at the floor, he shrugged and stood.

The clinic took the usual stupid amount of time, and in the end it was just the start. They measured him and took blood and asked questions and made an appointment for him to return. And after all that, he didn't want to go get a drink. So Mary walked him back to his co-op and then went to the tram station and took a 6 tram up the hill to Kirche Fluntern, past her old apartment and up the Zuriberg to her safe house. Bodyguards following her all the way, yes; she always gave them a nod when she saw them and then forgot about them, or tried to.

She sat there in the strange house feeling low. Something to celebrate, and no one to celebrate with. Even the beasts of the field had company of an evening. Ireland too; in the old days the men would go to the pub, the women gather in kitchens or hang out with the kids. A social species. Of course there were animals such as cats who were isolatoes most of the time. Some people were like that. But not her. She was like the marmots or the chamois. A group to chat with, or just to sit in a room with while reading or watching a screen; this was her preference. She wondered if she should move into something like Frank's co-op. The bedrooms there were private, small and cozy, and then all the rest of it was communal— the big kitchen, the dining room, the common room with its books and piano. People moved into places like that to have that experience.

Well, but she would probably just sit in her room anyway. And here she was. They probably wouldn't let her move. And she didn't like to move. All the moves, all the years; she was tired of it.

That night she slept uneasily, and the next morning it was a relief to walk down to work. This group of people who worked for her were serving as her family, she realized. That was fine, a lot of people functioned that way, maybe, although it felt a little odd when you considered it. Work colleagues did not usually function as family. Family, what was the saying? When you had to go there, they had to take you in. Probably that meant not the people you happened to work with.

Or maybe not. Because when she went into the office she found everyone crying, or palely grim-faced, sitting on desks with their heads in their hands. News had come just seconds before: Tatiana had been found dead outside her apartment, which turned out to have been in Zug. Shot. Her security team said a drone strike of some sort. "Ah no!" she exclaimed, and went to the nearest chair and collapsed in it, getting down before she fell. "No."

The police had put Tatiana in a safe house under full protection, including a big crew of bodyguards, but she had not been regular at following their instructions, they said. They had thought her gone to bed for the night, but she had stepped outside her door, no one knew why; they heard her fall to the ground, rushed out, there she was. Shot three times, no sound of gunfire. Bullets not yet recovered, they had gone right through her and had not yet been found.

Mary stopped asking questions when she heard that, feeling sick. The officers telling her regarded her warily. She would have to move again, they said. Go into better hiding, more complete.

"No!" she said instantly, wildly. "I'm not leaving! I won't leave Zurich. I want to keep working here. We can't cave to this kind of thing. You just do your jobs!"

They nodded, looking worried. Telecommuting, one suggested. For a while. From a secure location. Until they figured out more about how this had happened, and who might have done it.

They had some evidence already, of course. A little body camera that Tatiana had been wearing; Mary didn't want to look at that. What an ugly job these people had. Someone had to do it. The cameras on the building apparently didn't show anything. Shot from a distance. Little to go on. News came in while they were sitting there pondering her, remembering her. They had found two bullets nearby. American bullets. Not a definitive sign of anything. The investigation would continue. And so on.

Tatiana. Their tough one, their warrior. Her brother in arms. They kill the good ones, Mary thought bitterly, the leaders, the tough ones, and then dare the weaker ones to pick up the torch and carry on. Few would do it. The killers would prevail. This was how it always happened. This explained the world they lived in; the murderers were willing to kill to get their way. In a fight between sociopathic sick wounded angry fucked-up wicked people, and all the rest of them, not just the good and the brave but the ordinary and weak, the sheep who just wanted to get by, the fuckers always won. The few took power and wielded it like torturers, happy to tear the happiness away from the many. Oh sure everyone had their reasons. The killers always thought they were defending their race or their nation or their kids or their values. They looked through the mirror and threw their own ugliness onto the other, so they didn't see it in themselves. Always the other!

And of course she was doing it now. So it was another way the bad could infect the good, by making it furious, making it join the general badness. Fuck them, kill them back, jail them, lock them in some room of mirrors where they had to see themselves for what they were. That would be the punishment she would concoct for them, if she could: they would have to live in a mirror box, look at themselves all day every day for the rest of their lives. See what they saw. Narcissists could not look in the mirror, the myth had it backwards.

She tried to focus on her helpers, weeping as she looked at them. Swiss police. Switzerland. Think about how this little city-state of a country had gotten by in the world. In part it had been by accepting each other despite their differences. Some clever rules and a few mountain passes, both now irrelevant to power in the world; really it was just a system, a method. An old hoard and a way of getting along. The faces watching her now, their strange fairness, their insistence on some kind of justice for all. Some kind of enlightened self-interest, the notion that Switzerland was safest when the whole world was safe. Really very odd, this culture; and right now she wished with all her heart that it could conquer the world. No Zurcher would ever do anything like this. This thought made her weep harder than ever for a while.

Then she listened to these individual Swiss people advise her. Work from hiding for a while, please. It could be done. Until they knew what had happened, and knew it wouldn't happen to her too.

Can I live near Utoquai? Can I still go out swimming?

Probably not a good idea. It was well known she was a member of that swim club. Perhaps if she wanted very much to swim in the lake, they could go down the lakeshore and visit a private house, dive in from there.

Mary sighed. She saw immediately that her swimming was a social activity, that going out into the cold water was unpleasant enough to be a ritual that had to be shared by other sufferers to turn it pleasant. That and the shower and meal afterward, the kafi fertig. As with Tatiana, so many times. Their little ritual. She had been such a beauty, such a power.

Now Mary's eyes were burning, now she was furious. Her Swiss helpers were standing around looking uncomfortable. Other people's grief so awkward. One of the women officers sat down next to her, put a hand on her arm. She began to weep like a faucet.

When she stopped, she said, Get me Badim, please.

He was out of the office.

Get him on the phone! Now. And I'll need to be able to confer with him, even when I'm hiding.

So she would agree to hide?

Yes. But in Zurich only. Same safe house she was in now would be best.

They nodded uneasily. If you were to stay in it all the time, for a while, maybe.

She agreed to that. Then she thought of Frank. Oh hell, she said. Fuck.

She asked for a phone, they gave her one. She called him.

He picked up. "I'm sorry your friend was killed," he said immediately. "I heard about it on the news. I hope you're being careful."

"I am. Thank you. Did you find out anything from the doctors yet?"

"Well...They're still working on it."

"Come on, tell me."

"They're still working on it!"

As in, they found something wrong but I don't want to tell you now, when you're upset. Meaning it was bad.

"Tell me now," she said furiously, "or I'll think it's worse than it is."

He laughed.

"What!" she said, alarmed. He was laughing at the idea of there being something worse than what it was. "What is it!"

There was a silence. "Something in my head," he said at last. "As always. But this time it's visible on a scan. Some kind of tumor. They don't know what kind yet. Could be benign."

"Damn," she said.

"Yeah," he said. Briefly he blew air between his lips. "Oh well. It was always going to be something."

After another silence, she said, "I have to go into hiding for a while."

"Good. Hide."

"But I want to see you."

An even longer silence. She thought maybe he didn't approve. That he was letting her contemplate how strange she had become.

"We can talk like this," he said.

"When will you know more?" she asked.

"I go back tomorrow."

"All right. Good luck. I'll stay in touch. I'll come see you when I can."

"Stay safe," he told her. "If you aren't safe, I don't want to see you."

* * *

"Did you get Badim for me?" she asked her minders.

They nodded. She handed them the phone and they tapped at it.

Badim came on the line. Mary stood and left her minders, went into her office, shut the door.

"Can you talk?" she asked him.

"Yes."

"And is this phone private, do you think?"

"Yes."

"So—who killed her?"

"We don't know."

She said angrily, "What good is your fucking black wing if it can't figure out stuff like this?"

He let a silence stretch out, to make her hear how pointless and stupid her question was.

"I loved her too," he said. "We all loved her."

"I know."

After a grim pause, both of them lost in their own worlds, he said, "I'll tell you what I think, although most of it is obvious, stuff you probably already know. I think she was killed by Russians. Russia is really opaque at the top, but they've been making some bold moves lately, and I mean by that some good moves. Really important moves, both in the open and in the black. I think it's very likely Tatiana was part of those. She kept a lot of contacts there. So, whenever a government changes direction like that, it leaves some people behind. They're on the wrong side of the change, they're scared, they get angry. If some of those people who got caught on the wrong side thought that Tatiana was part of the change, maybe even directing it, then killing her could stop the turn, or at least exact a cost for it."

"Revenge."

"Yes, but also maybe an attempt to reverse the change. Serve as a warning to the people Tatiana was working with, and so on."

"All right. So that must mean just a few suspects."

"A few thousand."

"A few groups, I mean."

"A few dozen groups. Russia has been a kleptocracy for a long time.

The people who got rich after the fall of the Soviet Union, their kids have always felt like they ruled the world. So now it's as fractured and complex as anywhere else."

"Even at the top?"

"Especially at the top. This is a fight for how Russia operates."

Mary sighed. "None of this will ever end."

"That's true."

She thought about it for a while. "Listen, I'm not going to hide. I won't be forced into hiding by this sort of thing."

"It's dangerous right now."

"Are you in hiding?"

"Yes. I've got our whole team secured right now."

"But if it was a Russian matter, like you say, then they won't care about us."

"Maybe."

"I want to face up to them. To all of them, to everyone like them. Listen, this year's COP meeting was going to be in Zurich anyway. It's the sixth global stocktaking, so it's supposed to be big. We can use that. Make it as big as can be. A memorial to Tatiana. Everyone there, all of us. And we'll make a public accounting of where we are now. We have to rally the cause. We have to show people where we are, how far we've come, and how we can get the job done."

"It would be a security challenge."

"That's always true! So it's best to put the thing right out in the open and make it as big as possible. No more hiding from these fuckers."

"Maybe so."

"You know I'm right. Otherwise we're just caving in to terrorism. Do that and there'll just be more of it. There'll always be more anyway. Even if you identify and kill these particular murderers, there'll always be more. Meanwhile we have to live."

"All right."

"What?"

"I said, all right. You're the boss. The COP is three months from now. We'll put out the call and prepare for it, and the Swiss security services will go all in. After Tatiana, they'll be worried enough to lock the whole country down."

"Yes. They'll call up their army, I've no doubt. Bigger is better in this case, all round. The whole world watching."

"All right."

"And listen, your black wing—sic them on those bastards!"

"I already have. When we find them, we'll kill them."

"Good. I hope you can find them."

So she let the security people move her to a new safe house, near the Opera. She stayed in it all the time for a couple of weeks, communicating by phone and screen with her team and with everyone else she needed to talk to. They helped her to organize the coming COP as the regularly scheduled global stocktake, plus more: a full progress report from every country, every continent, every industry, every watershed. To that account of the good done, they would add a description of every outstanding problem, every obstacle to getting where they needed to be. The global situation was to be judged actor by actor. Rated, scored, judged; and if judged malingering, then penalized. Time was passing, patience was running out. The sheriff would have to be formed by a concoction of every sanction and penalty they had at hand. The general intellect. The world in their time. In all the blooming buzzing confusion of their moment, they would take stock and get some clarity on the situation, and act. And if Badim's black wing had anything to it, they would act there too. The hidden sheriff; she was ready for that now, that and the hidden prison. The guillotine for that matter. The gun in the night, the drone from nowhere. Whatever it took. Lose, lose, lose, lose, lose, lose, fuck it—win.

Meanwhile, however, as they approached the time of this conference, a lot of the news coming in was actually quite good. Real progress was being made on many fronts. The main point, or maybe what the financial sector would have called the index that stood for all the other factors involved—which was to say, the amount of carbon dioxide in the atmosphere—it really had dropped in the previous four years, and pretty sharply too. This was confirmed from multiple sources. And it had been leveling in the previous ten years before that, shifting up and down with the seasons as always, but staying more level than at any time since measurements had begun in the 1960s. That in itself had been celebrated, but it was now cycling downward between 450 and 445, they said, cycling seasonally as it did, with the current trend moving downward by about 5 parts per million per year, but that rate too was increasing, it looked like. This meant not only that they had stopped burning carbon to a large extent—not entirely, because that would not be possible in their lifetimes—but they were also drawing carbon down from the air in significant and measurable quantities, by way of all the carbon drawdown efforts in combination. There were discussions as to how much the oceans were still serving as a sink for carbon burned into the air, but now, in the Great Internet of Things, the Quantified World, the World as Data, all these aspects of the problem were being measured, and the ocean's uptake or drawdown was measured to within a fairly small margin of error; and the conclusion from the scientists involved was that since the ocean had already been quite saturated by the carbon it had absorbed in the previous three centuries, the drop they were seeing was only slightly explained by continuing ocean uptake. The majority was being drawn down by reforestation, biochar, agroforestry, kelp bed and other seaweed growth, regenerative agriculture, reduced and improved ranching, direct CO_2 capture from the air, and so on. All these efforts were paid for, or rather rewarded beyond the expense of doing them, in carbon coins, and these coins were trading strongly with all the other currencies in currency exchanges. In fact, it looked like there was a possibility that carbon coin might soon supersede the US dollar as the world's hegemonic currency, the ultimate guarantor of value. The US was a big backer of this complementary currency, of course, which was no doubt a factor in its success; in some senses the carbon coins were like dollars created by the sequestering of carbon. This made for a kind of double standard, or rather something finally to replace the lost gold standard; they had now a carbon standard, and also the dollar to use for exchanges. But the carboni, as more and more people called it, was also complementing the euro, the renminbi, and all the rest of the fiat currencies that had underwritten the new one.

As significantly, money itself was now almost completely blockchained, thus recorded unit by unit in the consolidated central banks and through the digital world, such that any real fiat money now traveled within a panopticon that was in itself a global state of sorts, unspoken as yet, emerging from the fact of money itself. Another brick in the controlocracy, some said of this recorded money; but if the public kept ultimate control of this new global state, by way of people power exerted by the ever more frequent strikes and non-compliances, then the people too would be seeing where all the money was and where it was going, move by move, so that it couldn't be shuffled into tax havens or otherwise hidden, without becoming inactivated by law. Digital distribution of the total blockchain record through YourLock and other sources meant there was a kind of emerging people's bank, a direct democracy of money. So now the various old private cryptocurrencies were only being used for criminal activities, and traded at fractions of a penny. Lots of investors who still held these worthless coins were looking for a moment to get out with a minimal haircut. Others had simply cut their losses and sold for mills on the million. Holders of the various cryptocoins owned trillions of them, but trillions multiplied by zero still came to zero, so they were mainly done for; their owners might as well have been holding piles of copper pennies, except copper was more valuable.

It was still argued by some that the Super Depression of the previous decade had created the greatest part of the carbon drop observed. They were burning less because the global economy had tanked. But this would only explain the flattening, not the reduction; and besides, even with the depression, the world's GWP, and even all the better measurements of human economic activity, showed activities worth nearly a hundred trillion US dollars a year. It was still big, but it didn't run on carbon anymore. It was this and the drawdown efforts that explained the drop.

So this was the financial and the carbon situation, what Mary thought of as the two macro signals, the global indexes that mattered. And at the meso- and micro-levels, the good projects that were being undertaken were so numerous they couldn't be assembled into a single list, although they tried. Regenerative ag, landscape restoration, wildlife stewardship, Mondragón-style co-ops, garden cities, universal basic income and services, job guarantees, refugee release and repatriation, climate justice and equity actions, first people support, all these tended to be regional or localized, but they were happening everywhere, and more than ever before. It was time to gather the world and let them see it.

Sick at heart, she was going to declare victory. Declare victory as if sticking a knife in the heart of her worst enemy, with a feeling not unlike posting a suicide note. And if the real truth was that in fact they had somehow lost, then she was going to try to see to it that the evil ones were winning a Pyrrhic victory. They were going to be the losers of a Pyrrhic victory; and the losing side of a Pyrrhic victory could be said to have won. They were therefore the winners of a Pyrrhic defeat. Because they were never going to give up, never never never. History was going to go like this: lose, lose, lose, lose, lose, lose, win. And the evil ones in the world could go down under the weight of their damned Pyrrhic victory. They could go fuck themselves, murdering cowardly bastards that they were.

90

Today we're here to discuss the question of whether technology drives history.

No.

I'm sorry, are you saying that technology doesn't drive history, or that you don't want to discuss it.

Both.

Well, let's focus on the first no instead of the second, and see if we can get some clarification on just why you would say that.

It's a ridiculous question.

And yet it has been the title of books, essays, seminars, conferences, and the like. We ourselves are poor forked radishes, to quote someone, unless we augment our poor powers with tools. We are *Homo faber*, man the maker, and our tools are the only thing that allow us to cope with the world. We even co-evolved with our tools; first stones were picked up and sharpened, then fires started on purpose, and these tools made us human, as it was our precursor species who invented them, and after that we evolved into ourselves, and then on from there. Clothes to keep warm. The moment we find bone needles in the archeological record, for instance, we see people moving twenty latitude lines farther north than before.

So what?

So what? Excuse me? The question is, where would we be without our tools?

We think of ways to do things. If one way doesn't work, we find another way.

But these are the ways we've found!

Path of least resistance. Dealing with the laws of physics. Picking up a rock for instance. It's just trial and error.

Well, say we call that technology then! Doesn't it follow that technology has been the driving force in history?

We are the driving force in history. We make do, and on it goes.

All right then. Enough of philosophy, I'm afraid I'm getting confused.

Yes.

Let's move on to some specific examples. Have you heard of those drones developed to shoot mangrove seeds into the mud flats, thus seeding hundreds of thousands of new trees in terrain difficult of human access?

Very nice, but that same drone could shoot a dart through your head as you walk out your door. So it illustrates my point, if you care to think about it. Our tools are expressions of our intentions, so what we want to do is the key driver.

I'll save that for our next foray into philosophy, which we will certainly schedule soon. Meanwhile, what do you think of these new bio-engineered amoebae that are now grown in vats to form our fuel, while also drawing carbon out of the atmosphere? Kind of a next-stage ethanol?

Useful.

And other amoebae grown in vats can be dried to make a very tasty flour, thus eliminating the need for lots of ag land while feeding us all.

Also useful.

What about this blockchain technology, identifying where all money is and where it's been on its way there?

Good idea. Money only works when you trust it. Tracking it might help with that. Get all that right and you might find money itself becoming irrelevant and going away.

Very good! And what about all the other recent transformations in that area that we've been seeing, the carbon coin, the guaranteed jobs, and so on? What you might call the social engineering, or the systems architecture?

These are the areas that matter! Our systems are what drive history, not our tools.

But aren't our systems just software, so to speak? And software is a technology. Without the software, the hardware is just lumps of stuff.

My point exactly. By that line of reasoning, you end up saying design is technology, law is technology, language is technology—even thinking is technology! At which point, QED—you've proved technology drives history, by defining everything we do as a technology.

But maybe it is! Maybe we need to remember that, and think about what technologies we want to develop and put to use.

Indeed.

So, but back to all the new innovations in our social systems we've been seeing recently. It really does seem like an unusual time. You were saying you think these changes are good things?

Yes. Strike while the iron is hot. Put the crisis to use. Change as much as you can as fast as you can.

Really?

Why not?

Won't so many changes at once lead to chaos?

It was chaos before. This is coping with chaos.

It's a bit of a case of inventing the parachute as you fall, isn't it?

Beats landing without one.

But will we have time?

Best work fast.

There are some who think we've already run out of time, that the hard landing is upon us now, which is what we're seeing in current events. Have you heard that the warming of the oceans means that the amount of omega-3 fatty acids in fish and thus available for human consumption may drop by as much as sixty percent? And that these fatty acids are crucial to signal transduction in the brain, so it's possible that our collective intelligence is now rapidly dropping because of an ocean-warming-caused diminishment in brain power?

That would explain a lot.

91

Zurich's Kongresshall had been built for meetings like this one, big as could be. And right down on the lakefront. Every day of this big conference, there would be speakers and display booths celebrating the emergent accomplishments. It was going to be hard to fit them all in, in fact there would have to be some compression, by nation or project type, to get a proper overview of the effort. Looking at the lists of people who were coming, Mary could begin to believe that they were actually gaining some traction, making some progress.

Then it was time to visit Frank.

He was there in his little apartment. He told her that the doctors had determined that he had a brain tumor, and had just recently identified it as a glioblastoma.

"That sounds bad," Mary said.

"Yes," he said. His mouth tightened. "I'm done for."

"But they must have treatments?"

"Average survival time from diagnosis is eighteen months. But mine is already pretty big."

"Why did it take so long to affect you?"

"It's not hitting anything too crucial yet."

She stared at him. He met her gaze unflinchingly. Finally it was she who shook her head and looked away, as she sat down at his side. He was on his apartment's little couch, looking slumped in an odd manner. She reached across the coffee table and put her hand on his knee. It was like the first moment they had met, when he had grabbed her arm and scared her, terrified her. Turnabout fair play.

"I'm sorry," she said. "It's shitty luck. You—you've had some really bad luck."

"Yes."

"Do they know why it happens?"

"Not really. Genetic maybe. They don't really know. Maybe it's too many bad thoughts. That's what I think."

"That's another bad thought."

"Doesn't matter now."

"I guess not. What will you do?"

"I'll take the treatments for as long as they tell me they're helping. Why not? Something might work."

She was encouraged to hear him say this. That must have showed, because he shook his head very slightly, as if to warn her to give up any hopes.

"Tell me about how your plans for this conference are coming," he said. And so she did.

After a while she could see he was tired. "I'll come back later," she said. He shook his head.

"I will," she insisted.

Actually that turned out to be hard to do. Not that she couldn't clear the time, because she could. And the safe house she was still living in was relatively nearby. Everywhere in Zurich was pretty close to everywhere else, really; it was a compact city. Although if you had to stay off the trams and streets, moving around in cars with tinted windows, it got harder.

But it wasn't the logistics. It was knowing what she would see when she got there. Frank May had never been an unguarded person, not since she had known him, anyway. The blaze in his eye, the set of his jaw, he had been easy to read, ever since that first night, when violent emotions had torn across his face like electric storms. Now he was closed. He had checked out. He was waiting for his time to come. And that was hard to watch. It was how she would be if she were in his situation, she guessed, but still. She kept finding reasons not to go.

But also she felt a duty to go. So eventually she would realize it had been ten days, two weeks, and she would send a message.

Can I drop by?

Sure.

She would be let in by his housemates, polite but distant people, and go in to see him. He looked bloated and unwell. He would look up at her, as if to say, See? Here I am, still fucked.

She told him about what was happening, talking away nervously to keep the silence at bay. Things were happening, as always. The big conference was looking good, coming soon. The Mondragón cooperative system was spreading through Europe, and it was reaching out to make connections elsewhere. Spain itself was slowest on the uptake, because in Madrid they didn't like the Basques having that much influence. But elsewhere it was catching fire, it was the latest thing, the obvious thing. Turned out each European nation had a tradition of working communally around their old commons, which had lasted until suppressed by Napoleon or other powers, but still there, if only as an idea, now put back in play.

"Good," Frank said.

Also, Mary went on a little nervously, the upcoming COP was going to propose a detailed refugee plan that used some of the principles of the Nansen passports of the 1920s. Some kind of global citizenship, given to all as a human right. Agreement had been signalled by all Paris signatory nations, which meant all the nations on Earth, to grant legal status to this global citizenship, and share the burden equally, with the historical disparities in carbon burn factored into the current assessments of the financial and human burden going forward. Some kind of climate justice, climate equity; a coming to terms at last with the imperial colonial period and its widespread exploitation and damage, never yet compensated, and still being lived by the refugees themselves.

"Good," Frank said.

Mary regarded him. "You don't seem very opinionated today."

He almost smiled. "No." He thought it over. "I've been losing my opinions," he concluded. "They seem to be going away."

"Ah." She didn't know what to say to that.

"It's interesting to see what goes first," he remarked, still looking inward. "I'm presuming it moves from less important to more."

"Seems likely."

Mary didn't know what to say, how to respond. She felt foolish and helpless. She wanted to stick with him, no matter where his line of thought took him; but it was hard to know how to do that when moments like this one came. Ask questions? Speculate? Sit there silently, uncomfortably?

"How are you feeling?" she asked.

"Sick," he said. "Weak. Fucked up."

"You still sound like yourself though?"

"Yes. As far as I can tell." He shook his head. "I can't remember what I felt like before. I'm betting I wouldn't score very high on any tests right now." He thought it over. "I don't know. I can still think. But I can't think why I should."

Mary shied away from understanding that, or tried to. But she couldn't; it was too obvious. A black weight in her stomach began to pull her down. She stifled a sigh, she wanted to get out of there. It was depressing.

But of course it was. She hadn't come there to get cheered up, but to do some cheering up. The difficulty of that was a given. That was why she avoided coming. But since she had come, it was a duty. But what could be said?

Nothing really. And he didn't really look like he expected any answer from her, or any encouraging words of any kind. He looked calm; desolate; a little sleepy. If she stayed quiet for a while, it looked like he would drop off. After which she could slip away.

He stirred. "Hey," he said, "I wanted to introduce you to someone who lives here. Let me see if he's in." He tapped at his phone, put it to his ear.

"Hey Art, can you drop by my room for a second? I'd like to introduce you to a friend of mine. Okay good. Thanks."

He put his phone down. "This guy is hardly ever here, but he's one of the original co-op members, and they let him keep his room. I just met him a while ago, and I like him. He flies an airship all over the world, following wildlife corridors and wilderness areas, basically looking for animals. He takes people along with him."

"What, like nature tours?"

"I think that's right, but there are only a few passengers. They do some citizen science and the like. And he gives what he charges the passengers to the World Wildlife Fund and other groups like that."

Mary tried to sound impressed.

A quiet tap came at Frank's door. "Come in!"

The door opened and a slight man entered, nodded at them shyly. Balding, beaky nose, blinking pale blue eyes, looking back and forth between Frank and Mary, attending to them with a diffident gaze.

Frank said, "Art, this is my friend Mary Murphy, she runs the UN's Ministry for the Future here in the city. Mary, Art here is the owner and pilot of *The Clipper of the Clouds*, a blimp—or is it a dirigible?"

"A dirigible," Art said with a little smile, "but you can call it a blimp if you want. Many people do. Actually airship seems to be becoming the usual term, to avoid that confusion."

"And you take people up to see wild animals."

"I do."

"Mary and I went up to the Alps a while ago and saw a little herd of chamois, and some marmots."

"Very nice," Art said. "That must have been lovely."

"It was," Mary said, trying to join in.

Art attended to her. "Do you go up to the Alps often?"

"Not really," she said. Not in ways I like, she didn't say. "I'd like to go more." As long as it isn't to hide from assassins, she didn't say.

This man seemed to be hearing her unspoken sentences, perhaps, or noticing the spaces they left in the air. He cocked his head to the side, then exchanged a few more pleasantries with Frank, and pulled up Frank's schedule for helpers that week, so he could add his name. Then he nodded to Mary and slipped out the door.

Frank and Mary sat there in silence.

"I like him," Frank said. "He's a good guy. He's been helping me, and when he's here he's always pitching in around the place."

"He seemed shy," Mary said.

"Yes, I think he is."

More silence.

Mary said, "Look, I have to go too. I'll drop by again. Next time I'll remember to bring you one of those orange slices dipped in chocolate."

"Ah good."

92

Word got around one morning that we were to be released. Did you hear? someone said, bursting into the dorm. They're letting us go!

Quickly everyone had heard the news, and then most people doubted it. Yet another camp rumor. They always flashed around at the speed of speech, word of mouth moving through the camp to full saturation within an hour, I would say. But then always the doubt. So many rumors had proved false. Almost all of them in fact. But occasionally it was just the news, that's all. And this time there was a meeting announced for right after breakfast. Or meetings; there was nowhere in the camp that would have held everyone in it at once. So each block would have a meeting in its dining hall. Request to keep the smallest children back in the dorms, so there would be room for almost all the adults at once.

Of course it would happen this way. No warning; ad hoc; an improvisation, just as it had been all along. They toy with us: cage us, release us, it's all made up moment to moment. That's history.

We assembled. There was a team from the camp administrators, a couple of them familiar faces, and a few more strangers. Swiss looking. Whatever else, the meeting was real.

They had said the meeting started at eleven, and we had learned well that the Swiss meant what they said when it came to schedules and times. The big digital clock on the wall went from 10:59:59 to 11:00:00, and one of our regular keepers went to the mike and said, "Hello everybody," in English, then "Guten Morgen" and then "Sabah alkhyr."

She continued in English, which I thought was odd, as about seventy or eighty percent of the people in this camp spoke Arabic, and about thirty percent of those didn't speak English, at least that was my impression. But then I got caught up in her message.

They were indeed releasing us. We were to be given world citizenship, meaning we had the right to live anywhere. We were warned there would be immigration quotas for many countries, and possibly we would have to get on a waiting list for some of them. But we could do that, and the quotas for all the countries put together added up to two hundred percent of the number of people who had been held in refugee camps for over two years, which was the criterion for this world citizenship. Citizenship would be in your name, non-transferable, with a global passport. Families would be wait-listed together. The requests for residency were to be coordinated all over the world, and the ones who had been in camps the longest would be the first allowed to choose. They could take their immediate families with them. The process of relocation would begin at the start of the next month. Combined with a worldwide universal job guarantee commitment, and transport and settlement subsidies, everyone should end up okay.

Switzerland had committed to taking twice as many people as were now being held in all the Swiss refugee camps. The housing for these new citizens had been built already or was being finished. It was distributed throughout the country, every canton taking a proportional share. The housing was to consist of apartment blocks conforming to ordinary Swiss building and housing codes. Employment would be offered according to need, the canton as employer of last resort. There was work to be done. Facilities for cooperative restaurants were already in place, ready for opening if the newcomers so desired. It was felt that food could be both a gathering place for the new residents and an outreach to the host community. So it had often been in the past.

This arrangement was not quite the same as open borders, they said. Countries would still have passports and immigration quotas. The hope was that many people would want to return home. Polling showed that many refugees felt that way, and would go back home if it could be done safely. The destabilized countries that had generated the most refugees would be helped to restabilize as much and as quickly as possible.

People had lots of questions, of course. That part shifted into Arabic for the most part, and other people on the stage answered, taking turns, or answering when a question matched their expertise.

Before it was over I left the meeting and went to the north perimeter, full of clashing thoughts. Anywhere! What did it mean? Where would I choose?

Most of us would be talking it over with family. Some of us would go back home. I could see the lure of that. Assuming it would be safe, why wouldn't you? But I didn't think it would be safe. I didn't trust any of it. Surely there had to be a catch. Surely, if this was possible, they would have done it long before.

But why think that? Things change.

I tried to convince myself that things change. It wasn't that easy. Do things change? I had lived the same day for 3,352 days. It seemed proof that things don't change.

But of course that was wrong. Nothing stays the same, not even life in the camp. We had formed study groups, classes, sports clubs, activity groups. We had made friends. We taught the children. People had been born, people had died. People had gotten married and divorced. Life had gone on in here. It wasn't the case that things hadn't changed, that time had stopped outright.

But there is change and there is change. Looking through the fence at the mountains, hazy in the late morning light, I felt a deep stab of fear at the idea that my life might really and truly change. A big change. New people. Strangers. A new life in a new city. After such giant changes, would I still be me? Of course I recalled the poem about how you can never escape yourself, every place is the same because you are the one moving to that place. No doubt true. I recalled also the old notion from psychotherapy that people fear change because it can only be change for the worse, in that you turn into a different person and are therefore no longer yourself. Thus change as death.

But death of habits. That's all it is, I told myself. Remember the poem; you can't help being yourself. You'll drag yourself with you all over the Earth, no matter how far you flee. You can't escape yourself even if you want to. If what you fear is losing yourself, rest easy.

No: the fear I was feeling was perhaps the fear that even if things changed, I would still be just as unhappy as before. Ah yes, that was a real fear!

Well, but I was always afraid. So this was no different.

Would I miss this place? The beautiful mountains, the beautiful faces...

No. I would not miss it. This I promised myself; and it seemed like a promise I could keep. Maybe that was my form of happiness.

93

Project Slowdown had been active for a decade, and the thirty largest glaciers on the planet, all of them in Antarctica and Greenland, had seen expeditions to their crux points where wells had been melted through their ice and the meltwater under them pumped to the surface and spread to refreeze as near the pumping wells as was convenient. Our team had been involved with the Weddell Sea area effort, which was particularly complicated, as a dispersed fan of glaciers and ice streams had fed into the Filchner Ice Shelf and the Ronne Ice Shelf in a way that was difficult to deal with. The landforms under the ice resembled a half bowl, not steep enough to easily identify the places upstream where glacial input was fastest. But we had done the best we could with that, and drilled 327 wells over a five-year period, focusing on the crux points we could find and hoping for the best.

It wouldn't have been possible without the navies of the United States, Russia, and England. They let a little village of their aircraft carriers freeze into the sea ice and overwinter in the Weddell Sea, and from these carriers we were able to keep the work going year round, and supply the land bases that were set on the ice of the Ronne and Filchner. Fleets of helicopters kept these camps supplied, and helped to move camps from drill site to drill site. Something like ten billion dollars was spent on the effort just in our zone alone. Such a deal, as Pete Griffen used to say. A lot of us had worked with him back in the day, and he was often remembered.

All good. Only four deaths, including his, all from accidents, and three of those accidents resulting from stupid decisions, including his. The other death, weather. Pretty good. Because Antarctica will kill you fast. And none of the deaths were people on our team, although we never said that of course. But it was a comfort, given what had happened to Pete. No one in my group wanted anything like that to happen to us.

So; ten years in Antarctica, with good work to do, and no more grant applications either. Papers got written, science got done, but mostly it was engineering the drills and pumps and dispersion technologies. There were papers to be had there too, even if it wasn't exactly what we had gone down there for. Actually the glaciologists were getting data like never before, especially structures of ice and flow histories, and most of all, bottom studies. For sure no one had ever had the kind of information about glacial ice/glacial bed interactions that we have now! If we had been doing that research only for its own sake, it would have taken centuries to learn what we've learned. But we had an ulterior motive, an overriding concern.

So, at the end of the season, we were flown into the middle of the Recovery Glacier, where we had drilled a double line of wells five years before. One of the lines was reporting that all its pumps had stopped.

Helo on up to a pretty dramatic campsite, on a flat section of the glacier between icefalls upstream and down, with the Shackleton Range bulking just to the north of us, forming the higher half of the glacier's sidewalls. Lateral shear at the glacier's margin was a shatter zone of turquoise seracs, so tall and violently sharded that it looked like a zone of broken glass skyscrapers. You never get used to helo rides in Antarctica. Not even the helo pilots get used to it.

Out on the flat we went to the wells that were reported as stopped. We had drilled these long before, back at the beginning, and now it was a familiar thing to check them out. Everything looked okay on the surface, and it wasn't the monitoring system. Very quickly the problem noticed by the automatic monitoring was confirmed, pump by pump, just by looking in the exit pipes and seeing nothing there. The closer to the center of the glacier the holes were, the less water they were pumping. Most were pumping nothing at all.

We were moving around on skis, and roped together, just in case the crevasse-free route between the wells had cracked in the years since someone had been there. There were no crevasses, so we flagged the new route, then got on the snowmobiles and tested the route to be sure. No fooling around in our team.

The wells were in the usual line cross-glacier. Tall pole with transponder and meteorological box, tattered red flag on top. Under that a squat orange insulated plywood box covering the wellhead, a very small shed in effect, heated by solar panels set next to it. The pipeline was lime green, crusted with gray rime. It pumped the water south, up to a hill beyond the south bank of the glacier, joining a big pipeline there, which took a feed from all the pumps in the area.

We got the door to open, and went into the hut covering the wellhead. Nice and warm in there. Dark even with the lights on, after the glare outside. Wind keening around the sides of the thing. Nice and cozy; it had to be kept above freezing. Checked the gauges; no water coming up. We opened the hatch on the well cap, fed a snake camera down the hole. The snake's reel was so big a snowmobile had had to haul it here on a sled of its own; two kilometers of snake on one big wheel.

Down went the camera. We stared at the screen. It was like doing a colonoscopy of an exceptionally simple colon. Or probably it's more like the cameras that plumbers use to check out a sewer line. No water in the hole, even two hundred meters down; this was a sign something was wrong, because when a hole is open from the bottom of a glacier to its top, the weight of the ice pushes water up the hole most of the way. But here we were looking far down the hole, and no water.

Got to be blocked, someone said.

Yes but where?

Eventually we got to the bottom of the hole; no water at any point along the way.

Hey you know what? This glacier has bottomed out. There's just no more water to pump!

So it will slow down now.

For sure.

How soon will we know?

Couple years. Although we should see it right away too. But we'll need a few years to be sure it's really happened.

Wow. So we did it.

Yep.

There would be maintenance drilling, of course. And the glaciers would still be sliding down into the sea under their own weight, at their old slower speed, so every decade or so they would have to be redrilled upstream a ways from the current holes. There were going to be lots of people working down here for the foreseeable future—maybe decades, maybe forever. A rather glorious prospect, we all agreed, after thawing out and getting into the dining hut, standing high on its big sled runners. Little windows on the south side of the hut gave a view of the Shackleton Range, oddly named, as he never got near this place. Possibly it was near where his proposed cross-Antarctic route would have gone, but when the *Endurance* got caught in the ice and crushed, all that plan had to be scrapped, and they had set about the very absorbing project of trying to survive. We toasted him that day, and promised his ghost we would try to do the same. Drop Plan A when the whole thing goes smash, enact Plan B, which was this: survive! You just do what you have to, in an ongoing improvisation, and survive if you can. We toasted his rugged black-cliffed mountains, rearing up into the low sky south of us. We were 650 meters above sea level, and ready for food and drink. Another great day in Antarctica, saving the world.

94

The 58th COP meeting of the Paris Agreement signatories, which included the sixth mandated global stocktake, concluded with a special supplementary two-day summing up of the previous decade and indeed the entire period of the Agreement's existence, which was looking more and more like a break point in the history of both humans and the Earth itself, the start of something new. Indeed it can never be emphasized enough how important the Paris Agreement had been; weak though it might have been at its start, it was perhaps like the moment the tide turns: first barely perceptible, then unstoppable. The greatest turning point in human history, what some called the first big spark of planetary mind. The birth of a good Anthropocene.

So the last two days of this meeting consisted of one day of people summarizing, listing, and celebrating various aspects of the positive changes made since the Agreement was signed. The second day was devoted to listing and describing some of the outstanding problems they had yet to solve if they were to secure the progress inherent or promised by the things mentioned on the first day. Both were very busy days.

On the first day, the day of celebration, Mary walked around the poster halls drinking them in and feeling first startled, then amazed. The big banner with the Keeling Curve stretching across it, with its continuous rise, then the leveling, then the recent downturn, stood over everything else like a flag. And under that, there was so much going on that she had never heard of. She felt again the power of the cognitive error called the availability heuristic, in which you feel that what is real is what you know. But there was so much more going on than any one person could know, reality was so much bigger than the self, that it was alarming to contemplate. This explained the error: one felt the vastness and shrunk in on oneself like a snail's horns, instinctively trying to protect one's mind.

And yet there was no real harm in it, this contact with the larger reality. Mary tried to feel that. She stuck her horns out and took it all in. The conference halls had the look of any other scientific meeting, poster after poster describing project after project. In this case all the posters, and the various organizational tables, and the panel discussions and plenary talks, described good things being done; but then again this was always true, at any scientific meeting ever held: they were utopian gatherings, spaces of hope. The difference here was that together the posters were describing a global situation that no one would have believed possible even ten years before; and forty years ago it had looked impossible.

First, powering everything in a most literal way, was the news that a lot of clean energy was being generated. Mary recalled Bob Wharton saying something to this effect many years before, back in the beginning, during that time that now seemed on the other side of a great mountain range, those invisible Alps over which there was a younger self she could barely remember: if we can generate lots of clean energy, Bob had said, lots of other good stuff becomes possible. Now they were doing that.

Also, crucially, even though they were creating more energy than ever before, they were burning far less CO_2 into the atmosphere, less per year than in any year since 1887. So Jevons Paradox appeared to be foxed at last; not in its central point, which stated that as more energy was created, more got used; but now that it was being generated cleanly, and for fewer people as the population began to drop, it didn't matter how much of it was being used. Since there was a surplus at most points of supply, most of the time, and created cleanly, they had simply outrun Jevons Paradox. And where excess energy was being lost by lack of completely effective storage methods, people were finding more ways to use it while they had it: for desalination, or more direct air carbon capture, or seawater pumped overland into certain dry basins, and so on. On and on and on it went. So clean energy, the crux of the challenge, had been met, or was being met.

Then also, another great poster: the Global Footprint Network had the world working at par in relation to the Earth's bioproduction and waste intake and processing. World civilization was no longer using up more of the biosphere's renewable resources than were being replaced by natural processes. What for many years had been true only for Cuba and Costa Rica had become true everywhere. Part of this achievement was due to the Half Earth projects; though this was not yet an achieved literal reality, because well more than half the Earth was still occupied and used by humans, nevertheless, broad swathes of each continent had been repurposed as wild land, and to a large extent emptied of people and their most disruptive structures, and left to the animals and plants. There were more wild animals alive on Earth than at any time in the past two centuries at least, and also there were fewer domestic beasts grown for human food, occupying far less land. Ecosystems on every continent were therefore returning to some new kind of health, just as the result of the planetary ecology doing its thing, living and dying under the sun. Most biomes were mongrels of one sort or another, but they were alive.

It was also the case, though not a single poster or panel referred to it directly on this day, that the global human replacement rate was now estimated to come to about 1.8 children per woman. As a level replacement rate was hit at about 2.15 children per woman, the total human population on Earth was therefore going down, slowly but surely. The idea that there was still a demographic surge ahead of them had gone away; demographers no longer predicted it. Some economic theorists worried that the economy couldn't handle such a diminishing population; others welcomed the change. But all of this was so new and controversial that at this conference, they left it unspoken. It was a topic for another day. The old attack on the environmentalist movement, that it was antihuman, still had enough force to make many in the scientific community wary of speaking on the subject. It was too hot to handle. But in this case, good news; so it was mentioned pretty often, it was one of those things being spread by word of mouth.

It was somewhat the same in any discussion of the recent Super Depression, and how the social and economic disruption of that had actually been a good thing in terms of carbon burn and biosphere health. That events which had caused suffering for millions of people might be good for the rest of the planet's life was again seen as a possible antihumanism. Best just to frame all this aspect of the general situation as managing disaster as best one could, making the best of a bad situation, etc. And in the flood of information being presented, some things were clearly being left to word of mouth, especially inferences and suppositions.

Over all of it, in the most literal sense because of the banner, and the air itself, the immense flux of information was often summed up well by what was being called the Big Index or the Big Number, meaning the parts per million of CO_2 in the atmosphere. This had now dropped 27 parts per million in the previous five years. It was down to 451 now, same as in the year 2032, and it was on a clear path to drop further, maybe even all the way to 350, the pre-industrial high point on the 280–350 ppm sine wave that had existed for the previous million years, marking shifts in the shape of Earth's orbit around the sun. 350 parts per million of CO_2, if they wanted that! The discussion now was how far down they wanted to take it. This was a very different kind of discussion than the one that had commanded the world's attention for the previous forty years.

In this same period, the Gini index figures for the world at large had flattened considerably. Every continent was showing improvement. The pay justice movements, the wage ratio movements, and the central banks' recommended tax plans, plus political movements everywhere supporting job guarantees and progressive taxation, sometimes under the rubric of "an end to the kleptocracy of the plutocrats," as one poster put it, had had powerful effects everywhere. Setting a generous definition of a universal necessary income, guaranteeing jobs to all, and capping personal annual income at ten times that minimum amount, as they had done in many countries, had immediately crushed Gini figures down. The EU had led the way, the US and China had followed, and then everywhere else had begun to leak their most educated young people to these flatter countries, until the countries losing educated people also instituted it. Guaranteed jobs, yes, but also universal basic services, and supported social reproduction, along with infrastructure and housing construction projects, had completed the rise out of poverty at the low end of the world income scale. Capping individual income and wealth had flattened the top of the scale. Of course many rich people had attempted to abscond to a safe haven with their riches, but currency controls, and the fact that all money was now blockchained and tracked, meant that all the old havens and shelters were being rooted out and eliminated. Money was now simply a number in the global banking system, so even if one shifted money into property, that property got listed, with an asset price on it, and then got taxed accordingly, and often therefore sold to avoid property taxes that had gone sharply progressive. Some land was surrendered in order to keep the owners solvent in the new tax regimes, which meant there was now more and more public land, defined as such and used as a commons. State-owned enterprises using a lot of big data and Red Plenty algorithms became less lumbering than they had been, avoiding the old bad inefficiencies, while keeping the good inefficiencies in ways that were important for resilience and justice.

A whole new economics was springing up to describe and analyze these new developments, and inevitably this new kind of economics included a lot of new measuring systems, because economics was above a system of quantified ethics and political power that depended on measurement. So now people were using older instruments like the Inclusive Prosperity Index, the Genuine Progress Indicator, the UN's Human Development Index Inequality Adjusted, and the Global Footprint Index. And they were also making up many new ones as well. All these new indexes for economic health were often now amalgamated to a new comprehensive index of indexes, called the Biosphere and Civilization Health Meta-Index. BCHMI.

The carbon coin had played its part in all of this. For a while, in its earliest days, it had looked like the creation of carbon coins would simply make the rich richer, as some of the largest fossil carbon companies declared their intention to sequester the carbon they owned, and took the corresponding pay-out in carbon coins, and then traded most of those coins for US dollars and other currencies, and then made investments in other capital assets, in particular property, thus becoming richer than ever—as if all their future profits were going to be paid to them at once, at a hundred cents on the dollar, even though their assets were now stranded as toxic to the biosphere and thus to human beings.

But the central banks had worked out a scheme to deal with this. The fossil fuel companies were being paid, yes, and even at par, if that meant one carbon coin per ton of carbon sequestered, as certified by the Ministry for the Future's certification teams, just like any other entity doing the same sequestration. But pay-outs above a certain amount were being amortized over time, and would be paid out, when the time came, at zero interest; zero interest, but not negative interest; and with guarantees, thus becoming a kind of bond. And then the companies were required by law and international treaty to do carbon-negative work with the initial use of the carbon coins they were given, in order to keep qualifying for their pay-outs, because if they merely invested in other biosphere-destroying production, especially carbon-burning production, then they wouldn't be sequestering carbon at all in the larger scheme of things. The upshot of these policy implementation decisions was that the oil companies and petro-states were being paid in proportion to their stranded assets, but over time, and only for doing carbon-negative work, as defined and measured by the Paris Agreement standards and certification teams. The young staffs of the central banks were all quite proud of this arrangement, which they had concocted over the years in an effort save the carbon coin, and then watched as their bosses approved and implemented it. Those staff reunion parties were raucous to the point of almost scandalizing staid old Zurich.

What the success of the carbon coin meant was a huge amount of money was now going to landscape restoration, regenerative ag, reforestation, biochar and kelp beds, direct air capture and storage, and all the rest of the efforts described elsewhere in the hall. A banner over one of the rooms put it this way:

"Revolution comes; not the expected one, but another, always another."

The people under the banner who had put it up told Mary it was a phrase from one Mario Praz, which had been quoted by a John P. Farrell, who had then been quoted by a Christopher Palmer. They were pleased with it, these Swiss staffers.

The second day, their final day, devoted to "outstanding problems," was a sobering reminder that they were still in the thick of it. And yet, given what had been revealed and celebrated the day before, there was a sense in the hall that these problems, as wicked as they were, were impediments to a general movement, to history itself, and thus susceptible to being overwhelmed, or solved piecemeal, or worked around, or put off to a later time when even more momentum would be available for deployment against them.

That day's banner over the entryway proclaimed the motto:

"Overcome difficulties by multiplying them."

This the staffers attributed to Walter Benjamin, who apparently when he wrote it had added that this was "an old dialectical maxim," though no one could find it anywhere earlier in the relevant literatures. Probably he had made it up and given it a false provenance, although someone remarked that that wouldn't have been like him, as he had been a very diligent historian and archivist. There were people trying to find the true source on the internet, but again this only demonstrated that the internet, although huge, was not even close to comprehensive when it came to the material traces of the past. A drop in the bucket, in fact.

Anyway, Mary felt there were more pressing matters than finding the real source for this quote. She wandered the halls as she had the day before. Overcome obstacles by multiplying them: easy to do! The outstanding problems were on this day being discussed, evaluated, rated, even put into a hierarchy of urgency; then compiled into an index somewhat like the Association of Atomic Scientists' "minutes to midnight" clock, which indeed still stood in the hallway, set at about twenty minutes to midnight, where it had been stuck for decades, making Mary wonder if it had any validity, or rather carried any weight. Stuck clocks; not the best image for a deadly serious danger. For all the most pressing dangers they faced, she felt that the clock was always ticking. But when she mentioned this to Badim, he shook his head; that atomic clock was simply saying the nuclear danger has never gone away, he said. We act like it has, but it hasn't. So it's a way of saying we are good at ignoring existential dangers. Really it would be better to do something about that one. Shoot it in the head. You could disarm all those weapons in under five years. Fission materials used as energy fuel, burned down to nothing and the residues buried. It's stupid.

After that, she wandered the halls feeling that maybe she was in fact no good at trying to rate danger. The Arctic petro-nations still had a not-so-secret affection for climate change, for instance; was this a danger? Recently it had seemed that keeping the Arctic sea ice thick and robust was now Russian policy. So Mary wasn't so concerned. They had the fleet, and they would keep staining the open Arctic Ocean yellow, to keep sunlight from penetrating deep into the water and cooking them all; same with the icing drones now sent out every long dark winter night, to spray the sea ice with added layers of frozen mist, and seal unwanted polynas and open areas and so on. No; if the Arctic was an outstanding problem, it was at least not intractable. Russia would do its part, she judged.

Down the hall, she was thrust back into the realm of nuclear weapons and nuclear waste. Badim wasn't the only one worried; there was a whole slate of panels devoted to it. Could these problematic materials be turned into nuclear power somehow, burned down to a concentrate that could be safely cached, or slung into outer space? No one could make a compelling case either way on this, Mary judged. It was indeed an outstanding problem.

Then the thirty poorest countries. The bad thirty, the sad thirty, the weak thirty; she heard all these names. These thirty included at least ten so-called failed states, and some of those had been failed for decades, immiserating their people. Wicked problems, in the technical sense of the term, were problems that not only could not be solved, but dragged other situations down into them; they were contagious, in effect. So it seemed that these countries suffering wicked problems needed to have interventions made by their neighbor countries, meaning the whole world; in effect, to be put into regional or international receivership. But sovereignty disregarded for one nation implied it could be disregarded for any nation, if the political winds blew against it; so no nation liked to tamper with sovereignty. And it was usually the prosperous old imperial powers that were insisting most strongly that the various post-colonial disasters be put back into subaltern positions, which insistence never looked good, even if the intentions were benign. The American empire had been mostly economic and undeclared, and so had never been understood or acknowledged to be an empire by Americans themselves, despite their eight hundred military bases around the world, and the fact that their military budget was larger than all the other nations' on Earth combined. So you could get things like the Washington Consensus, in which the World Bank and the IMF and even the WTO had been used as instruments of American hegemony, forcing small poor countries to join the world as a new kind of colony, American colonies in all but name, or else suffer even worse fates. Even the Chinese with their Belt and Road Initiative, and their local power in Asia, were not as bad as the Americans when it came to imperial self-regard pretending to be charity, as in the structural adjustment procedures at the end of the twentieth century, which had wrecked local subsistence so that entire countries could become cash crop providers for American markets. No: Mary would include American stupidity and hubris, and the assumption of being the world's sole superpower, as one of their outstanding problems; but there wasn't a panel or even a poster given over to that idea, no, of course not. Another word of mouth issue. It would take the whole world combined and coordinated to attack that problem, and of course the other countries could never agree among themselves to the extent of being very good at that, especially since most of them were beholden to the US, and more than a few bought lock stock and barrel.

Then the continuing poisoning of the biosphere by pollution, pesticides, plastics, and other wastes and residues of civilization was growing in people's perceptions as the CO_2 problem began to look like it might be receding. The biosphere was robust, sure; but taking in and processing poison was something that any living thing could only do up to a point, after which it was simply poisoned, and in trouble.

Then came a hall devoted to discussing the mistreatment of women, individually and collectively, often in the same countries that rated lowest in all the other categories of well-being and political representation and functionality. That was no coincidence, of course. The status of women was not just an indicator, but fundamental to the success of any culture. But many of the old forms of patriarchy were not dead, and so among the worst of the outstanding wicked problems were patriarchy and misogyny. Mary sighed as she looked in at that hall, where of course there were lots of women, and fewer men by proportion than elsewhere. In a room of a hundred people she counted five white men, twenty men of color. Even if it were demonstrated to be the worst problem left, still there would be few men willing to take a look that way. It wouldn't be men who solved this one, but women; laws crafted by women and rammed through by women. So it was indeed an outstanding problem, a wicked problem indeed. Although to be categorized like that, as a problem alongside pollution and nuclear arms and carbon and misgovernance and the like—it was galling, it was another aspect of the problem itself. Women as

Other—when would that stop, them being as they were the majority of the species by many millions?

On then to other intractables. Resource issues. Women as resources, Mary thought, no, forget about that. Water. But with copious clean energy, they could desalinate. Soil: regenerative ag was the hope there, biology itself. The biosphere generally: loss of habitat, of safe habitat corridors, of wildlife numbers. Extinctions. Invasive biology problems. Watershed health. Insect loss, including bee loss. Where to store the CO_2 they were drawing out of the atmosphere. Even with the progress made, these were still acute problems.

Ocean health. They could do nothing about ocean acidification, nor the heating of the ocean that was baked in by the previous century's carbon burn, nor the deoxygenation. Thus die-offs were happening, and presumably extinctions they didn't even know about, that might have catastrophic cascading results. Ocean health would be an outstanding problem for centuries to come, and little to nothing they could do about it, except to leave big parts of the ocean, half of it at least, alone, so that its biomes and creatures could adapt as best they might. Coral reefs and beaches and coastal wetlands of course were also a big part of that, and almost equally beyond what humans could do to help. Stand back, get away, keep out; maybe try fishing for plastic rather than fish, at least in the big areas left alone; or even in the fishing zones. Set new foundations for coral reefs. And so on. It would be a wicked problem for the rest of their lives.

Oh yes, it went on like that, all day and all over the Kongresshall. And it didn't help that many of these problems were incommensurate, that it was offensive to have women's welfare put on the same card as the welfare of coral reefs or nuclear stockpiles. To hell with these anthologies of outstanding problems! Lists like these were in some senses useless, she felt. Perhaps better to have ended the conference the day before, with the celebration of progress they already made and were still making. That had felt good, this felt bad. Possibly the anger generated on this day could be put to use, but she wasn't sure. A lot of shocked or depressed-looking young people were wandering Kongresshall, especially young women. Mary stopped some of them when they were in groups that looked as if they were talking things over, and she tried to encourage them to fight on, to go out there and kick ass, as they had been. Some nodded, some didn't.

A mixed day, therefore. And then she got a call from the clinic where Frank was being helped. He had collapsed and was doing poorly.

They had him in a room of his own, a small room almost filled by his hospital bed and the monitoring and life support equipment, and three chairs. His bed was tilted so that he was sitting up. He wore a hospital gown and had an IV port in the back of one hand, tube running up to an IV bag on a stand. Monitor had his pulse bumping a graph, pretty fast she thought. His face was white and swollen, dark circles under his eyes. Hair cropped short; she noticed his receding hairline.

He was asleep, or at least had his eyes closed, and she sat down on one of the chairs and decided not to wake him up, to wait for that.

He looked ill. The instrumentation hummed, his pulse bumped the scrolling graph over his head. Faint smell of starch and sweat and soap. Ah yes: she knew this world.

She sighed and sat back. Hospice, in effect. Even if they were still making efforts to save him or give him time, this was still hospice. She knew this place. The halfway house between this world and no world.

It was a quiet place, much attenuated. Much had already gone away. Remaining was water, some food, food as fuel, which she had seen be refused, to speed the process along; also painkillers, also removal of wastes. Catheter tube running out from under sheet to a plastic bag hanging from the bed frame. She had sometimes brought music into these places, as the one thing that need not fall away, that some part of the failing mind might recognize and enjoy, or at least be distracted by. The boredom inherent in the situation was as bad for the dying person as for those visiting; or worse. Left with time to think, as in a night of insomnia that never relented. Except sleep did come. The drugs helped that, and simple exhaustion. Failing functions in the brain. Sleep came in to fill the emptied spaces, and as so often in life it was a blessing, for even if one woke uneasy and unrefreshed, some time had been passed out of awareness, out of pain. Sleep that knits the ravelled sleeve of care; not exactly, in her painfully unravelled insomniac experience, but it was a beautiful phrase nonetheless. The rocking of the vowel sounds was like the cradle itself. Shakespeare's gift for phrasing. A true poet. The great poet, erratic though the plays always seemed to her. Hit or miss, shots in the dark. Garbled messes with perfect knots of tense confrontation. She recalled once in the Abbey Theatre, Falstaff and Hal confronting each other in a scene that must have lasted an hour, the two fencing with each other, enjoying the contest of wits, but with something dangerous at stake too, something deep. Their friendship, unstable on the great disparity of their situations. Maybe she and Frank had been a bit like that. That hour would never leave her.

Nor this. Well, this was hospice. There was too much time to think. Let the mind roam. She could pull out her phone and read or check her mail, sure. She could listen to music on earbuds, or bring in a little speaker box if he wanted to listen too. But also, she could just sit there. Rest. Think. Run over in her mind the conference that had ended, and everything else: their last hike in the Alps; those wild creatures up there living their lives. Those creatures too must die in pain, with some of their kin huddled round them seeing them out, presumably. Or just alone out there. Tatiana. Martin, the hospice he had died in. There was a tendency to range widely in these situations, she had found, as if sitting on the edge of a cliff, with a view out to the edge of the world, like at the Cliffs of Moher but even higher. The sidewalk over the abyss, as Virginia Woolf had put it: sitting on the edge of that sidewalk with one's feet kicking in the empty air, staring down into the abyss or out at the horizon, or back at the sidewalk that had seemed so important as they walked it, now revealed as a gossamer strip through careless air. No, life—what was it? So deep and important and full of feeling, so crucial, then suddenly just a blink, a mayfly moment and gone. Nothing really, in the grander scheme; and no grander scheme either. No. A vertiginous perch, the bedside chair in a hospice.

Certainly it was impossible to avoid the realization that this was going to happen to her too, someday. Usually one could dodge that thought; it was a distant thing, one could avoid thinking of it. Take it on, get used to it in small theoretical doses, forget about it again. Live on as if that would always be the case. But in a room like this one the real situation obtruded as if out of another dimension. Thus the vertigo, the sense of scale. A great reckoning in a little room, indeed. This phrase was supposed to be Shakespeare's reference to Marlowe's death, which was generally understood to have resulted from a drunken fight over a pub bill. Sudden and stupid eruption of hospice reality into mundane reality.

That first visit right after the conference, he never woke. She left with a little guilty sensation of relief. She knew very well it would be hard to talk with him in any comfortable way. He had not ever been one for mincing words or papering over a situation with some kind of comforting nostrum. He had not ever been comfortable with her visiting him. It had been like punishing him and she had done it anyway.

The next time she visited, she found his ex-partner and her child in the room. The girl was now a young woman and looked miserable. Youth was no protection from the exposure of a room like this, indeed for youth it might even have more of an impact, being perhaps new, or at least unusual; the calluses had not yet formed, the reality of death was more shocking, less buffered.

His ex looked unhappy as well. They sat on the chairs upright, twisting their hands together, looking anywhere but at him. He regarded them with what Mary thought was a kindly look, full of regret and love. Seeing it was like a needle stuck in her heart; she had thought he was beyond feeling, that he had dismissed them somehow. But of course not. A mammal never forgets a hurt; and they all were mammals. And here these two were. They were afraid of him, she saw. Not just afraid of death, but of him. She wasn't sure he saw that. Maybe he did and loved them for coming anyway. The young woman would remember this for the rest of her life. All her anger would have to include this too. The one who lives longest wins. But then has to carry the burden of that victory, that horrid feeling of triumph. It would never be good to feel that.

I'll come back later, Mary said.

We were just leaving, the mother said. The daughter nodded gratefully. We were just leaving. He may be tired.

Frank's little smile belied that, Mary thought. Thanks for coming, he said. I appreciate it. I'm sorry you had to come. Sorry for everything—you know.

No no, the woman said, tears spilling down her cheeks all of a sudden. We'll come back again. It's not far.

Thank you, Frank said. He reached out a hand and she took it and squeezed it. Then his daughter lunged forward and put her hand on the back of her mother's. For a moment they held hands like that. Then the young woman rushed out of the room, and, weeping, the mother followed.

Sorry about that, Mary said to Frank. I didn't mean to intrude.

That's all right. They wanted to leave.

No.

It's all right. They were here a while.

I can come back later.

Not necessary. Sit for a while if you want. Maybe you could hit the nurse button. I'd like to get them to bring some juice.

Sure.

They sat there in silence. After a while a nurse came by with apple juice in a small cup with a sealed cap and a straw protruding from it. Hospital equipment, disabled people accommodated as well as could be; he drained it in a single suck.

I can probably find a water fountain and refill that, Mary offered.

Please.

She wandered the hall, found a water fountain near the restrooms. Sealed cap hard to get off, but she could do it. Cup too small. Trying to keep patients from choking, from overhydrating? She didn't get it.

She dragged her heels getting back to the room, but there was no avoiding it.

Back on the chair. He avoided meeting her eye, she thought. She would have to exert herself to make this work.

We had a good meeting, she said.

Yeah? Tell me about it.

So she did. It was interesting to try to summarize the whole event in a way he would appreciate. The day of accomplishments, the day of outstanding problems. The difficulty in reconciling the two. The difficulty in giving up any sense of influence on the process, much less control. When you ride a tiger it's hard to get off. The Chinese had known this feeling for a long time.

How were the Chinese? he asked, curious. Are they on board with all this?

Yes, I think so. I get the impression that they feel confident they're at the center of the story, or one of the main players. For them, it's not just the United States anymore. They have rivalries with Russia and India that are also collaborations. They have contacts everywhere. I think they know they're crucial. I don't see the Party beating the drum about the

century of humiliation anymore, or not as much. It doesn't make sense to
the Chinese alive now. Including the leadership. So they seem to be relaxing into a feeling of confidence, of being taken seriously. No one can bully
them anymore, not even the US. It would be stupid to try. And they can
see that everyone's beginning to do things more like them. I mean state-
owned enterprises. Everyone's taking over money and energy and even
land, they're all seen as public trusts now, and that's how the Chinese have
always treated them. So the containment of the market, of finance—they
must feel they led the way on that, or gave everyone an example of how
to do it.

So it's really America that is the main problem, Frank said.

Mary sighed. I suppose. It's so easy to blame you for everything, you
lead with your chin, but I'm never comfortable with that. There's so much
good along with the bad. The country of countries, that kind of thing.

I wonder if people said that about Britain when it was the world power.

I don't know.

Not in Ireland.

Well that's true! She laughed. Although it has to be said, there was
some good in the Brits and their empire, even in Ireland.

I bet you don't say that when you're there.

No, I don't.

Suddenly he winced. His forehead popped sweat.

Are you all right? she asked.

He didn't reply. He buzzed the nurse, which was a reply of sorts. When
one showed up, he asked her for pain relief. Mary felt her stomach clench.
Of course. Pain. Enough to cause sweat to pop out of you, to cause your
face to go gray. Break-out pain, they called it. She had seen this before,
but it had been a long time.

Maybe I'll come back later, she said.

Okay, he said.

95

I am a thing. I am alive and I am dead. I am conscious and unconscious. Sentient but not. A multiplicity and a whole. A polity of some sextillions of citizens.

I spiral a god that is not a god, and I am not a god. I am not a mother, though I am many mothers. I keep you alive. I will kill you someday, or I won't and something else will, and then, either way, I will take you in. Someday soon.

You know what I am. Now find me out.

96

In the weeks that followed she began to take her pad with her to the clinic where Frank was being cared for. A period of surgeries and interventions had given way to a routine of palliative care. He could get out of bed, and with help shuffle out into the clinic courtyard, a pleasant walled space dominated by a big shade tree, a linden. He would sit there looking up into the leaves and the sky. There were flowerbeds, well-tended, but he never seemed to look at these. As far as Mary could tell, his ex and her daughter never came back to visit again. Once she asked him about it and he frowned and said he thought they had come by once or twice, but he couldn't be sure when it had been. She even asked one of the nurses about it, and was told it was not information they could share, that she would have to ask him.

It didn't matter. Acquaintances from the apartment co-op he had so briefly occupied came by, and friends from jail. So he said. Whenever she came by he was alone and seemed like he had been that way for the whole of that day, no matter when she came. It could have just been his manner, which was getting more and more withdrawn, but she began to think her impression was right; he was seldom visited. As his condition worsened, and he was more and more confined to his room and even his bed, on an IV drip of pain meds and who knew what else, she began to spend more and more time there. She realized that she believed, as much as she believed anything, that when someone was dying, it wasn't right that they be left alone, stuck in a bed, attended only sporadically by nurses and doctors. That wasn't proper; it wasn't human; it should never happen.

And so she began to make his room her office. She brought in a music box; found a small chair she could borrow from the clinic to use as a footstool; added a pillow to the chair she sat in, to give her back more support.

After that she began every day with a quick breakfast in her safe house, filled a thermos with coffee, then went to the office to check in, then continued on to Frank's room. There she settled into her chair with her pad, started Miles Davis's *Kind of Blue* on the music box to announce her arrival, and got to work on her pad. If she had to make a call she stepped out into the hall and spoke as briefly and quietly as she could. Her bodyguard detail, almost always Thomas and Sibilla, got comfortable in their own ways out in the clinic reception area. Their job had to be boring, she judged, but they did not complain. When she mentioned it they just shrugged. We like it that way, they said. Better that way. Hope it stays that way.

These days, Frank spent most of his time asleep. This was a relief to both of them. When he woke, he would stir, groan, blink and rub his eyes, look around red-eyed and confused. His face was swollen. He would see Mary and say "Ah." Sometimes that was all, for minutes at a time. Other times he would ask how she was doing, or what was happening, and she would reply with a quick description of the latest news, especially if it pertained to the refugee situation. If it was about Switzerland in particular, she read to him off her pad, so he could get as much information as there was. Otherwise she gave him her impressions.

Most of the time he slept, uneasily, fitfully. Drugged. Sometimes he lay still, but often he shifted restlessly, trying to find a more comfortable position.

Sometimes he would come to suddenly and seem fully awake, although eyeing her from a great distance. Once when he was like this, he said out of the blue, "Now you've got me kidnapped."

She growled at that, a little nonplussed. "A captive audience is never very satisfying," she replied at last, trying to keep it light.

"You could help me escape."

"That's what I'm doing."

"You're not very good at it."

"Well, you're in maximum security here."

"Still visiting me in jail then."

" 'Fraid so."

Another time he woke and stared at her, then knew her, and where he was. He said quietly, "I'll be sorry not to see what happens next. It sounds like things are getting interesting."

"I think so. But, you know. No one will live long enough to see an end to it."

"More trouble coming?"

"For sure." She looked at her in-box; she would have to scroll down for a couple of minutes to get to the bottom of it. "Something this big is going to go on for years and years."

"Centuries."

"Exactly."

He thought it over. "Even so. The crux, you called it once. The crux is a crux. You might see an end to that anyway."

She nodded, watching him. Instinctively she always shied away when he talked about his death. She recognized that fear in her—that some barrier would crack and they would fall together into an unbearable space. But she had learned to stay quiet and let him go where he would. There was no point in keeping someone company if you wouldn't follow them where they wanted to go.

This time, he fell asleep while still formulating his next thought. Another time when she walked in he was already awake, sitting up and agitated. He saw her and reached out for her so convulsively she thought he might fall off the bed.

"I just jumped through the ceiling," he exclaimed, wild-eyed. "I woke up and I was standing on this bed, and then I jumped up through the ceiling, right up there!" Pointing up. "But then I still couldn't get away. I tried to but I couldn't. I fell back down and then I found myself here again. But I jumped right through the ceiling!"

"Wow," Mary said.

"What does it mean?" he cried, transfixing her with his look, his face vivid with dismay and astonishment. "What does it mean."

"I don't know," she said immediately. She reached out and touched his hand, both twining her fingers with his and shifting him back toward the middle of the bed. "Sounds like you had a vision. You were trying to get out of here."

"I was trying to get out of here," he agreed.

She let go of him and sat in her chair. "It's not time yet," she ventured.

"Damn," he said.

"You're a very strong person."

"So I should be able to do it," he objected.

She hesitated. "Well," she said. "It cuts both ways, I guess. It wasn't your time yet."

He stared at her, still completely rattled. Of course, to have a real vision—to hallucinate—to try to fly out of this world—it was bound to be upsetting.

She didn't know what to say. Now he was weeping, looking right at her still, tears rolling down his cheeks. Seeing it she felt her eyes go hot and tears well up. Something leaping the gap from face to face, some kind of telepathy, some primate language older than language. It was like seeing someone yawn and then yawning yourself. What could you say?

She tapped on the music box and got *Kind of Blue* going. Their theme music now, this album, flowing along in its intelligent conversation. She sat back in her chair, let the familiar riffs flow over them together. She reached out and they held hands for a while. He clutched her hand from time to time. After a while he relaxed, fell asleep, and was deeply out the rest of the day.

Another day, struggling unconscious on his bed, writhing even, he suddenly came up from under, as if to breathe, and saw her there and turned his head aside, pained somehow. He was drugged, confused, only semi-conscious if that. Out of that condition she heard him mutter, "It's only fate. It's only fate."

She stared at him. His face was sweaty, both bloated and drawn at the same time. His breathing was labored; he pulled in air with desperate heaves, as if he could never get as much as he needed. When she was sure he was fully out, she said, "My friend, there is no such thing as fate."

Then one morning she came in and there were two hospice nurses in there, women attending to him—but no; cleaning up the scene.

One looked up and saw her and said, "I'm sorry, he's gone."

"No!" Mary said.

That one nodded at her, the other shook her head.

The one who had spoken said, "They often slip away when no one's around. Seems like some of them want it to be that way. A kind of privacy, you know."

She was not upset, as far as Mary could see. Not even alarmed. This was her job. She helped people at this point of life, helped them get to the end with a minimum of pain and distress. Now this one was gone.

Mary nodded absently, regarding Frank's still face. He looked to her like he had when he was sleeping. She had been coming by for two months. Now he was very still. She took in a big breath, felt herself breathing. Felt her heart beating. She was confused; she had thought there would be a struggle, a final clutch at life. She had thought it always went that way. As if she knew anything about it. It had been a long time since she had attended a death; and there hadn't been that many.

"We can take care of him now."

Mary nodded. "Give me a moment with him," she said.

"Of course."

They left. Mary arranged his stiff curled hands on his chest. They were cold; his chest was still warm. She leaned over and kissed him very lightly on the forehead. Then she picked up her pad, put it in her bag with the rest of her things, and left him. She walked out of there and wandered over to Bahnhofstrasse, and turned toward the lake.

She walked the handsome prosperous streets of Zurich both sightlessly and seeing things she hadn't noticed in years. Mind skittering, feelings blank. The chipped heavy stone blocks that formed the buildings flanking Bahnhofstrasse. They were amazingly regular geometrical objects, not perfect, lightly pocked and nobbled to give the faces of the buildings texture, but regular in that too, and set so perfectly that it was hard to imagine the process that would manage to accomplish it. In the end it had been the human eye, the human mind. Swiss precision. Buildings from a time when stone masons still did most of this kind of construction by hand. Artists of a very meticulous aesthetic, maybe even fanatics. Monomaniacs of cubical form. Stolid. Permanent. Many of these buildings had been here since 1400. Their stonework repaired probably in the nineteenth or twentieth century, but maybe not. Maybe set in stone for good.

Unlike a human life. A mayfly thing, a wisp of smoke. Here then gone. Frank May was gone. Well, now she would never wake up one night to find herself being murdered by him. She shook off that thought, shocked by it. The one who lives longest wins. No. No. Never again his spiky rebukes. The pleasure in an Alpine day, the rare moment of peace, the dark brooding anger, the ceaseless, useless remorse. He was released from that at last. Thirty years carrying that burden, manifesting whenever he let down his guard.

A PTSD sufferer. If that was really the way to think of it. Weren't they all post-traumatic in the end? So that it was just a way of pathologizing being human? Martin had died on her just like Frank—hospice bed, pain-killers, suffering through the final breakdown of his body's functions, the end of his life, at age twenty-eight, when they had only been married five years; wasn't that trauma? It most definitely was! She could see all that as if it were yesterday, and of course sitting with Frank had brought it all back in ways she hadn't felt for years. So wasn't she post-traumatic too?

Yes. But PTSD meant someone whose trauma had been—brutal? But death was so often brutal. Violent? That too. Shocking, bloody, prema-ture, evil? Something cruel and unusual, so that the person who survived it couldn't get it out of their minds, kept having flashbacks to the point of reliving the experience, as in some nightmare of eternal recurrence? Yes.

Maybe it was a matter of degree. Everyone was post-traumatic, it was universal, it was being human, you couldn't escape it. Some people had it worse, that was all it came down to. They were haunted by it, stricken, disabled. Sometimes it was bad enough they killed themselves to get free of it. Not uncommon at all.

What a thing. Death and memory. Martin had been young, he had struggled against death with a kind of furious resistance, with a sense of injustice. Right to the end he was never reconciled to it. Long after he had lost consciousness for the last time his body had fought on, the lizard brain in the cerebellum rallying every last cellular spasm to the cause. Those last hours of labored gasping breaths, what used to be called the death rattle, would never leave her. It had gone on too long. She seldom thought of that, she had learned to forget most of the time. This was the key, maybe, that ability to forget; but she dreamed of it sometimes, woke gasping and then remembered, as of course she always would. One didn't forget but rather repressed. Some kind of boxing up or compartmentalization; what that meant inside the brain and the mind she had no idea. Somehow they managed not to think of certain things. Maybe that was what PTSD was—the inability to do the work of forgetting, or of not recalling.

Not working at all for her now, she had to admit. The trigger in her brain had been pulled, and she was shot. Poor Martin. She wandered in an abreaction, through this handsome stone city that she quite loved. Remembering Martin. It wasn't so hard when she let herself. In fact it was easy. How she had loved him. Ah Zuri Zuri my town, my town. Some old poem from her German class. This was her town. Martin and she had lived in London, in Dublin, in Paris, in Berlin. Never in Zurich or anywhere in Switzerland. She loved it for that. Really she was very fond of this town. She even loved it. The way they could make her laugh with their Swissness. Their stoicism, their insistence on order suffused by intense feelings of enthusiasm and melancholy. That peculiar unnameable combination that was a national affect, a national style. It suited her. She was a little Swiss herself, maybe. Now aching with old pain, heartsick at the loss of someone gone now forty-four years.

She wandered the narrow medieval streets around Peterskirche and the Zeughauskeller. There was the candy shop where Frank had bought them candied oranges, proud of them, how good they were, how much an example of Swiss art at its finest.

Down to the lake. She headed toward the park with the tiny marina below it, intent on visiting the statue of Ganymede and the eagle. Ganymede perhaps asking Zeus for a ride to Olympus. It wouldn't be good when he got there, but he didn't know that. The gods were godlike, humans never prospered among them. But Ganymede wanted to find out. That moment when you asked life to come through for you.

It was so hard to imagine that a mind could be gone. All those thoughts that you never tell anyone, all those dreams, all that entire pocket universe: gone. A character unlike any other character, a consciousness. It didn't seem possible. She saw why people might believe in souls. Souls popping in and out of beings, in and out, in and out. Well, why not. Anything might be true. All things remain in God. Some saint's line, then Yeats, then Van Morrison, the way she knew it best. All things remain in God. Even if there was no God. All things remain in something or other. Some kind of eternity outside time.

As she stood there above the little marina she heard a roar, saw smoke across the lake to the left. Ah yes: it was Sechseläuten, the third Monday of April. She had completely forgotten. *Sachsilüüte*, to put it in Schwyzerdüütsch. The guilds had marched in their parade earlier, and now a tall tower erected in the Sechseläutenplatz had been set on fire at its bottom. Stuck on top of the tower would be a cloth figure of the Böögg, the Swiss German bogeyman, his head stuffed with fireworks that would explode when the fire reached them. The time it took for this to happen would predict whether they would have a sunny summer or a rainy one; the shorter it took, the nicer the weather would be.

Mary hurried across the Quaibrücke to Bürkliplatz, past the squeak and squeal of the trams over their tracks. If the Böögg went fast she wouldn't get there in time. Had to hope for a bad summer if she wanted to see the fireworks burst out of its head.

She got there sooner than she thought it would take. The platz was jammed with people, as always. The cleared circle they kept around the burning pyre was smaller than any other people would have kept it; the Swiss were strangely casual about fireworks. Their independence day in August was like a war zone. Some kind of wanton pleasure in fireworks. In this case they were at least going to go off well overhead, as opposed to August 1 when they were shot off by the crowd into the crowd.

The tower in the center of Sechseläutenplatz was about twenty meters high, a flammable stacking of wood and paper. On top, the humanoid big-headed figure of the Böögg, ready to ignite. The crowd was thick to the point of impenetrability, Mary was as close as she was going to get.

Then the Böögg went off. A fairly modest explosion of colored sparks bursting from out of the head of winter's monster. Some booms, then fireworks pale in the late afternoon light, then a lot of white smoke. Giant cheer from the crowd.

The smoke drifted off to the east. She walked over to the lakeshore, just a few blocks from her schwimmbad. It was near sunset. She could see three ridges to the south; first the low green rim of the lake, then the higher, darker green ridge between them and Zug; then in the distance, far to the south, higher than the world, the big triangular snow-splashed peaks of the Alps proper, now yellow in the late light. Alpenglow. This moment. Zurich.

97

There are about sixty billion birds alive on Earth. They've been quicker than anyone to inhabit the rewilded land and thrive. Recall they are all descended from theropods. They are dinosaurs, still alive among us. Sixty billion is a good number, a healthy number.

The great north tundra did not melt enough to stop the return of the caribou herds to their annual migrations. Animals moved out from the refuge of the Arctic National Wildlife Refuge in Alaska and repopulated the top of the world. In Siberia they're establishing a Pleistocene Park, and re-introducing a resuscitated version of the woolly mammoth; this has been a problematic project, but meanwhile the reindeer have been coming back there, along with all the rest of the Siberian creatures, musk oxen, elk, bears, wolves, even Siberian tigers.

In the boreal forests to the south of the tundra and taiga, wolves and grizzlies have prowled outward from the Canadian Rockies. This is the biggest forest on Earth, wrapping the world around the sixtieth latitude north, and all of it is now being returned to health.

It's the same right down to the equator, and then past the equator, south to the southern ends of the world. Certain inhospitable areas everywhere had been almost empty of humans all along; now these have been connected up by habitat corridors, and the animals living on these emptier lands are protected and nourished as needed. Often this means just leaving them alone. Many are now tagged, and more all the time. There is coming into being a kind of Internet of Animals, whatever that means. Better perhaps to say they are citizens now, and have citizens' rights, and therefore a census is being taken. Watersheds upstream from cities are adopted by urban dwellers who observe their fellow citizens from afar, along with making occasional visits in person. Wild animals' lives and deaths are being noted by people. They mean something, they are part of a meaning. Not since the Paleolithic have animals meant so much to humans, been regarded so closely and fondly by we their cousins. The land that supports these animals also supports our farms and cities as well, in a big network of networks.

What's good is what's good for the land.

There are fewer humans than before. The demographic peak is in the past, we are a little fewer than we were before, and on a trajectory for that to continue. People speak now of an optimum number of humans; some say two billion, others four; no one really knows. It will be an experiment. All of us in balance, we the people, meaning we the living beings, in a single ecosystem which is the planet. Fewer people, more wild animals. Right now that feels like coming back from a time of illness. Like healing, like getting healthy. The structure of feeling in our time. Population dynamics in play, as always. Maybe that makes us all living together in this biosphere some kind of supra-organism, who can say.

In a high meadow, wild bighorn sheep. Their lambs gambol. When you see that gamboling with your own eyes, you'll know something you didn't know before. What will you know? Hard to say, but something like this: whether life means anything or not, joy is real. Life lives, life is living.

98

Notes for Mary's last meeting, taken for Badim again. All in attendance except Estevan, who is in Chile. Also Tatiana of course. Someone took away her chair.

All stand as Mary comes in late. Mary shakes head. People, please. Sit. Let's get down to business, there's things to do.

We laugh at her. First pleasure, then business. Retirement cake on coffee table, etc. A little party, quick and subdued, as Mary obviously doesn't like it. She adjusts to the idea as people toast her with coffee. Changing of the guard, she says. The winter general, handing things over for the spring charge. Etc. Toasts. Awkward.

Then everyone to their usual places around the table. Mary convenes meeting, looking relieved.

M: All right, thanks for that. Yes, I'm retiring. Going emerita. Very happy Badim has agreed to take over. The secretary general and all other relevant parties have agreed to this and have named him acting minister. I'll be trying to see to it that they make it a permanent appointment. That would be good for all.

Badim thanks her. Looks at her as he always does: mongoose regarding cobra. She likes that he regards her as a problem to solve. Still.

M: Want to stay involved but without messing things up. Ambassador for the ministry, agent at large, that kind of thing. Available for whatever.

Janus Athena: Staying on as minister?

People laugh. J-A smiles briefly at Badim to indicate joking, but several are nodding. Badim among them. He says, At least go to the San Francisco meeting for us. I think you need to do that one.

M: I'm done. It's your turn now. Everyone has to do what they can in their time. Then it's time to pass it along. All of you will get to this point sooner or later. Hanging on too long never good. I may have done that already. But you're all young. Me, I'm done. I'm around if you need me. I'll stay in Zurich. Professor at the ETH and so on.

Badim: You'll always be welcome. In fact, we need you.

Mary smiles. I doubt that. But it's all right. It's time.

99

We're here today to discuss whether any of the so-called totalizing solutions to our current problems will serve to do the job.

No.

I suppose I have to ask, do you mean no to the question or no to the topic.

No to the question. There is no single solution adequate to the task.

And so what can we expect to see?

Failure.

But assuming success, just for discussion's sake, what shape might that take?

The shape of failure.

Expand on that please? A success made of failures?

Yes. A cobbling-together from less-than-satisfactory parts. A slurry, a bricolage. An unholy mess.

Will this in itself create problems?

Of course.

Such as?

Such as the way like-minded people working to solve the same problem will engage in continuous civil war with each other over methods, thus destroying their chances of success.

Why does that happen, do you think?

The narcissism of small differences.

That's an odd name.

It's Freud's name. Means more regard for yourself than for your allies or the problems you both face.

Well, but sometimes the differences aren't so small, right?

The front is broad.

But don't you think there's a real difference in for instance how people regard the market?

There's no such thing as the market.

Really! I'm surprised to hear you say this, what can you mean?

There's no more of a real market behind what we now call the market than there is gold behind what we call money. Old words obscure new situations.

You think this happens often?

Yes.

Give us another example.

Revolutions don't involve guillotines anymore. Alas.

You think revolutions are less visible now?

Exactly. Invisible revolutions, technical revolutions, legal revolutions. Quite possibly one could claim the benefits of a revolution without having to go through one.

But doesn't already-existing power resist revolutionary changes?

Of course, but they fail! Because who holds power? No one knows anymore. Political power is itself one of those fossil words, behind which lies an unknown.

I would have thought oligarchies were pretty known.

Oligarchic power is the usual answer given, but if it exists at all, it's so concentrated that it's weak.

How so? I must say you amaze me.

Brittle. Fragile. Susceptible to decapitation. By which I mean not the guillotine type of decapitation, but the systemic kind, the removal from power of a small elite. Their situation is very unstable and tenuous. It's highly possible to shift capital away from them, either legally or extra-judicially.

Just capital?

Everything relies on capital! Please don't be stupid. Who has capital, how it gets distributed, that's always our question.

And how does it get distributed?

People decide how it gets distributed by way of laws. So change could happen by changing the laws, as I've been saying all along. Or you could just shift some account numbers, as happened in Switzerland.

Ah yes. The banks. That reminds me of a fine story. Do you remember what the bank robber Willie Sutton said when a reporter asked him why he did what he did?

I do! Good of you to ask. And good of that reporter too.

The reporter said, Why do you rob banks?

And Sutton replied, Because that's where the money is.

100

She took an overnight train to Montpellier, slept the sleep of the blessed. Her ship left that evening, so that day she wandered the city's big old plaza, then the line of new Doric columns running from the plaza toward the harbor. Then onto an ocean clipper, sleek, seven-masted, looking like a cross between a schooner and a rocket ship laid on its side. On board to sleep again.

When she woke they were at sea. Every surface of this ship was photovoltaic or piezoelectric or both. Its passage through the waves, its very existence in the sun, generated power which got sent to the props. With a good wind filling the big sails, and the kites pulling from far overhead, tethered to the bow, they could fly on the thing's hydroplanes. A hundred kilometers an hour felt really fast.

Next morning they surged through the Pillars of Hercules and out into the Atlantic. Some vague memory of Tennyson's poem "Ulysses," a kind of Victorian-Homeric ode to retirement: To live, to fight, to strive, to something, and not to yield. Brit love of heroic death; charge of the Light Brigade, Scott in Antarctica, World War One. A sentimentalism very far from Irish, although the Irish had their own sentimentalities, God knew. Into the open sea, leave behind the sight of land.

The blue plate of the ocean. Sea and sky, clouds. Pink at dawn, orange at sunset. Winds pushing and pulling them, the sun, the waves. The glorious glide, crest to trough, trough to crest, long rollers of mid-ocean. How had they forgotten this? She recalled her last flight from London to San Francisco, passing over Greenland at midday, no clouds below them, the great ice expanse as alien as Callisto or Titan, and everyone with their window shades pulled down so they could watch their movies. She had looked out her window and then around at her fellow passengers, feeling they were doomed. They were too stupid to live. Darwin Prize, grand winner. The road to dusty death.

Here, now, she stood at the taffrail of a seven-masted schooner, a craft that could maybe be sailed solo, or by the ship's AI. AI-assisted design was continuously working up better ships, as with everything, and solutions were sometimes as counterintuitive as could be (kites? masts curving forward?), but of course human intuition was so often wrong. Foxing their own cognitive errors might be one of the greatest accomplishments of contemporary science, if they could really do it. As with everything, if they could get through this tight spot, they might sail right off into something grand.

A stop at Havana, handsome seaside city. Beautiful monument to the communist idea. Then to Panama, through the canal and up the sunny Pacific to San Francisco. They sailed under the Golden Gate Bridge on a cold cloudy day, the marine layer so low that the orange bridge was invisible, like a return to the bridgeless time. Then into their appointed wharf and onto land again, feeling its unsteady steadiness. Walking in hilly San Francisco, the most beautiful city in the world. Time for a last bit of work.

She had agreed to Badim's request to represent them one last time at the CCCB. They met again at the top of the Big Tower, and again Mary was distracted by the city below, and Mount Tamalpais, and the Farallons poking blackly over the western horizon.

Most of the same people were there, including the Chinese finance minister. Again Mary found her cheerful and articulate. One of the most powerful people in China. Hoping, she said when introduced, to find more that could be done to crank the Great Turn. Dynastic succession for the whole world, she suggested with a smile.

Jane Yablonski asked her what she had in mind.

Chan spoke of equity, getting better in China and the world, but still far from achieved. She spoke of income floors and ceilings, of land taxes and habitat corridors. Of the world as a commons, one ecosphere, one planet, a living thing they were all part of. Looking at the central bankers listening attentively to her, Mary saw it again; these people were as close to rulers of the world as existed. If they were now using their power to protect the biosphere and increase equity, the world could very well tack onto a new heading and take a good course. Bankers! It was enough to make her laugh, or cry. And yet by their own criteria, so pinched and narrow, they were doing the necessary things. They were securing money's value, they still told themselves; which in this moment of history required that the world get saved.

She had to smile, she couldn't help it. Saved by fucking bankers. Of course the whole world was making them do it. Now they were discussing other new ideas, experiments beyond anything she had ever dreamed of. Minister Chan was now smiling, sweetly but slyly—it seemed because she had looked over at Mary and seen Mary's little smile. The two of them were complicit in their amusement; both were amused that Chan herself would be taking the helm here, and heading off in new directions. It was so amusing that when someone asked Mary what she and her ministry might think of all these new ideas for reform that Madame Chan was proposing to them, Mary stood to extend her hand toward the young Chinese woman, and smile at her, and say, I yield the floor, I pass the torch, I like all of these ideas. I say, be as bold as you can dare to be!

101

What did we teach Beijing, you ask? We taught them a police state doesn't work! They thought it could, and they tried for fifty years to bring Hong Kong to heel, using every tool that came along—buying people, using CCTV cameras and facial recognition, propaganda, phalanxes of police and army, drone surveillance, drone strikes—and all of that just made the people of Hong Kong more resolute to keep what we had.

—Why do you say that! Of course what we had was real, because hegemony is real. That's a feeling, and feelings are something your culture explains to you. We in Hong Kong have a very particular culture and feeling. We lived as servants of the British, and we know very well what it felt like to be subaltern to a hegemonic power. That only lasted a few generations, but it set a particular feeling here, right in people's hearts: never again.

So when the British turned us over to Beijing, fine. We are Chinese, Beijing is Chinese. But we are also Hong Kong. It creates a kind of dual loyalty. Part of that is we speak Cantonese while Beijing speaks putonghua, Mandarin to some Westerners, although that's actually an elite or written version of putonghua, but never mind, we were different. We speak Cantonese, we are Hong Kong.

—Yes, of course there were Hong Kong people in favor of joining Beijing completely! These people often got money from Beijing, but I'm sure it was a genuine feeling for some of them. But most of us were for one country two systems, just as the saying had it. Our system, many of us call it rule of law. Laws in Hong Kong were written and passed by the legislature, enforced by the police, and ruled on by the courts. That's why the world trusted us with their money! Beijing didn't have that. They only had the Party. What got decided behind the closed doors of their standing committee became their equivalent of the law, but it was a rule only, rule without law, and it couldn't be challenged. It was arbitrary. That's why when Beijing tried to build Shanghai as their own finance center, to counter us, it didn't work. The world didn't trust Shanghai the way they trusted Hong Kong. So we in Hong Kong fought for it, we fought for the rule of law. All through the years between 1997 and 2047 we fought.

—Why 2047? The deal was that on July 1, 2047, we were to be folded entirely into the Republic of China. That was the British kicking the can down the road. They were not the worst empire by any means, but they were definitely an empire, and all empires are bad. So they struck that deal with Beijing, which had it that for fifty years we were to be one country two systems. And during those fifty years we in Hong Kong got used to fighting for our rights. Part of that meant going out into the streets and demonstrating. Over the years we saw what worked and refined our methods. Violence didn't work. Numbers did. That's the secret, in case you are looking for the secret to resisting an imperial power, which was what we were doing through those years. Non-violent resistance of the total population, or as much of it as you can get. That's what works.

—Yes, of course Beijing could have crushed us! They could have killed every person in Hong Kong, and repopulated the city with people from mainland China who didn't know any better, and would have been happy to take over such a nice infrastructure. Not that they would have known how to operate it! Still, this joke that a Western acquaintance once told me, about the government dismissing the people and electing another one, that was no joke to us. Because Beijing could have done that. It's kind of what they did to Tibet.

But there were constraints on Beijing too. For one thing they were always trying to entice Taiwan back into the fold. One part of that effort was to say to Taiwan, you'll be fine if you join us—we'll treat you just like we treat Hong Kong! One country two systems, and if you come back to us, three systems! Let a hundred flowers bloom! But that argument would only work if they were indeed treating Hong Kong well.

—Yes, there were other reasons that weren't so convincing to Taiwan, of course. There are always multiple causes, always. In this case there was always June 4. Tiananmen Square, 1989. Now known also as May 35th, or

April 66th, or all the way around the calendar, although the joke gets old, and even hard to calculate, and besides every one of those dates is now blocked on the mainland internet, of course. Because Beijing wants to erase that day from history. And though it seems like that can't really happen, in fact it is more possible than you might think. At least on the mainland. The world remembers it for China, and Beijing definitely did not want another such incident in the twenty-first century, when the whole thing would have been recorded by every phone and broadcast worldwide. No. Murdering the citizens of Hong Kong was not an option.

—Sure, the billion's occupation of Beijing helped us. No doubt about it! That was huge, and of course part of why that happened was that the mainland's illegal internal migrant population saw how we did it in Hong Kong, and decided to do it themselves, right in Beijing. And of course the Party was terrified, how could they not be? The people without proper hukou, without residency papers for where they lived, did all the dirty work in China, and there were about four hundred million of them when the occupation happened. That's a lot of people without any feeling of representation or belonging. So yes, the Party had to deal with that, or lose everything. In that struggle Hong Kong became a smaller matter and had some wiggle room, you might say. Which we used very smartly. It was never about independence, you must understand that. It was only for one country two systems. For the rule of law that we had here to persist past 2047.

—Yes, of course Beijing had other big problems. And as I said, they couldn't just kill us off. That left only talk. The discursive battle. And happily, we people of Hong Kong recognized this, and banded together. Solidarity—there's no feeling like it. People talk about it, they use the word, they write about it, they try to invoke it. Naturally. But to really feel it? You have to be part of a wave in history. You can't get it just by wanting it, you can't call for it and make it come. You can't choose it—it chooses you! It arrives like a wave picking you up! It's a feeling—how can I say it? It's as if everyone in your city becomes a family member, known to you as such even when you have never seen their face before and never will again. Mass action, yes, but the mass is suddenly family, they are all on the same side, doing something important.

—How did that play out in reality? What, didn't you see? Have you forgotten? We took to the streets every Saturday for thirty straight years!

—Of course sometimes it was more intense and other times less so. Often we let the young people do it, the idealism of youth is very good at persisting in such things, young people want something to believe in, of course! Everyone does, but the young aren't yet used to not getting it, so they persist. And they can handle the physical stresses of it better. But whenever push came to shove, we older people would come back out onto the streets also. As July 1 drew ever closer, we got back out onto the streets such that on some Saturdays the entire population of Hong Kong was out there. Those were stupendous events.

—Yes, there were other things we had to do also. Of course. These were not so exciting, in fact they were often tedious, but they needed to be done. Eventually you have to recognize that many necessary things are boring, but also, quite a few things are both boring and interesting at the same time. So we went to meetings, we joined the neighborhood councils, we went to the HK Legislature and did all the things that being part of a citizenry requires. It wasn't just the demonstrations, although that was part of it. There was all kinds of work to be done, and we did it. You have to pace yourself for the long haul.

—No, they didn't give it to us! Don't put it that way. Your questions are kind of offensive, I'm not sure if you are aware of that or not, but whatever, I will answer you politely, because I know the difference. So, eventually what I think happened is we just wore them down. They couldn't beat us; they had not the hegemony to do that. In fact, now some people say we beat them. These are the Tail Wags Dog people, who think our sterling example will eventually transform all mainland China into one big Hong Kong.

—No, I myself don't think that's right. It goes too far. China is too big, and the party elites are too convinced they are right. I'm more in the camp that gets called Tail Wags Dog's Butt. This is more realistic, even to the image itself; when a dog wags its tail, even when it is most excited, it's only the butt that also moves with the tail, not the whole dog. That wouldn't make sense, just in the physical sense. You can see that when you look at any dog, even one in a frenzy of happiness: only the butt wags with the tail. The head and chest stay steady. So, it's the same with China. The Cantonese-speaking part of China is in the south. It's Guangdong, a very big and prosperous province in south China, centered on the city Guangzhou, used to be spelled Canton in English, and there are a hundred million of us who speak Cantonese, and it's an older dialect than Mandarin. And most of the Chinese who live elsewhere in the world speak it, and we in Hong Kong speak it. Also in Shenzhen, the Special Economic Zone where Beijing tried to piggyback on Hong Kong's success in the world. So Beijing made a big mistake when they tried to suppress Cantonese as a language, which they did for many years, because that meant all of Guangdong didn't believe in Beijing either—they were more with Hong Kong than with Beijing, even if they never did much to show it. But language is family. Language is the real family.

—What has been the upshot? Well it's a work in progress, but since you ask so politely, I will say that since July 1, having agreed to keep one country two systems in Hong Kong, Beijing also has had to grant more and more rights to Guangdong. And they definitely don't mess with the Cantonese language anymore! Of course they said this change of policy was made just to integrate south China better into the country at large, but really it was done to turn what had been a defeat into a victory, or at least put it to use, which Beijing, it has to be said, is very good at doing. They cross the river by feeling the stones. Still, in this case, the tail wagged so vigorously that it wagged the dog's butt too, that I will grant. You can see how that works just by telling your dog it's time to leave the apartment and go for a run!

102

Meeting in San Francisco over, retired in full, Mary considered how best to get home to Zurich. There was no hurry. She looked into it online, and to her surprise found that Arthur Nolan, the airship pilot Frank had introduced to her in his co-op, was flying into San Francisco the following week, as part of a voyage around the world. This tour of his was headed to the Arctic, then down over Europe and the east side of Africa to Antarctica.

Mary contacted him and asked if she could join the trip, and he texted back and said yes, of course. Happy to have her.

On the airship they called him Captain Art. He met her on a pad on the side of Mount Tamalpais where his craft was tied to a mast, and ushered her up the jetway and along to the craft's viewing chamber, which was located at the bow of the airship's living quarters, a long gallery that extended under much of the length of the airship's body, like a big keel. The gondola, they called it. A little group of passengers were already in this clear-walled and clear-floored room, eating appetizers and chatting. Nature cruise. Mary tried to keep an open mind about that, tried to remember names as she was introduced. About a dozen people, mostly Scandinavian.

At the end of the introductions, Captain Art told them they would stop next in a particular Sierra meadow, to see a wolverine that had been spotted, an animal he obviously considered special. They were happy to hear it.

Shortly thereafter they took off. This felt strange, lofting up over the bay, bouncing a little on the wind, not like a jet, not like a helicopter. Strange but interesting. Dynamic lift; the electric motors, on sidecars up the sides of the bag, could get them to about two hundred kilometers an hour over the land, depending on the winds.

East over the bay and that part of the city. Then the delta. It reminded Mary of the model of northern California that she had been shown long ago, but this time it was real, and vast. The delta an endless tule marsh below them, cut into patterns by lines of salt-tolerant trees, remnants of the old islands and channels. Blond-tipped green grasses, lines of trees, open water channels, the V wakes of a pair of animals swimming along— beavers, Art told them. The viewing chamber had spotting scopes, and what they showed when one looked through them was that the delta was dense with wildlife. Mostly unvisited by humans now, they were told. Part of California's contribution to the Half Earth project. Mount Diablo, rising behind them to the southwest, gave them a sense of the size of the delta; it was immense. They could still just see the Farallons marking the sea on the western horizon; to the north stood the little black bump of Mount Shasta; to the south the coastal range walled the central valley on the right, the Sierra Nevada walled it on the left. Huge expanse of land. It looked like California would have an easy time meeting the Half Earth goal.

Over the central valley. Habitat corridors looked like wide hedgerows, separating giant rectangles of crops and orchards. A green and yellow checkerboard. Farther east hills erupted out of orchards; the first rise to the Sierra, now a dark wall ahead. The airship rose with the land, floating over wild oak forests and then evergreen forests, with steep-sided canyons etched deeply through the hills. Snow ahead on the highest peaks.

Art brought the airship down onto Tuolumne Meadows, a high expanse of snow and trees, punctuated by clean granite domes. The roads to it were still closed, they seemed to have it entirely to themselves. Along with a family of wolverines.

The airship attached itself to a mast sticking out of the snow near Lambert Dome. The two members of the crew maneuvered the craft down and secured it to anchor bollards. A ramp extended from a door in the side of the gondola, and they walked down into chill still air, onto hard white snow.

What do they eat in winter? the passengers asked Art, looking around at the white folds of snow, the steep granite faces; no obvious signs of food, unless you could eat pine trees. Did they hibernate?

They did not. They were fine in winter, Art told them. Dense fur, feet like snowshoes. In winter they ate little mammals dug up from under the snow, but mainly bigger creatures found dead. Carrion eaters. Not uncommon in winter to find creatures that had died.

They followed him over the hard snow. He was looking at his phone as he walked, his other hand balancing a spotting scope on a monopole over his shoulder. Then he stopped, phone extended to point. They all froze. From a knot of trees emerged three black creatures, galumping over the snow. They looked somewhat like dogs with very short legs. A mama and two kits, it appeared. Broad backs, reddish black fur. Bands of lighter fur around the sides of their bodies. The mother had a buff band across her forehead.

She stopped and suddenly began digging hard in the snow. Not far beyond them, a steaming hot springs had melted an open patch of ground, muddy and crusted at the edge with brown ice. Possibly a water hole in wintertime. Then the mama wolverine stuck her head in the hole she had dug and began to tug upward. Ah: a dead deer, buried under the snow near the spring. Patiently, with some more pawing and many hard tugs, the mother pulled it up out of the snow. Serious strength in that small body. She began to tear at the corpse, and her kits flopped around her trying to do the same.

One old name for the wolverine, Art said quietly as they watched, was the glutton. They always ate with great enthusiasm, it seemed, and usually tore their prey to bits and ate every part of it, including bones. They had a tooth at the back of their mouths that made it easier for them to tear flesh, and their jaws were so strong they could break any bone they came on. *Gulo gulo* was their Latin name, referencing this supposed gluttony.

"We're lucky to see this," Art said quietly. "Wolverines are still rare here. There weren't any at all in the Sierra from about 1940 to the early 2000s. Then a few began to show up on night cameras near Lake Tahoe, but they were wanderers, and it didn't look like there were any breeding pairs. Now they're being seen all up and down the range. There were some re-introductions to help that along. Now it seems like they're back."

"Lots of deer up here?" Mary guessed.

"Sure. Like everywhere. Although at least here the deer have some predators. Mountain lions, coyotes."

They settled in and shared time at the spotting scope. They watched the wolverine family eat. It was a somewhat grisly business. The kits were playful in the usual style of youngsters. This was their first year, Art said. Next year the mom would shoo them off. They weren't graceful, being so low and foursquare. They reminded Mary of otters she had seen in zoos, the way otters moved on land; but otters were very graceful underwater. For wolverines, this was it. Not graceful. But of course this was a human perspective; they were also obviously capable, confident, happy on the snow. Unafraid. Wild creatures at home, back at home, after a century gone. Knitting up the world.

Mary moved away, stood watching. For a time she was distracted, thinking of Frank and the chamois and marmots. Then these animals brought her back. The kits harassed their mom, played in the usual style of the young. Such a deep part of what mammals were, playing when young. Did baby salamanders play? She couldn't remember ever playing, her childhood was so far behind her—but no, there it was, she remembered. Kicking a ball in the yard and so on. Sure.

Then also the careless tolerance of the mom, ignoring her youngsters as they clambered on her, wrestled with each other, fell over themselves. They snuffed and worried at her underside, she knocked them away with a flick of her foreleg. They did have big feet, clawed pads like snowshoes, broad and long. Lords of winter. Nothing up here could harm them, nothing scared them. Art said people had seen them chasing away bears, mountain lions, wolves. Masters of all they surveyed.

Captain Art regarded them with a fixation Mary found pleasing. He was lost in it. They were in no hurry; this was the place to be. Again she thought of Frank in the meadow above Flims, but now it was all right, she could be grateful he had taken her up there, that he had introduced her to this man. It was getting cold, she felt the first pinch of hunger, she had to pee. But there before her, wolverines. It was a blessing.

Only when the sun dropped into the treetops did Art stir and lead them back to the airship. By then they were really cold, and warming up in the gondola was a kind of party. They flew east on the wind, looking down at the pink alpenglow suffusing the range of light.

★   ★   ★

North and east above desert, the Rockies, flat prairie, and then tundra. The border between the great boreal forest and the tundra looked ramshackle and weird; a lot of permafrost here had melted, Art said, creating what was called a drunken forest, trees tilted this way and that. Then lakes everywhere under them, more lakes than land. Flying over this huge wet expanse it looked like the Half Earth goal would be easy to reach, or was even already accomplished. Which was not the case, but one always judged by what stood before one's eyes. In fact they infested the planet like locusts. No, that too was wrong. In the cities it looked like that, but not here. There were many realities on a planet this big.

The new port city on the Arctic Ocean, called Mackenzie Prime, looked like an old industrial site. A single dock six kilometers long, studded with cranes for handling container ships. The opening up of the Arctic Ocean to ships had made for one of the odd zones of the Anthropocene. Traffic was mostly container ships refurbished as autopiloted solar-powered freighters, slow but steady. Carbon-neutral transport on a great circle route, and as such not much to complain about. Also there were few to complain, at least in terms of locals; the total population on the coasts of the Arctic Ocean still numbered less than a million people: Inuit, Sami, Athapaskan, Inupiat, Yakut; Russian, American, Canadian, Scandinavian.

The great shock of their arrival was to see that the ocean, clear of ice to the northern horizon, was yellow. Naturally this looked awful, like some vast toxic spill; in fact it was geoengineering, no doubt the most visible act of geoengineering ever, and as such widely reviled. But the solar heating of the Arctic Ocean when there was no ice covering it might be enough all by itself to tip the world irrevocably into jungle planet. All the models were in agreement on this, so the decision to try to forestall that result had been made according to Paris Agreement protocols, and the color dye released. Yellow water didn't allow sunlight to penetrate it, and even bounced some sunlight back into space. Relatively small quantities of dye could color a large area of ocean. Both the artificial and natural dyes they were using broke down over a summer season, and could be renewed or not the following year. Petroleum-based dyes were cheap to manufacture, and only mildly carcinogenic; natural dyes, made of oak and mulberry bark, were non-petroleum-based, and only a little bit poisonous. The two could be alternated as they learned more about them. The energy and heat savings in terms of albedo were huge—the albedo went from 0.06 for open water (where 1 was total reflection and 0 total absorption) to 0.47 for yellow water. The amount of energy thus bounced back out into space was simply stupendous, the benefit-to-cost ratio off the charts.

Geoengineering? Yes. Ugly? Very much so. Dangerous? Possibly.

Necessary? Yes. Or put it this way; the international community had decided through their international treaty system to do it. Yet another intervention, yet another experiment in managing the Earth system, in finessing Gaia. Geobegging.

Mary looked down at the ungodly sight from the airship's gondola and sighed. It was a funny world. "Why did you bring us up here?" she asked Art. "Was it to see this?"

He shook his head, looking mildly shocked at the suggestion. "For the animals," he said. "As always."

And a few hours later they were flying over a herd of caribou that covered the tundra from horizon to horizon. Art admitted he had brought the airship down to the right altitude to create this effect; they were about five hundred feet above the ground. From this height there seemed to be millions of animals, covering the whole world. These were migrating west, in loose lines like banners or ribbons, which bunched whenever they were crossing a stream. It was stunning to see.

South over Greenland.

As they flew they saw a lot of other airships. Giant robot freighters, circular sky villages under rings of balloons, actual clippers of the clouds sporting sails or pulled by kites, hot-air balloons in their usual rainbow array. There had not yet been any regularization of shapes and sizes; Art said they were still in the Cambrian explosion moment of airship design. Many people were moving up into the sky, and traffic lanes and altitudes had been established, as with jets in the old days. Airspace was humanized and therefore also bureaucratized. And carbon neutral.

As they flew, Mary spent more and more time listening to Art talk to his passengers, his clients or guests or customers. He had lived most of his life on this airship, he told them. He was about sixty, Mary reckoned, so the "most of his life" seemed a bit premature, a statement of intent as much as a history. She liked him. A slight man, angular face, hooked nose, balding. Startling pale eyes, a distinguished look, a sweet shy smile. He looked like the photo of Joyce Cary that her father had kept on his bookshelf, next to a row of Cary's novels. Despite his job as ship captain and chief naturalist, he seemed to her a shy man. He spoke mostly of animals and geography. Which given their position made sense, but days passed and she never learned a thing about him except what she might deduce. Irish; eventually, she even had to ask, she learned he was from Belfast, his dad Protestant, mother Catholic.

Something had sent him aloft, she thought as she watched him. It had been an escape, perhaps. A refuge. An ascension into solitude. Then, after years had passed, perhaps, he had gotten lonely, and begun running these tour cruises. This was her theory. Now he liked to share the pleasures of his life aloft, and he gained some company by it, some conversation. And he had an expertise he could teach to people, the various joys and fascinations of a bird's life. An Arctic tern he was, back and forth, pole to pole. A few years before he had hired an events coordinator in London, who booked his tours and helped him arrange their various ports of call.

So: nature cruise. Mary was still very dubious. It was not her kind of thing. She doubted she would do it again. Still, for now, the other passengers were pleasant; some Norwegians, a few Chinese, a family from Sri Lanka. They were all interested to see the world from the air, in particular the world's animals.

Earth was big. At this height, at this speed, that immensity was becoming clearer and clearer. Of course scale was so variable. Pale blue dot, mote of dust in the sunlight, true enough; but from this vantage it was beyond enormous. You could walk your whole life and never cover more than a small fraction of it. Now they lofted like an eagle over it.

"We're so stupid," she said to Art one night.

He looked at her, startled. It was late, they were alone in the viewing room, the others had gone to bed. This had already happened once or twice before; it was beginning to look like a habit, a little conspiracy to chat.

"I don't think so," he said.

"Sure you do," she replied. "Why else are you up here?"

Again he was startled. His other guests didn't speak to him like this, she saw.

"Didn't something drive you up here?" she pressed.

"Oh," he said, "let's not talk about that."

She relented, feeling she had gone too fast, hit a wall. "You like the beauty," she said. "I know. And it is beautiful."

"It is," he agreed quickly. "I never get over it."

She smiled. "You're lucky."

"It's true." And he added: "Especially tonight."

She laughed at that.

He was still young enough to blush. She knew that kind of fair skin very well; her grandmother had blushed furiously right into her nineties.

After that conversation, the habit was set. They stayed for a nightcap in the viewing chamber after the others had retired. There they had the view of everything below. When he dimmed the room lights, the world below them became visible. This was especially true when the moon was up; then the land and ocean became eldritch things, glittery and dark, distinct in their forms.

The airship also had a tiny viewing chamber on top of its big body, there among the solar panels, so that Art and his guests could see the stars when the moon was down. In its earliest phase, after the thin glowing crescent of the moon set, he took guests up through the body of the ship to this chamber to observe the starbowl. One night at new moon he led Mary up there after the others had retired. Milky Way low in the west, Orion climbing up over the eastern horizon, all of this very far from cities, and at five thousand feet, it was simply amazing how many stars they could see. It was a whole different sky, primal and alive. Art knew the constellations, and some of the stories behind them. He had a telescope in that bubble set with a tracking motor that kept it fixed wherever he aimed it, but on that night he left it alone. He taught Mary to see a galaxy visible to the naked eye, in the north near Cassiopeia.

But mostly they stayed in what he called the understudy, looking down at the Earth. As they flew down the Atlantic, over Iceland, then the Hebrides, then Ireland—this last part for her, and for him too, perhaps—then over the Bay of Biscay—they would say good night to everyone, then she would go to her cabin, go to the bathroom, change clothes perhaps, and slip down the private stairs he had taught her to find, using the key code he had taught her to use, back to the viewing room at the bow, now locked and empty, except for them.

One night they watched the Pillars of Hercules float by below them, framing the Strait of Gibraltar. The little lumps of Gibraltar Rock and Jebel Musa stood like sentinels over the black water. Art told Mary the story of the flooding of the Mediterranean; it had been a dry low plain between Europe and Africa, then as an ice age had ended and sea level rose, the Atlantic had spilled through this strait into what had been flat playas. Two years of flow, he said, at a thousand times the rate of the Amazon, moving at forty meters a second, and carving a channel a thousand feet deep, until the Med was filled and the two bodies of water equalized in elevation.

"When did that happen?"

"About five million years ago, they say. There isn't total agreement."

"There never is."

She watched him closely. A flood, a sudden breakthrough. Now he was talking about the end of the last ice age, fifteen thousand years ago, when enormous lakes of meltwater on top of the great ice sheet had broken through ice dams and poured down into the ocean in stupendous floods, changing the climate of the whole world. Then the Mediterranean had risen high enough to flood through the hills of the Bosporus, filling the Black Sea's area in just a few years' time, flooding land that had been occupied by humans, giving rise to the legend of Noah's flood.

He was nattering on. He was perhaps a little nervous. Was he a dry plain himself, she wondered, a space waiting to be flooded? Was she the Atlantic, he the Mediterranean? And she? Was she rising? Would she pour over into him and fill him up?

There was no way to know, no rush to decide. They were headed for Antarctica, and they hadn't even reached the equator yet. There was time. She could enjoy the idea of it, mull it over in mind and body. When she got up to go to her cabin, at the end of that night, she leaned over and gave him a quick kiss on the top of the head.

Across the Atlas Mountains, east over the Sahel. Here there were new salt lakes and marshes being created by water pumped up from the Atlantic or the Mediterranean. Salt seas in dry basins, an interesting experiment. They definitely changed things. Here in the Sahel, the dust storms that used to fly off these desert basins over the Atlantic were much diminished, and certain kinds of plankton out to sea were going hungry. Unexpected consequences—no, unforeseen consequences. Because now they were expected, even when they couldn't be predicted.

For now, the desert below them was dotted by long lakes. Green, brown, sky blue, cobalt. Cat's paws. Little towns hugged their shores, or stood on outcrops nearby. Irrigated fields formed circles on the land, circles of green and yellow like quilting art. Local culture was said to be thriving, Art said. Polls indicated most residents loved their new lakes, especially younger people. Without them we would have left, they said. The land was dying, the world had killed it. Now it would live.

A red dawn, punctuated by two black masses rising up higher than they were: Ethiopian highlands to their left, Mounts Kenya and Kilimanjaro to their right. As they flew through this immense gap, Art told them about Jules Verne's first hit novel, *Five Weeks in a Balloon*. Also about his later works *The Mysterious Island* and *The Clipper of the Clouds*, both describing balloon and airship travels, as did of course a big part of *Around the World in Eighty Days*. Art also told them about Verne's *Invasion of the Sea*, which told the story of pumping seawater onto Saharan deserts to create lakes, just as they had seen during the previous few days. Verne's books had bewitched him as a youth, he said. An idea of how to live. He had taught himself French in order to read them in the original, said that Verne's prose was far better than people usually supposed when judging by the wretched early translations.

"And so we're here," one of them said, "with our own Captain Nemo!"

"Yes," Art replied easily. "But without his brooding, or so I hope." This said with a lightning glance Mary's way. "I hope I'm more like Passepartout. Passing by all, you know, with the least amount of difficulty."

The green and gray masses of Mounts Kenya and Kilimanjaro loomed over them to the south, one very flat-topped, the other a little flat-topped. Neither had glaciers, nor any sign of snow. No such thing as the snows of Kilimanjaro. Something they could only hope for in distant times to come.

But the great plains of east Africa were still populated by animals. Yes, they were now doing a safari from the air. Elephants, giraffes, antelopes, great herds of all these, migrating from river to river. Some of the streams' water was now piped in, Art said quietly. Desalinated at the seashore and then piped up to the headwaters and released to keep the streams flowing, the herds alive. They were in their twelfth straight year of drought.

Then Madagascar. The reforestation of that big island had been happening for well over a generation, and such was the fecundity of life that its rugged hillsides already looked densely forested, dark and wild. It had changed during Art's time aloft, he said, and now the people of Madagascar were joining with Cubans and other island nations to help similar restoration efforts all over the world. Indonesia, Brazil, and west Africa were teammates in this effort. Rewilding, Art called it. They were rewilding down there.

That night when Mary joined Art in his understudy, Madagascar was already behind them, but the air still seemed to carry its spicy scent. Art sat looking back at the island, bulking like a sea creature with thick napped fur. He seemed content. They sipped their whiskies for a while, enjoying a companionable silence. Then they talked about other voyages they had made. He asked about her escape on foot over the Alps, apparently a famous story of her professional life, and she made it brief, talked about the Oeschinensee, and Thomas and Sibilla, and the Fründenjoch. Have you flown your *Clipper* over the Alps? she asked.

Once or twice, he said. It's a bit much. They're too high. And the weather is just so variable.

I love the Alps, she said. They've caught my affection.

He regarded her with a little smile. A shy Irishman. She knew that type and liked it. She had always liked those men who kept to themselves, who had only a sidelong look for her. Probably there was something back there in his past, some event or situation that had made him so aloof; but the island world he had made for himself was one that she was coming to like. Or she could see why he liked it. He was younger than she was, but old enough that they were in that temporal space that felt roughly contemporaneous.

These were fleeting thoughts. Mostly she just watched the ocean and the great black island behind them, slowly receding. But they were thoughts that led in a certain direction. She tried to track them as if they were shy animals. Desire stirring in her, maybe that was like a tracker's curiosity. On the hunt. A hope for contact. Then in the midst of her musing he stood, saying he was feeling tired. Time for bed. He led the way up to the main gallery, said good night and turned toward his cabin.

Not a mind reader, she thought. Recalling that in her youth she had seemed to be able to reach out telepathically. Or maybe it had been a matter of looks, of pheromones. Animals in heat. Not at their age. But there was no rush.

Nothing more than that happened between them for the rest of that voyage. They still met on some nights in the viewing chamber to chat, but no more Madagascars; they were too far south.

Over the endless ocean, angling west to fight the great shove of the westerlies. Their pushback caused the *Clipper* to tremble and rock more than earlier in their voyage. Then one morning she woke and went to the viewing chamber, and there to the south lay Antarctica. Everyone was standing at the forward window to see it. The ocean was distinctly darker than before, almost black, which itself was strange, and faintly ominous; then to the south a white land like a low wall, white flecked with a black blacker than the strangely black sea. This great escarpment of ice and rock extended from east to west for as far as they could see.

Antarctica. It was early autumn, the sea ice minimal, although as they flew south they saw that there were icebergs everywhere. These made no pattern, just white chunks on black water. An occasional misshapen iceberg of jade or turquoise hue. Flocks, or it seemed rather shoals, of tiny penguins dotted some of these icebergs. Once they flew over a pod of orcas, sleek-backed and ominous. On tabular bergs they sometimes saw Weddell seals, looking like slugs splayed on the ice, often with smaller slugs attached to their sides like leeches. Mother and child. Their cousins, down here thriving on ice. If they were thriving.

Then Antarctica itself, white and foreboding. Ice Planet.

It was surreal in that icy desolation to come on six giant aircraft carriers, arranged in a rough hexagon like a coven of city-states, surrounded by smaller craft—icebreakers, tugboats, shore craft—it was hard to tell what the smaller craft might be, they were so small.

Apparently aircraft carriers made excellent polar stations, being nuclear powered, and outweighing ordinary icebreakers by a thousand times or more. Sea ice stood no chance against such behemoths, they were icebreakers from God and could leave anytime they wanted to; but they didn't. They made a little floating city, anchored by the shore of Antarctica, supporting various inland encampments, all of which had been airlifted upcountry from here.

They passed over the carrier city and landed on the snowy surface of the continent itself. Out of the *Clipper* onto flat snow. Very bright and very cold, although no colder than Zurich on a windy winter's day.

Around them cloth-walled huts, blue-glassed boxes. The camp managers were happy to meet Mary, seemed to consider her their patroness, which made her laugh. And they were well-acquainted with Captain Art, a frequent visitor.

The glacier slowdown operation had been a success. Ice fields had therefore also slowed. All this would have been impossible without the navies helping. Aircraft carriers were mobile towns. Deploying them like this was a chance to make use of the huge amounts of money that had been spent building them. Swords into plowshares kind of thing.

"Like the Swiss," Mary observed. "Can we see a pumping site?"

Of course. Already scheduled. For sure a feature of interest.

All of Art's passengers occupied only a corner of one of their giant helicopters. Helmets on, sit sideways looking out through a small window, then up, straight up, not like the *Clipper*'s rise. Then over endless white snow, listening to the pilot and her crew discuss things in a language she didn't recognize. Black sea behind them receding from view.

"Why is the ocean black down here?" she asked into her helmet's microphone.

"No one knows."

"I heard one guy say it's because the water is so clean down here, and the bottom so deep starting from right offshore, that you're seeing down to the dark part of the ocean where the sunlight doesn't penetrate. So you're seeing through super-clear water to the black of the deeps."

"Can that be right?"

"I heard the plankton down here are black, and they color the water."

"Lots of days it looks as blue as anywhere, I think."

"No way!"

Then they were descending. Snow or ice as far as they could see. Then a cluster of black dots. Around the dots black threads, like a broken spider web. These dots and lines held civilization suspended over the abyss.

"How many stations are there like this one?" she asked.

"Five or six hundred."

"And how many people does that add up to?"

"The stations are mostly automated. Maintenance and repair crews fly in as needed. There are caretakers in some of them. But mainly it's construction crews, moving around at need. I don't know, twenty thousand people? It fluctuates. There were more a few years ago."

The helo landed with a foursquare thump. They unbelted, stood awkwardly, filed down the narrow gap between seats and wall and down metal steps to the ice.

Cold. Bright. Windy. Cold.

Light blasted around the edges of her sunglasses and blinded her. Tears were blown off her eyes onto her sunglasses, where they froze in smears. She tried to see through all that. Blinking hard, she followed the others toward this settlement's main hut, like a blue-walled motor home on stilts.

"Wait, I don't want to go inside yet," she protested. "I want to see."

A couple of their hosts stayed out with her and walked her to a pump, which stood inside a little heated hut of its own. Not very heated, as the floor inside was ice, with the black housing of the pump plunging right into it. The saving of civilization, right there before her. A piece of plumbing.

They went back outside and followed one of the pipelines up a gentle gradient. It stretched across the land from black box to black box, lying right on the ice. Mary stopped to look around. The snow seemed to her like a lake surface that had flash-frozen, all its little waves caught mid-break. Glowing in the light. Her guides explained things to her. She liked their enthusiasm. They were happy to be here not because they were saving the world, but because they were in Antarctica. If you like it, one told her when she asked, you like it a lot. It gets into you, until nowhere else seems as good.

A white plane under a blue dome. Some cirrus clouds over them looked close enough to touch.

"It's like another planet," Mary said.

Yes, they said. But actually just Earth.

"Thank you," she said to them. "Now I'm ready to go back. I'm glad to have seen this, it's just amazing. Thank you for showing it to me. But now let's go back."

Because I too have a place I love.

They flew north up the Atlantic, to see St. Helena and Ascension. Before Art dropped Mary off in Lisbon, where she would train home, she joined him in his understudy one last time. When they were sitting in their usual spots, sipping their drams, she said, "Will we meet again?"

He looked uncertain. "I hope so!"

She regarded him. A shy man. Some animals are reclusive.

"Why do you do this?" she said.

"I like it."

"What do you do when you're on the ground?"

"I resupply."

"Aren't there any places you like to walk around?"

He considered this. "I like Venice. And London. New York. Hong Kong, if it isn't too hot."

She stared at him for a while. He shifted his gaze down, clearly uncomfortable. Finally he said, "Mostly I just like being here. I like the sky people. The sky villages are a lot of fun to visit. I like the way they look. And the people in them. Everyone's on a voyage. Did you ever read *The Twenty-one Balloons*? It's an old children's book about a sky village."

"Like your Jules Verne."

"Yes, but for kids."

Verne is for kids, Mary didn't say.

"Anyway I read it when I was about five. Actually my mom read it to me."

"Is your mom still alive?"

"No. She died five years ago."

"Sorry to hear."

"Is your mom still alive?"

"No. My parents both died young."

They sat there for a while. Mary saw that he was unsettled. Rejecting all the fashionable diagnostics of their time, knowing him to be fond of her, maybe, she pondered it. So, he was quiet. Perhaps he was shy. Perhaps he played a part for people: Captain Art, doing his best to get by.

She was not quiet, nor was she shy. A bossy forward girl, one teacher had said of her at school; and that was true. So she could only guess at him. But this was always the case, with everyone. And it seemed to her they got along. His silence was restful. As if he were content. She wasn't content, and she wasn't sure she had ever met anyone who was, so it was a hard thing for her to recognize. Maybe she was wrong. No one was content. She was projecting onto his silence. But from what, and onto what? Oh it was all such a muddle, such a swamp of guesswork and feeling.

"I like you," she said. "And you like me."

"I do," he said firmly, and then waved a hand, as if to push that aside. "I don't mean to be intrusive."

"Please," Mary said. "I'm about to disembark here."

"True."

"And so?"

"And so what?"

Mary sighed. She was going to have to do the work here. "So—maybe we can meet again."

"I'd like that."

After a pause during which Mary watched him, making him go on, if he would, he said, "You could come with me again. Be my celebrity guide. We could make a tour of all the greatest landscape restoration sites, or geoengineering projects."

"God spare me."

He laughed. "Or whatever you like. Your favorite cities. You could be a guest curator or whatnot."

"I'd rather just be your girlfriend."

His eyebrows rose at that. As if it were an entirely new idea.

She sighed. "I'll think about it. One nature cruise may be enough for me. But some ideas might come to me."

He took a deep breath, held it, let it out in a long sigh. Now he looked really content. He glanced at her, met her eye, did not look away. Smiled.

"I always come back to Zurich. I have my room there."

She nodded, thinking it over. Say it took years to get to know this man; what else did she have to do? "I'll want you to talk a bit more than you have," she warned him. "I'll want to know things about you."

"I'll try," he said. "I might have some things to say."

She laughed at that, knocked back the whisky in her shot glass. It was late.

"Good," she said. She stood and kissed him on the top of the head, ignoring his flinch away. "Maybe you can tell me when you're in town, and we can get together. Fasnacht is at the end of the winter, that's a party I like. We could do the town on Fasnacht."

He frowned. "I'll be out on another trip that month. I'm not sure I'll be back by then."

Mary stopped herself from sighing, from saying anything sharp. This was not going to be anything quick, or even normal. "We'll figure it out," she said. "Now I'm off to bed."

103

I don't think anyone ever figured out who organized it. Whoever they were, they wanted to stay out of the way and have it look self-organized. Have it emerge out of the Zeitgeist. And maybe it did, I mean ultimately we all did do it together. It was already a feeling everyone had. I think something like three billion people tapped their phones to say they had taken part.

It was sort of like New Year's Eve, except it was agreed it should be a simultaneous moment all over the Earth. Near the spring equinox in the northern hemisphere, like Narooz or Easter. Having it be the same very moment for all seemed right, it was important to feel the connection with everyone and everything else, as a kind of vibe. *Kulike*, in Hawaiian, means harmony. Or *la ʻoluʻolu*, harmony day. Evoke the noösphere, call it into existence by everyone thinking of it at the same time—that's not a time-delayed thing, it has to be simultaneous. So we in Hawaii kind of got the short end of the stick, time-wise. The timing was presented as a given, which I think means that someone somewhere had to be doing it in terms of organization, but anyway east Asia got the late night, then going west they went down through the time zone hours until western Europe got noon, then across the Atlantic it got earlier and earlier across the Americas, to a dawn patrol kind of thing on the west coast, so we in Hawaii were looking at 3 AM I think it was. Fine, whatever, an excuse to stay up all night and party, and it has to be admitted that it was still nice and warm for us even in the middle of the night, so we could go to Diamond Head and look out over the ocean as we partied. And the moon was full that night, no coincidence I'm sure. So it was nice. Down in the concert bowl bands played through the night, and we sat on the ridge talking and drinking and watching the ocean by moonlight, good south swell too, so that a lot of us were talking about going to Point Panic at sun-up to catch some waves, great way to finish this event, back in Mother Ocean where we all began. Slight offshore wind too.

So the time came and we listened to the voices on our phones. We are the children of this planet, we are going to sing its praises all together, all at once, now is the time to express our love, to take the responsibilities that come with being stewards of this earth, devotees of this sacred space, one planet, one planet, on and on it went, it seemed clear to me that the original had been written in some other language, that we were listening to a translation into English, and in fact you could tap around and hear what was being said in other languages; Gupta insisted on listening to it in Sanskrit, which he admits he doesn't understand when spoken, though he reads it, but he claimed that what we were hearing had to have been written or thought originally in Sanskrit, maybe even thousands of years ago, and in fact the Sanskrit version did sound very primal, which made me curious and I clicked around and found a version in Proto-IndoEuropean, why not? It sounded like Spanish. I switched to Basque, supposedly a living fossil of a language, and it too sounded like Spanish. Actually both sounded quite a bit stranger than Spanish, older than Spanish, odd harsh primal sounds, but no more so than Dutch or many another language that isn't Hawaiian, you always hear all the same sounds, and no matter which language I tapped on, I kept hearing *mamma Gaia*. Yes of course *mamma* would be one of the oldest words, maybe the first word, invented over and over by babies trying their best to talk but having limited control of their mouths, and yet always trying to say the same thing, to beseech or celebrate that great goddess filling their sight, the fountain and source of all food, warmth, touch, love, and eye contact—mamma! I cried out that night on the ridge, seeing the why of it for the first time, the why of everything, of course it's a category error on my part to genderize the planet in that crass way, but we were high that night on the worldwide lovefest, and since everyone else was singing and cheering and hooting as after having caught a great ride on a great wave, I just kept shouting Mamma Mia! Mamma Mia! Because of course, being human, the other first word we speak is always me, mine, me me me, and God bless the Italians and whoever else in the Romance languages for holding fast to that very first Ur phrase, the same in all the languages, I checked Proto-IndoEuropean and sure enough it was the same there too, Mamma Mia! Mamma Mia! Genius of a language!

Yes, I was a little drunk, a little high. A little giddy. I mean think about it, a worldwide moment during which all sentient beings aware of the project were to sing praise together to the one planet we stood on, to perform the noösphere created by this so-vast and complex biosphere, while standing on the lithosphere and contemplating the hydrosphere and circulating the atmosphere in and out of us, breath after breath—it's great, but it's a little hypothetical too, right? It's hard to know how to feel it. What could we do in that moment but try? As a linguist I naturally think of the words involved, but there was more to it than that, so I tried all that too, I drank and I looked around at the faces of the other people on the ridge, all of them also trying, and many of them had their dogs with them there, the dogs too were trying, trying to understand it, very aware that something unusual was happening, such that some of them barked or howled, which some guys instantly took up, of course it was time to howl, howl at the moon like wolves. What a great language! And besides we *were* like wolves! We turned wolves into dogs and they turned us into humans—we were something like orangutans before, solitaries who didn't know how to work together, it was the wolves who taught us that, who taught us the idea of friendship and cooperation. So we howled at the moon and hugged the people around us, if they were hugging types, and the dogs, and I kept looking at all the faces, so vivid and real, and I kept coming back to saying Mamma Mia. As one does when in awe. Hugging Gracie in particular, as always. We're lucky that way.

This went on for about fifteen minutes. Then we quieted down. Time to get back down in the bowl of the concert space and dance to the music for what was left of the night. Had we done it right? Had we joined with every sentient being on the planet, brought into existence a new Earth religion that would change everything? Were we all brother and sister now, as they were always telling us we should be? Hard to tell. It felt like a lark. But larks are beautiful. All these bird and animal names we use for our moods and actions, of course they're always perfectly apt. We are all family, as the new religion was telling us, and as every living thing on Earth shares a crucial 938 base pairs of DNA, I guess it's really true. So yeah, we went down there and danced all night long, feeling very high, and when the sky lightened and dawn approached, and they sent us out to greet the day and go home and do whatever we were going to do that day, they played Bruddah Iz Israel Kamakawiwoʻole's medley of "Somewhere Over the Rainbow" and "What a Wonderful World," a great piece of Hawaiian schmaltz that we could sing along with on our way out, and hum the rest of the day. Later I read that people said they really felt it, that moment when everyone worldwide sang the same song of praise and devotion, it was said to be like an electric pulse filling you or like that. I must admit I didn't feel anything like that in the moment itself, maybe I was too drunk or too aware of Gracie's hand on my ass, but that morning I caught the longest left of my life at Point Panic, with Bruddah Iz's pair of tunes running in my head the whole time, what a wonderful world indeed, and I flew out of that wave just before it closed out, and up there in the air over the wave, suspended weightless with the offshore spray and not seeing the ehukai rainbow because I was right there in it, yes that's when I felt it. Of course it doesn't come on command or to a schedule, grace isn't like that, it touches down on you in unexpected moments, a matter of accidents, but you have to be open to it too, so maybe that was it for me, a slight delay after the sacred ceremony, which maybe was the riding of that wave anyway—anyway I felt it then, hanging in the air, then I crashed back down onto the white hissing backside and swam head up, laughing out loud. Yes, it worked. Mamma Mia!

104

Back in Zurich, Mary told the Swiss that she was going to move out of her safe house. Retirement had changed her security status, she didn't need to occupy a safe house that was probably needed for others, and so on. They didn't object.

They didn't want her to move back to her apartment on Hochstrasse, but neither did she. That faded blue apartment block was now part of her past, it wouldn't come back. It was time to move on. And she had an idea anyway. There were housing cooperatives all over Zurich. She didn't want to move into the one Frank had lived in; that too was the past. Anyway it was Art's place too, and so for multiple reasons she felt it wasn't a good idea. But there were other housing co-ops in Zurich, a lot of them it turned out, and so she spent some time visiting a few.

During these visits around the city, she realized that she liked her neighborhood. Fluntern, it was called, there on the lower slope of the Zuriberg. She liked it there. It was her neighborhood. So, in a less strongly felt way, was the district behind the Utoquai schwimmbad. She liked that part of town too. So she focused on those two neighborhoods, and then the area between them; they weren't that far apart. Of course the whole city was compact.

There were some cooperatives that were all right, but most didn't quite suit, and almost all of them had waiting lists that would take quite a while to work through to her. The more she looked, the more she realized how much she had liked her old place. But she needed to change.

Finally she found a place because Badim knew she was looking for one. Someone who had been working for him in the ministry had to move back to Ticino to care for her father, and when she heard from Badim that Mary was looking for a place, she wanted Mary to take it. It could be an informal arrangement, she said, a kind of sublease; the board had heard her appeal concerning this plan, and liked the idea of having Mary Murphy among them, and so had approved it. It was just several blocks from her old place, south past her usual tram stop at Kirche Fluntern and along Bergstrasse to an odd three-streeted corner where the co-op took up one wedge of the intersection. A little twenty-apartment co-op, four stories high, well-maintained like every other building in Zurich, all except for the old wreck across from the tram stop one down from Kirche Fluntern, which was some kind of special thing.

The woman subletting it to her met her at the door of the place. Trudi Maggiore, she said.

"Mary," Mary said, shaking her hand. "I've seen you around the office."

The woman nodded. "I worked two buildings down, but I took notes for Badim at a lot of your meetings. I sat against the wall with the other assistants. And I went with you on a trip to India."

"Ah yes, I remember now."

Trudi led her up the broad stairs of their stairwell. On the top floor she unlocked the door of her place. "It used to be the attic," she explained as she opened it. "I hope you don't mind. You get used to it."

It had been a very small and low attic, Mary saw at once. It was a single room, tucked under the big roofbeam of the building such that only under the roofbeam itself could you stand upright. To left and right of it the ceiling sloped down until the walls to left and right were only about two feet high. The left side of the room did have an interior wall sticking out into the room, and a door in that wall opened on the bathroom, which likewise sloped down to a low wall. It was hyperclean, like any Swiss bathroom, of course, but as with the rest of the place, about half its volume was far less than head high. The toilet was located past the sink; kind of a woman's apartment in that sense, in that you could sit on it, but if you stood before it you would have to duck.

"I like it," Mary said. "It's funny."

Trudi looked pleased. "I like it too. I'm sorry I have to leave. But I'm glad you'll be the one in it. I admire what you did."

"Thank you," Mary said.

She walked up and down the midline. Past the bathroom the room widened again to the left, and there lay the bed, set right on the floor. To lie down on it, it would be easiest to sit first on a short chair set next to it, and then roll on. Once in bed it didn't matter how low the ceiling was, as long as you didn't leap to your feet in a dream or something like that.

The kitchen was back against the wall next to the front door, just a counter with a sink in it, and to the left a stove and a short refrigerator. All very functional.

"I'll take it, of course," she said. "I really appreciate it."

"Me too," Trudi said.

After that they went out for coffee in a bakery next door, and traded a bit of their stories. Trudi watched her curiously, as if trying to correlate her with the minister she had known in the office. Mary resisted an impulse to explain herself.

So she had a place. Her Swiss security team helped her to move, checking out her room as they did so. Priska and Sibilla were none too impressed. Thomas and Jurg thought it was funny.

Once settled in, she tried to form a new rhythm for her days. No more going to work; she didn't want to intrude. She hoped they might ask her for help in some capacity or other, but in fact, now that she had seen how things had gone at the meeting in San Francisco, she realized that there was probably little help she could give. The authority of her position had been a big part of her effectiveness. That was a little chastening, but no doubt true. Now that she was a private citizen, it was an open question what she could do to help them, or anyone.

Well, she could always get her habits back. Up in the morning, tram down to Utoquai, walk to the schwimmbad, into the locker room, go to her locker, into her swimsuit, kissing the ghost of Tatiana, so beautiful, feeling the pain of her death and then encysting it again, chilling it off by stepping down the metal steps into the water, brr! and out into the lake. Thinking of someone gone kept them just that tiny bit alive, maybe. The Zurichsee, blue and calm, cool and silky. Offshore swimming freestyle, until she could look back and see the near shore properly; then a few circles of breaststroke, to see the whole city, now looking so low and far away. It was a big lake. If she felt strong enough, she could join the swim across the lake at the end of the summer, see what that felt like. It felt good to be out there. Of course this part of her habit set would only work from

May or June through October. In the other months both water and air were just too cold. But as a summer thing, a beautiful way to start the day.

Then back up to her place, join the communal lunch in the co-op, talk to people, if they wanted to, but it was important not to impose English on them, and as the talk around her was usually in Schwyzerdüütch, guttural and sing-songy, she didn't start conversations much. She liked being in the midst of their talk without being part of the conversation, it was soothing. She could feel her body, relaxed from her swim, slope down into her chair like a cat, satisfied just to be among people, unconcerned with the content of their chatter.

Later that year she began to go to the UN Refugee Agency, the UNHCR, which was headquartered in Geneva but had a small office in Zurich. Issuing UN passports to refugees, closing the camps, or rather opening them, emptying them, had made for a lot of work. Naturally the Swiss were intent to get it done, so when she presented herself at the Zurich office they were happy to give her things to do. In fact they wanted to use her fame to rally more volunteers, and she agreed to that, but only if she could do some basic work too. The local work also kept her carbon burn low. Because that too was a project she wanted to pursue.

Almost everyone in the co-op was part of the 2,000 Watt Society, so staying low-burn was not that difficult. The communal meals were vegetarian for the most part, and calculations were made for everything they did, so that she could keep a personal count easily enough, and always had people around to answer her questions. If she stayed in Zurich, if she traveled in Switzerland, in Europe, even around the world—they had all these rated for energy costs and also carbon burn, though the latter was getting lower and lower, especially if you stayed in Switzerland and used public transport. Together the occupants of the place owned one electric car, and there was a sign-up sheet for it, often almost empty, but not always. Most of the people living in the co-op actually traveled quite a bit in Europe, but factoring that in they still came to the end of the year having used well less than the amount of kilowatt hours the 2,000 Watt Society was calling for. The whole country was getting closer to hitting that usage goal; the world would then have a model to follow. Her housemates were sure other countries would then match them.

Mary was not so sure, but did not argue. She just lived the life. Quickly her habits clicked into place, day after day the same. She lived through the week trying to feel if she liked it, trying to figure out how to do more at the UNHCR, and so on. Day after day, week after week. Never had she been so immersed in Swissness. Before she had been an international person living an international life. Now she was a foreign-born Zurcher, living in Zurich.

Recognizing this change, she added a German class to her days. Turned out that would be easiest in the evenings. The city offered free classes, the people who signed up for them were from all over the world. She joined a class that met nearby on Monday nights. The language was wicked, the teacher kindly: Oskar Pfenninger, a white-haired man who had lived in Japan and Korea, and knew English among other languages, but would not speak any of them with his students, at least not in class; nothing but German in class. So they blundered together through the hours of his lessons, and went out afterward for pizza. There they spoke English. As the months went on they even tried speaking German to each other, shyly and with much laughter. Turned out they were all taking the class to fit better in Switzerland.

The days grew short, the air chill. The leaves on the lindens turned yellow and the west wind swept them away. The fire maples on the lanes running down to the ETH went incandescent red. Up on the Zuriberg the views got longer as the trees got bare-limbed and the air chilled and grew clear. She took walks up there around sunset, trudging up the hill, wandering the paths, then clumping or floating back down, depending on her mood. The nude concrete woman holding up her green loops of garden hose always regarded the weather stoically. Mary liked the way she held firm her position. I too will be a concrete woman, she said to her as she passed.

The year turned. She got through Christmas and New Year's without going to Ireland, and without thinking much about anything. The ministry's team invited her to their holiday parties, and she went to them.

Once at one of these she stood out on a balcony with Badim. They looked down on the lights of the night city, elbows on the rail.

How's it going? she said.

He considered it. Pretty well, I guess.

I had a good time on Gaia day.

He laughed. That wasn't us. But I did too. Where did you go for it?

I was out on the lake, swimming with some friends from my club. We made a ring and held hands.

He smiled. Did you feel the moment?

No. It was too cold.

Me neither. But people seemed to like it. I think it was worth supporting. I still think we need a religion. If others feel the same, who knows.

I think it was good. And I'm sure you're following on.

We are. But it's something that has to come from inside.

Mary regarded him curiously. Even now she knew so little about him, really. Her man from Nepal. She had heard things recently, not to her face but around the internet, rumors to the effect that the Ministry for the Future had been thousands strong and had waged a savage war against the carbon oligarchy, murdering hundreds and tipping the balance of history in a new direction. Bollocks, no doubt, but people dearly loved such stories. The idea that it all happened in the light of day was too frightening, history being as obviously out of control as it was—better to have secret plots ordering things, in a realm without witnesses. Not that she completely disbelieved this particular tale. Her man had a look that could freeze your blood, and a lot of money had disappeared into his division without explanation.

Do you have someone now like I had you? she asked him curiously.

He looked over the side of the balcony. Bare linden branches below. Someone to do the dirty work, you mean?

Yes.

He laughed. No, he said. No one I trust the way you trusted me. I don't know how you did it.

I don't either. Actually, you made me do it. Right? I mean what choice did I have?

You could have fired me.

I suppose. But I never even thought of that. I'm not that stupid.

He laughed again. Or you were entirely deceived.

I don't think so. But now, for you—it must be a problem. You need someone you can trust.

He nodded. I know. It's my problem. But I don't know. Maybe the need isn't as great now? Or I'm doing both, and I don't let my right hand know what my left hand is doing?

She shook her head. I don't think that's possible.

No, I suppose not.

What about your team? I mean in the darker side of things. There must be some people there who could do what you did.

I suppose. I'll have to think about that. I don't know. What I think now is that what you did was way harder than I thought it was.

What do you mean?

You trusted me.

She regarded him. She wondered if it were true. Maybe it was.

Sometimes you have to, she said finally. You just throw yourself out there. Throw yourself off the cliff and start making the parachute.

Or start to fly, he suggested.

She nodded dubiously. She didn't think they could fly.

Let me know if I can help, she said.

I will. But he was shaking his head, very slightly. No one could help him with this. As with so many things.

They went back into the party. As they crossed the threshold from balcony to room he touched her arm.

Thanks, Mary.

105

After we were given our passports we put our names on a few lists, and waited some more. Of course those were the days that felt longest of all. Finally we got lucky; our names came up on one of the Swiss cantonal lists—for Canton Bern, in fact, where we had been all along. We were invited to move to Kandersteg. A village up in the Oberland. On a train line, where the line ran into one end of a tunnel which cut under the mountain south to the Valais. Said to be quiet. A bit of a backwater. Room in a hostel for us, and apartments being built. We said yes. My daughter, her husband, their two little ones.

Once there we moved into the hostel, and got on the waiting list for an apartment building soon to be finished. Kandersteg turned out to be a very Swiss village, like a set from a movie, a rather clichéd movie, but fine, we were there. A family from Syria. There were five other refugee families living there, from Jordan, Iran, Libya, Somalia, and Mauritania. We said hello to them cautiously.

Of course we all knew about the SVP, the Schweizerische Volkspartei, the Swiss People's Party. They do well in the mountain cantons, and they don't like immigrants. People in the Fremdenkontrolle and the SEM, the State Secretariat for Migration, were often only marginally helpful to us, or even unfriendly, but the SVP are actively hostile. It's best to avoid attention. Part of that means not gathering in groups with other refugees, looking together so dark and strange. *Unheimlich*. We all knew that. When we met, at first, it tended to be privately.

So, a day came when I stood outside the doorway of our hostel. Green alps rising all around, gray mountains above them leaping to the sky. Like living in the bottom of an immense roofless room, or even at the bottom of a well. But a quick creek in a drainage channel ran right through the middle of the village, making a cheerful sound. The air was clean and cold, the sunlight on the rocks above a thick yellow. A real place, despite its unreal look. And we were here. Here, after twelve years in a Turkish camp, two years on the move trying to get to Germany, very crazy years, very hard; then fourteen more years in the Swiss camp north of Bern. Now we are finally somewhere.

And yet now I find I am seventy-one years old. My life has passed. I will not say it was wasted, that's not right. We took care of each other, and we taught the children. They got a good education in that camp. We did what we could with what we had. It was the life we could make.

And now we're here. And the SEM awarded us a small lump sum based on how long we had stayed in the camp, and that being so long for us, it wasn't all that small. We could combine it with the savings of the Jordanian family new to Kandersteg, and together we rented an empty space in a building on the main street between the train station and the cable car terminus. The space had been used as a bakery, so it was not too difficult or expensive to change it into a little restaurant. Middle Eastern food, we were told to call it. Kebabs and falafel, and other things people already knew about; then, when we got them into our place, we could have them try dishes more interesting. Six tables and we would be full. It seemed possible, and it was interesting to try. Well, who am I kidding—it was exciting to try.

Of course I am old now, but there is no changing that, except by death. At least I have this day, and these days. All that happened before seems now to have happened to someone else. It's like remembering a previous incarnation. Especially home itself. I remember when I left Damascus I looked around and promised myself I would someday come back. Damascus isn't like any other city, it's old, the oldest capital left on Earth, and you can tell that when you're there, it's in the streets and the way it feels at night. And when we were released from the camp, I had the chance to return. I even got a plane ticket. I went to Kloten thinking I'll just go back and see it. The family didn't want to go, but I did. But then in Kloten I had a kind of a, I don't know what—a kind of a breakdown I guess. Who was this going back, and why? I tried to put it together, all the pieces of my life, and I couldn't do it. I concluded that the person who thought of going back was not me, that I was no longer that person. The years in the camp had taken me, day by day, the same day every day, to another person. So at the last minute I said no to myself, and went back down to the train station under Kloten and returned to the camp. My family greeted me curiously, unaware of this shift, this fact that a different person had returned to them. Are you okay? they asked, and I said yes, I'm okay. I just don't want to go anymore. I didn't understand it, so how could I explain it? Who can tell the riddle of their own true self?

So all right, a new person. Old but new. I think about what I have now, as this new person in her life, not quite my life, it seems, but I'm trying to get my head around it. We work all day to prepare a meal. It's a fixed menu for those who want the whole supper, and we take reservations, which sometimes happen and sometimes not, but by eight or nine the restaurant is mostly full. Easy with the six tables. It's almost like hosting a dinner party at home, except instead of friends coming over, it's strangers. Or let's call them acquaintances. Many are there for the first time, but some have been before and come back. We always greet those ones with a smile, and they often talk to each other. Swiss German is such a funny language, it is sometimes hard not to smile. It's maybe like the sound of their medieval life, chopping wood and clanging cowbells and the nasal toot of their alphorns, and maybe rocks falling off the sides of their awesome mountains. This compared to the fluid birdsong of Arabic; it would be funny to have both in the same room at once, but we don't usually speak Arabic around them, we speak high German, *Hochdeutsch*, and they speak it back to us, slowly and clearly, with what I am told is a strong Swiss accent, but I don't hear that, it's the only high German I know. It was a tourist from Berlin who told me that, he said, in Berlin you would be taken for a Swiss woman, your German is that good, but with the Swiss accent. If it weren't for the color of your skin of course, you know what I mean. I agreed that I did, with a smile.

So what we have now, I would say, is not money (very short), nor freedom (we are still registered as *Ausländer*), but dignity. And this is what I think everyone needs. After the basics of food and shelter that we need just as animals, first thing after that: dignity. Everyone needs and deserves this, just as part of being human. And yet this is a very undignified world. And so we struggle. You see how it is. And yes, dignity is something you get from other people, it's in their eyes, it's a kind of regard. If you don't get it, the anger rises in you. This I know very well. That anger can kill you. Those young men blowing things up, they're angry because they don't have dignity. Which is something other people give you, so it's tricky. I mean you have to deserve it, but ultimately it's something other people give you. So the angriest of our young men blow things up because they aren't given it, and mostly they blow up their own people's chances in this world.

Take the Chinese. Chinese tourists who come in tell me, in English of course, that for a century they were oppressed by European countries, they were humiliated. They had no dignity anywhere on Earth, even at home. But who can imagine that now? The Chinese are so powerful now, no one can criticize them. And they forced that to happen by standing up for themselves. They didn't do it by killing strangers at random. That is so wrong I can't even express it. No, if it's going to happen, it has to be done like the Chinese did it. Possibly Arabia with its new regime will change, and the wars end and the rest of our suffering countries change in ways that force the rest of the world to give us the respect we deserve. It will take changes all around. It will take the young to do it.

Meanwhile we fill our restaurant, night after night. We are legal permanent residents of Switzerland. The years here will pass faster than in the camp, that's for sure. That the boredom of the camp made time go slow, so that I must have lived a very long life by that protraction, is an irony that I don't find all that funny. Better to have it all go by in a tearing rush. That I am sure of.

To get by here in this country, I've become a different person, and more than once. But this new person standing here now is not so bad. And there are things about the Swiss you have to admire. They are so *punktlich*, so punctual—this is funny at first, but what is it but a regard for the other person? You are saying to the other person, your time is as valuable as mine, so I will not waste yours by being late. Let us agree we are all equally important and so everyone has to be on time, in order to respect each other. Once we had the restaurant reserved by a single group, we decided to do it on Monday, our usual night off, so as not to inconvenience any of our regulars. So we were cooking away, fixed menu, pretty easy but had to be done right, and my daughter looked out the door and laughed. Look, she said, the invitation is for eight but some of them got here at quarter till, so they are waiting outside until it turns eight. Here, look at the clock, you'll see I'm right. And at eight there was a knock on our door. We greeted them with huge smiles, I'm sure they thought we were a little tipsy. Then also in the train stations, this I like to watch, the clocks over the platforms show the time, and whatever your train's departure time is, if you look out the window of the train right before, you'll see the conductor of the train also has his or her head out the window, looking at the clock; and when the clock hits the very minute and second of departure, the train jerks and off you go. That's the Swiss.

These people will accept us, if we aren't too many. If we are too many, they will get nervous, that's pretty clear. I think it's the same in Hungary or in any of these little European countries. They're prosperous, yes, but there are only a few million of them in each country. Seven million Swiss, I think, and three million *Ausländer* among them; that's a lot. And it's not just the sense of the nation, but the language. This I think is the crux. Say only five million people on Earth speak your language. That's already far less than many cities hold. Then another five million come to live with you and everyone speaks English to understand each other. Pretty soon your kids speak English, pretty soon everyone speaks English, and then your language is gone. That would be a big loss, a crushing loss. So people get protective of that. The most important thing, therefore, is to learn the language. Not just English, but the local language, the native language. The mother tongue. Their culture doesn't matter so much, just the language. That I find is the great connector. You speak their language and even when you're messing it up like crazy, they get a look on their face: in that moment they want to help you. They see you are human, also that their language is a hard one, a strange one. But you've taken the trouble. The Swiss are very good about that. Their language classes are free, and besides they have four languages among themselves, which they hack to bits with each other every day. Take the tunnel under the mountain from this town to the town at the other end of the tunnel, and they don't speak the same language they speak here! You go from weird German to weird French, and really, we speak German better than many Swiss who live just twenty kilometers from here. It makes them more tolerant, maybe. They joke about each other in that regard.

When everyone jokes like the Swiss do about each other, when everyone in the world has their dignity, we will be all right. In the mean-time, here I am, an old woman, my life mostly lived in refugee camps, out on the street outside our little restaurant, early sunset as always here, in the shade at 3 PM. It's a shady town so deep in its hole, a calm town, a sleepy town. Whatever happened in the past, whatever happens after this, today is today. In a little while I'll go back inside.

106

The day before Fasnacht, Mary got a message from Arthur Nolan. He was getting back to Zurich in time for it, could he still join her?

Yes, she messaged back. The rest of that day she thought about him, wondered what it meant that he was back. She was pleased he would be there. Had he cut short a tour?

She went out to see *The Clipper of the Clouds* descend onto the big new airship flughafen in Dübendorf. When he emerged from their little Jetway he saw her and smiled. A slight man.

She accompanied him to his co-op, looked around curiously as he put away his small bag of stuff. Frank's last place. Who remembered him now? Seemed like it might already be down to her. Maybe his parents were still alive. If so they would be so sad. Horrible the way mental illness spread its pain around, cut people off. Her Frank, she had done her best; and he had been a friend anyway, she had loved him in her way. Nothing to be done.

They trammed up to her place, and he laughed to see it. You took a place sized for me, he joked as he walked down the length it, farther to the left than she could have gone.

They dined in a nearby trattoria. Art told her where he had gone on his last trip: central Asia, mostly, circling the lower slopes of the various mountain ranges, where animals were doing very well. The Caucasus, the Pamirs, the Karakorums, the Altai, the Hindu Kush, the Himalaya. There was a Lenin Peak in the Pamirs, and Tajikistan was almost all a wilderness reserve, imperfect but real. They had seen a snow leopard, and black-faced langurs, and many other creatures. People had inhabited these mountain ranges for thousands of years, but the nature of the land meant it was a bit like Switzerland, only more so; some terrain was just too wild to make much of. His friends Tobias and Jesse were helping to create what they called the Anthropocene wilderness, a composite thing that was like the wilder wing of the Half Earth movement, and many of the governments there were cooperating in creating a vast integrated park and corridor system that included and supported the local indigenous human populations, as park keepers or simply local residents, part of the land doing their thing.

It sounds great, Mary said. I'd like to go on that one.

Would you? Because I'm going to do it again.

If I go, I'd like to spend more time on the ground, she confessed. Just stay in one place for a while, see what happens.

We could drop you off and pick you up again.

That sounds good.

After dinner he gave her another hug and headed off to the tram.

The next day was Fasnacht in Zurich. Shrove Tuesday, falling on February 14 of this year. Art came up to her place to meet her, and when she opened the door she found him wearing a silver lamé jumpsuit with a plastic red hat. You're going to freeze in that, she warned him. She herself had on a long black cape and carried a Venetian domino she could put on when she wanted, a beautiful cat face, which restricted her vision too much to keep on all the time, but looked nice. She put it on to show him and he said, Oh I love cats.

I know you do, she said. Do you want to borrow a coat?

I'll be all right.

They went out into the darkness of early evening. As so often, Fasnacht was going to be cold. On this night the air was particularly chill, temperature already well below freezing. This had a peculiar effect on the festival, because many Zurchers were like Art, dressed in costumes not really appropriate for such cold. But the Swiss were pretty cold-hardened people, and apparently Art was too. As they walked down Rämistrasse arm in arm, they saw people in grass skirts, Hawaiian short-sleeved shirts, bikinis and the like, also fur coats, band uniforms, national costumes from many nations, and every possible type of kitschy cantonal costume. And almost every person out there promenading carried a musical instrument. Fasnacht in Zurich was a musical evening. On every street corner, one or more musical groups were playing for small crowds surrounding them. For a while Mary and Art listened to a steel drum band banging away metallically at some spritely tune from Trinidad. Right behind the band, a fountain was gushing into the air, its water plashing down in time to the music. Bulbous ice knuckles made a thick white verge around the edges of the fountain's basin.

Lower on Rämistrasse they strolled slowly by the luxury shops, looking at window displays. The shop that sold Alpine curiosities held them for a long time: polished facets of stone, geodes, burls and cubes of wood, all enlivened by a small menagerie of stuffed Alpine animals. Also fur pelts, stretched out like artworks against the walls to right and left. Art stuck his nose to the glass to see better.

What are they? Mary asked.

I'm not sure. I mean the stuffed ones are easy, that's a fox, and a weasel. I'm not really sure about the pelts though.

Kind of sad, no?

I don't know, once they're dead I think stuffing some of them is okay. And keeping their fur. Once I came on a dead owl that was perfectly intact, a huge thing, and I took it to a taxidermist and had it stuffed. It was beautiful, I had it for years.

What happened to it?

I don't know, I was about ten.

Down the street to the next corner, where an Andean band in serapes played their pan pipes and guitars. They at least were appropriately dressed for the cold. They sang in tight harmonies, not in Spanish—maybe it was Quechua. These were professionals, or at least professional street musicians, and Mary and Art stayed and listened for a long time—so long that Mary got cold, and steered Art down into the Niederdorf.

Here they found that Zurich had put its lions out for the evening, a fact which caused Art to exclaim happily, time after time as they passed the little prides. Mary told Art what she knew about them, which she had just read in the paper the week before; they were fiberglass lions, life-sized, molded in ten or a dozen different postures, then painted different colors by different groups, and placed all over the city to celebrate its two thousandth anniversary, back in 1987. Turicum, Art interjected, a Roman city. Mary agreed. After the city's yearlong celebration of its two thousandth year, she went on, most of the lions had been auctioned off, but the city had kept a hundred or two in storage at one of the bus garages, ready to be redeployed, and this year's Fasnacht had been declared special for

some reason or other. So now they passed lions painted like alpine meadows, like flames, like the blue and white Zurich flag; like tram tickets and
zebras and sea serpents and the British flag (they booed it); like Art Deco
lamps or granite or brick; and as for their various postures, Art identified
most of them to Mary as they passed them: That's couchant, that's rampant, that one's assaultant; that one's at gaze, that one's accolé. That head is
caboshed, if you can believe it.

You liked heraldry as a child, Mary guessed.

I did! It was all about animals, it seemed to me.

Did you read Gerald Durrell?

I *loved* Gerald Durrell. That one's passant, the one next to it is trippant,
this one is saliant.

She steered him toward the Casa Bar. As they approached it he laughed
aloud at a group of lions outside its door, painted in uniforms of some
psychedelic Sergeant Pepper kind. He said, Do you know Fourier, Charles
Fourier, the French utopian?

No, Mary said. Tell me.

He was a utopian, he had followers in France and America, they started
communes based on his ideas, and in his books he went into great detail
about everything. Verne loved his work, he's a kind of secret influence on
Verne. And for him the animals were very important—they were going to
join us, he said, and become a big part of civilization. So at one point he
says, The mail will be delivered by lions.

By lions! Mary exclaimed.

That's right. The mail will be delivered by lions!

She laughed with him. They staggered down an alleyway laughing
helplessly. I'd like to see that, she said. I'm looking forward to that one.

They crashed into the Casa Bar still laughing. The drinks will be
served by kangaroos, Mary predicted. The usual house band was playing traditional jazz. The star of this band was the clarinet player, fluid and
long-winded beyond human belief. They drank Irish coffees as they listened, and prodded by Mary, Art told her more about his animalist youth.
It turned out they had grown up about a hundred miles apart, but he had
been a country boy in summer, a child of County Down, and Ireland's
back country still supported a fair population of small wild creatures, all
hounded relentlessly by the young Art, it sounded like.

They downed the last of their coffees and went back out into the night. As they threaded the crowded dark alleys it sounded like someone was playing an organ in the Grossmünster, but as they chased the sound they found it was coming from a single accordion player, sitting on a gold lion in a concrete and glass box on the western river walk. Bach's Toccata and Fugue in D Minor, in fact, all coming from a single man squeezing his big black box in and out, and fingering fast. Perfect pacing, articulation, volume. Mary had never heard any orchestra finish it so well. Afterward Art approached with the few other listeners to find out who this virtuoso was, while Mary stayed outside the box to listen to the general cacophony. A Russian, Art reported to Mary when he returned, due to play the next night in Tonhalle. Just passing the time tonight, not even rehearsing, just joining the party. He had played in subway stations in Moscow when he was young, and still liked it. So he had said.

To make such beauty out of a silly squeezebox! Mary marvelled.

Anything is possible on Fasnacht, Art said. Let's go find the guggen-musikplatz, I like those bands.

Guggenmusik?

You know, brass bands. They're mostly school band reunions, and they play really loudly and out of tune.

On purpose?

Yes. It's a Swiss thing, I think. On festival night you're supposed to go wild, so for them that means playing your French horn out of tune!

Again they laughed. They crossed the Limmat on the Münsterbrücke, into the little old streets between the river and Bahnhofstrasse. The candy store was open for Fasnacht, and Mary treated Art to a slice of dried orange half-dipped in dark chocolate. Bands were playing on every corner: a quintet of saxes, some west African pop, a tango ensemble blazing through some Piazzolla. Finally they found the guggenmusikplatz, which was indeed brass bands playing loudly and out of tune, all at once, different tunes. All a mix and a roar, the Zurchers costumed and flush-faced in the frigid air. It was cold! A choir of alpenhorn players blew into alpenhorns of different lengths, making them more versatile in terms of tones, so that Copland's *Fanfare for the Common Man* sounded very fine; the final movement of Beethoven's *Appassionata* sonata, on the other hand, proved to be a bridge too far when adapted for alphorn, guggenmusik indeed, to the point of falling apart into a giant blaring of all the alphorns at once. The crowd cheered and dispersed, Mary and Art among them. After a few more stops to listen to other groups, they took refuge in the Zeughauskeller and ordered crème brûlée and kafi fertig.

I like the mix of caffeine and alcohol, Mary confessed.

He nodded. I like the warmth.

A group of men dressed as American cheerleaders burst into the big room and leaped up onto the tables. The band accompanying them was Swiss traditional, and yet with the cheerleaders on the tables they began to play marching songs from America, John Philip Sousa no doubt. On the Swiss instruments this didn't work very well, nor did the men in drag manage to coordinate their can-can dancing with any precision, but everyone cheered them anyway; it was guggenmusik again, with guggendanse too, moustached bankers in pleated skirts and cashmere sweaters, arm in arm, high-kicking precariously overhead, it was too ludicrous not to cheer. Art shouted in Mary's ear that all this was sign of a syndrome, that when an orderly culture like Switzerland finally let loose it was inevitably even wilder than more relaxed cultures. It's a matter of venting, he said loudly so she could hear him. Lots of pressure through a small aperture kind of thing.

Like a French horn mouthpiece, Mary joked.

Yes exactly!

I'm like that myself, Mary shouted, she didn't know why.

Art grinned. Me too!

Let's get out of here before someone gets hurt! Mary said.

Sounds good!

They wandered on though the crowded streets. From the Tonhalle's opened doors they could hear the city's orchestra charging through the end of Brahms's Second Symphony, trombones leading the way. Brahms's own guggenmusik, the best of all. After that the orchestra was going to do the finale of Beethoven's Fifth, it wasn't a night to be subtle.

At midnight there would be fireworks over the lake.

How will people get home after the fireworks? Mary wondered. The trams stop at midnight no matter what day it is.

We can walk up to your place, right?

Yeah sure. I'd like that. Work off some energy.

And warm up a little.

So they made their way slowly through the streets and alleys toward the lake. A string quartet was screeching out something by Ligeti, or maybe it was Stockhausen, transfixing a crowd around them, except for the people singing along with it, or shouting abuse. A passing clown gave Mary and Art slide whistles, hers small and high, his bigger and lower, and Art made her stop by a pair of green and orange lions, assurgent affrontée, Art said, and in the colors of the Irish flag, so they could stand on one and become their own band, him leading her through "Raglan Road." The Gaelic name for the tune, Art told her before they started, meant "Dawning of the Day." Very apt, she replied, trying to focus on the music. Slide whistles were simple but hard, as she was now remembering from her childhood; every micrometer up or down the slide changed the tone considerably, it was impossible to hit the notes right, one had to adjust mid-note with minute adjustments, so it was guggenmusik again for sure. Possibly it was the same for trombones, which was why they liked to play out of tune on this night, being out of practice and thus making a virtue of necessity. But Art could carry a tune on his bigger whistle, and she did her best, and some people stopped to listen; Mary saw all of a sudden that they had stopped because no one on this night was to be allowed to play without an audience. This was so touching that Mary almost cried, she blew hard into her little whistle and squealed a looping descant that almost worked, guggenmusik indeed. Sloppy Irish song, the anthem of St. Patrick's Day, on Ragland Road in November, I saw her first and knew. When they finished she took Art by the arm and dragged him off, saying, Come on we'll be puncturing their eardrums with these things. He only laughed.

Many of the corner bands were also school reunions, people pulling instruments and jackets out of the closet, happy to be reimmersed in their mates and their old songs. The Swiss never gave up on music, Art shouted, everyone still learns to play something in school, and even if they stop doing it afterward, tonight's the night. Mary nodded as she looked around; people were red-faced and ecstatic with the joy of playing music together. Playing: music was adults at play.

The sound spheres in this part of the city overlapped, but as long as that didn't confuse the players, or even if it did, the listeners took it as part of the experience. Down near the lake it got more and more crowded, it was the place to be, and here they could always hear at least three tunes at once, often more. Cacophony, she shouted to Art.

Polyphony! he shouted back, grinning.

Mary nodded, smiling back at him.

Midnight was approaching. Ash Wednesday would soon be here, with its discipline and fasting. It was time to let loose. Fat Tuesday. Even more than New Year's Eve, it was time to let loose: winter's end was in sight, and though it wasn't spring, it was the promise of spring. Spring would come. That knowledge was always the real festival night.

They reached the lake. The Zurichsee, her summer schwimmbad. Mary recalled the big conference in Kongresshall just down the shore, the clot of trends, the Gordian knot of the world; but then she thought, No, not tonight. She had spent her entire life tugging on that Gordian knot, and at most had only loosened a snag, the big knot remaining intact despite a lifetime of ceaseless work. She shuddered in the cold at the thought of that, grabbed Art's arm, guided him to the little lakeside park with the statue of Ganymede and the eagle.

Have you seen this before? she asked him.

The statue? Sure. But who was Ganymede, I never really knew. And what's with the bird?

She said, It's kind of mysterious really. It's said to be Ganymede and Zeus, Zeus in the form of an eagle. So, but look at him. What is he saying to that eagle, do you think? What does it mean? I mean—what does it mean?

Art regarded it. Naked bronze man, arms outstretched, neatly balanced, one arm back and high, the other forward and low—as if offering something to the bird, as in falconry. But the eagle was almost waist high to him.

That's a really big bird, Art said. And there's something wrong with its wings.

A phoenix, Mary said as it occurred to her. Maybe it's a phoenix.

The man is offering it his life, Art guessed.

Mary stared at it. I don't know, she confessed. I can't get it.

It's some kind of offering, Art insisted. It's a gesture of offering. He's us, right? So he's us, offering the world back to the animals!

Maybe so.

He was definitely saying something. That we could become something magnificent, or at least interesting. That we began as we still are now, child geniuses. That there is no other home for us than here. That we will cope no matter how stupid things get. That all couples are odd couples. That the only catastrophe that can't be undone is extinction. That we can make a good place. That people can take their fate in their hands. That there is no such thing as fate.

Her lake extended blackly to the low hills in the distance, the Vorder-Alps, the forward alps. Black sky above, spangled with stars. Orion, the winter god, looking like a starry version of the Ganymede before them.

It has to mean something, Mary said.

Does it? Art asked.

I think it does.

That's Jupiter there to the west, Art said, pointing to the brightest star. So if your big bird is Zeus, that's where he comes from, right?

Maybe so, Mary said.

She tried to put that together with the burbling roar of the crowd, the overlapping music, the lake and the sky; it was too big. She tried to take it in anyway, feeling the world balloon inside her, oceans of clouds in her chest, this town, these people, this friend, the Alps—the future—all too much. She clutched his arm hard. We will keep going, she said to him in her head—to everyone she knew or had ever known, all those people so tangled inside her, living or dead, we will keep going, she reassured them all, but mostly herself, if she could; we will keep going, we will keep going, because there is no such thing as fate. Because we never really come to the end.

ACKNOWLEDGMENTS

My thanks for very generous help from:

Tom Athanasiou, Jürgen Atzgendorfer, Eric Berlow, Terry Bisson, Michael Blumlein (in memory), Dick Bryan, Federica Carugati, Amy Chan, Delton Chen, Joshua Clover, Oisín Fagan, Banning Garrett, Laurie Glover, Dan Gluesenkamp, Hilary Gordon, Casey Handmer, Fritz Heidorn, Jurg Hoigné (in memory), Tim Holman, Joe Holtz, Arlene Hopkins, Drew Keeling, Kimon Keramidas, Jonathan Lethem, Margaret Levi, Robert Markley, Tobias Menely, Ashwin Jacob Mathew, Chris McKay, Colin Milburn, Miguel Nogués, Lisa Nowell, Oskar Pfenninger (in memory), Kavita Philip, Armando Quintero, Carter Scholz, Mark Schwartz, Anasuya Sengupta, Slawek Tulaczyk, José Luis de Vicente, and K. Y. Wong